Kari Stenman
Karolina Hołda

Finnish Fighter Aces Their Planes and Units 1939–1945

Published in Poland in 2022
by STRATUS sp.j.
Żeromskiego 4
27-600 Sandomierz
e-mail: office@mmpbooks.biz
www.stratusbooks.pl

for
MMPBooks,
e-mail: office@mmpbooks.biz
© 2022 MMPBooks.
http://www.mmpbooks.biz

ISBN
978-83-66549-59-3

Editor in chief
Roger Wallsgrove

Editorial Team
Bartłomiej Belcarz
Robert Pęczkowski
Artur Juszczak

Colour Plates
Karolina Hołda
Artur Juszczak
Andrzej M. Olejniczak
Janusz Światłoń

Text
Kari Stenman

Proofreading
Roger Wallsgrove

DTP
Bartłomiej Belcarz

Printed by
Wydawnictwo
Diecezjalne i Drukarnia
w Sandomierzu
www.wds.pl

PRINTED IN POLAND

Table of contents

Acknowledgements

Most of the facts presented in this book have been drawn from the Finnish National Archives, from its Sörnainen division, which was previously the Finnish War Archives. The documents consulted were those of the Air Force Headquarters dealing with the combat reports, squadron log books, orders of the day and individual aircraft files. Another valuable source was the personal service records of each pilot introduced here.

I am indebted to the staff of this organization, having fulfilled my enquiries by delivering masses of archive files for my use, during the past fifty years.

Thank you is also due to the staffs of the Finnish Air Force Museum, Finnish Aviation Museum and Finnish War Museum, for a practically free access to their collections, whether material, documents or photos.

The bulk of the photos and especially the high quality ones came from the Finnish Defence Force collections, shortly abbreviated as SA-kuva. Thank you for making them available to the public.

I also wish to extend my gratitude to many fighter pilots, who shared their personal log books and photo albums. I would especially like to mention the late MajGen Gustav Magnusson, late Col Lauri Bremer, late Col Pauli Ervi, late Col Raoul Harju-Jeanty, late Col Aimo Huhtala, late Col Jorma Karhunen, late LtCol Aulis Bremer, late LtCol Olli Puhakka and late LtCol Jorma Sarvanto.

Photos were also received from the following wartime members of the Finnish Air Force, (or their families): Klaus Alakoski, Lars Bergman, Carl-Erik Bruun, Jaakko Dahl, Anssi Hartiala, Erkki Heinilä, Risto Hiltunen, Ilmari Juutilainen, Aarno Juurinen, Kyösti Karhila, Veikko Karu, Nils Katajainen, Esko Laiho, Vilppu Lakio, Reino Lampelto, Ossi Marttila, Pauli Massinen, Uolevi Nurmi, Onni Paronen, Jaakko Puolakkainen, Olavi Puro, Osmo Rantala, Olli Riekki, Esko Rinne, Ragnar Rosenberg, Joel Savonen, Sulo Suikkanen, Kaarlo Temmes, Jouko Timonen, Kauko Tuomikoski, Oiva Tuominen, Onni Utti, Kullervo Virtanen, Toivo Vuorinen, Erkki Wegelius and Hans Wind. Thank you all.

The support of my fellow researchers late Eino Ritaranta, late Klaus Niska, Mikael Forslund (Sweden), Lassi Eskola, Carl-Fredrik Geust, Pentti Manninen and Kyösti Partonen is duly acknowledged. All expressed no limitations to my wishes concerning their collections, photos, drawings or otherwise.

Last but not least I wish to thank Karolina Hołda, the main illustrator of this book. She has shown exceptional talent in the faithful reproduction of the colour profiles. With a keen eye to the smallest detail, she was also open to all my instructions. The end result being a series of the finest Finnish fighter profiles that have ever appeared in print.

The ace of aces, Warrant Officer Ilmari Juutilainen of 1/HLeLv 34 pictured at Taipalsaari on 30 June 1944, in the middle of the heat of the major Russian summer offensive. In repelling the Soviets, his contribution on this day was downing six enemy aircraft in one mission flown before noon, flying Mersu MT-457. (SA-kuva)

Abbreviation	Finnish	English
AOK	Aliupseeriohjaajakurssi	NCO pilot course
EK	Saksan Rautaristi (Eisernes Kreutz)	German Iron Cross
Er.LLv	Erillinen Lentolaivue	Detached squadron
EsikLtue	Esikuntalentue	Staff Flight
HLeLv	Hävittäjälentolaivue (14.2.44 >)	Fighter squadron (14.2.44 >)
IlmavE	Ilmavoimien Esikunta	Air Force Head quarters
IlmavVK	Ilmavoimien Viestikoulu	Air Force Communications School
IlmK	Ilmailukoulu (> 31.12.37)	Aviation School (> 31.12.38)
IlmVar	Ilmavoimien Varikko	Air Force Depot
IPAK	Ilmapuolustusaluekeskus	Area air defence centre
IRUK	Ilmavoimien Reserviupseerikurssi	Air Force reserve officer school
ISK	Ilmasotakoulu (1.1.38–2.1.41)	Air Fighting School (1.1.38–2.1.41)
IVAK	Ilmavalvonta-aluekeskus	Area air surveillance centre
KadK	Kadettikoulu	Cadet School
KoeL	Koelentue	Test Flight
LejKoulK	Lentojoukkojen Koulutuskeskus	Air corps training centre
LeLv	Lentolaivue (3.5.42 >)	Squadron (3.5.42 >)
LeR	Lentorykmentti	Air Regiment
LeSK	Lentosotakoulu (3.1.41 >)	Air Fighting School (3.1.41 >)
LeV	Lentovarikko	Air Depot
LLv	Lentolaivue (> 3.5.42)	Squadron (> 3.5.42)
MaaSK	Maasotakoulu (Kadettikoulu)	Cadet School
Ltue	Lentue	Flight
MHR	Mannerheim-risti	Mannerheim Cross
PLeLv	Pommituslentolaivue (14.2.44 >)	Bomber Squadron (14.2.44 >)
Os	Osasto	Detachment
RAOK	Reservin aliupseeriohjaajakurssi	Reserve NCO pilot course
SOK	Sotaohjaajakurssi	Combat pilot course
TLeLv	Tiedustelulentolaivue (14.2.44 >)	Reconnaissance squadron (14.2.44 >)
T-LeLv	Täydennyslentolaivue	Advanced training squadron
T-LentoR	Täydennyslentorykmentti	Advanced training regiment
UK	Upseeriohjaajakurssi	Officer pilot course
VL	Valtion Lentokonetehdas	State Aircraft Factory
VM	Vapauden Mitali	Medal of Freedom (for NCOs)
VR	Vapauden Risti	Cross of Freedom (for officers)
YhtLtue	Yhteyslentue	Liaison Flight

Ranks

Finnish	Abbreviated	British	American	Abbreviated
Kenraaliluutnantti	kenr.luutn	Air Marshal	Lieutenant General	LtGen
Kenraalimajuri	kenr.maj	Air Vice Marshal	Major General	MajGen
Eversti	ev	Group Captain	Colonel	Col
Everstiluutnantti	ev.luutn	Wing Commander	Lieutenant Colonel	LtCol
Majuri	maj	Squadron Leader	Major	Maj
Kapteeni	kapt	Flight Lieutenant	Captain	Capt
Luutnantti	luutn	Flying Officer	1st Lieutenant	1Lt
Vänrikki	vänr	Pilot Officer	2nd Lieutenant	2Lt
Lentomestari	ltm	Warrant Officer	Warrant Officer	WO
Vääpeli	vääp	Colour Sergeant	Master Sergeant	MSgt
Ylikersantti	ylik	Flight Sergeant	Staff Sergeant	SSgt
Kersantti	kers	Sergeant	Sergeant	Sgt
Alikersantti	alik	Corporal	Corporal	Cpl

Soviet Aviation Units

Abbreviated	English
ADD	Long Range Bomber Aviation
AP	Aviation Regiment
BAD	Bomber Aviation Division
BAP	Bomber Aviation Regiment
G	Guard's
IAD	Fighter Aviation Division
IAP	Fighter Aviation Regiment
K	Red Banner
KBF	Baltic Fleet Air Force
LBAP	Light Bomber Aviation Regiment
MTAP	Maritime Torpedo Aviation Regiment
NBAP	Night Bomber Aviation Regiment
O	Detached
PVO	Air Defence Aviation
RAP	Reconnaissance Aviation Regiment
PBAP	Dive Bomber Aviation Regiment
SAD	Mixed Aviation Regiment
SBAP	Fast Bomber Aviation Regiment
ShAD	Ground Attack Aviation Division
ShAP	Ground Attack Aviation Regiment
TAP	Transport Aviation Regiment

Introduction

Aerial Victories

According to air force headquarters statistics made in 1940, during the Winter War Finnish pilots claimed 281 air victories, of which 85 were unconfirmed. In the corresponding total from 1945, the Continuation War produced 1,621 claims containing 140 unconfirmed cases. The headquarters also kept a monthly statistic of about a dozen of top scorers, but just a small part of these lists, intended for internal use only, have been found in the archives, the latest one dated in October 1943. In addition to the headquarters, regiments and squadrons also compiled their own kill lists.

The only known official victory list made by the headquarters was published on 12 September 1944 in circular No. 3129/Ye.2/9. It was signed by the chief of staff Colonel Risto Pajari and verified by the head of dept 2 Major Oskari Haaki. The title of the circular was "List of fighter pilots who have shot down most enemy aircraft by 1.9.44" and it was the following:

The Air Force chief of staff LtCol Risto Pajari seen on 10 September 1943. He had the victory list "Top 18" compiled. (SA-kuva)

Successful fighter pilots (eight aces) of LeLv 32 at Nurmoila on 12 July 1942. From left Sgt Aaro Kiljunen, 2Lt Esko Ruotsila, Sgt Niilo Erkinheimo, 2Lt Kalevi Tervo, SSgt Jaakko Tiivola, Capt Aulis Bremer, Sgt Eero Tähtö, Maj Olavi Ehrnrooth, WO Eino Koskinen, 1Lt Aimo Euramo and 2Lt Kyösti Karhila. (SA-kuva)

No.	Rank	Name	Squadron	Winter War	Continuation War	Total	Book
1	WO	Juutilainen	HLELv 34	2	91	93	94
2	Captain	Wind	HLeLv 24	-	78	78	74½
3	Major	Luukkanen	HLeLv 34	2½	51½	54	54
4	WO	Lehtovaara	HLeLv 34	1	43	44	41½
5	WO	Tuominen	HLeLv 34	8	35	43	47
6	Captain	Puhakka	HLeLv 34	4	39	43	46
7	MSgt	Katajainen	HLeLv 24	-	36	36	34½
8	1Lt	Puro	HLeLv 24	-	35	35	33
9	1Lt	Nissinen †	HLeLv 24	4	28½	32½	30½
10	Major	Karhunen	HLeLv 24	5	26	31	31½
11	1Lt	Karhila	HLeLv 30	-	29½	29½	32½
12	MSgt	Vesa	HLeLv 24	-	28½	28½	30½
13	MSgt	Järvi	HLeLv 24	-	28	28	25½
14	SSgt	Alakoski	HLeLv 34	-	26	26	28
15	WO	Kinnunen †	HLeLv 24	3½	19	22½	22½
16	1Lt	Saarinen †	HLeLv 24	-	22	22	23
17	MSgt	Tani	HLeLv 34	-	20½	20½	20½
18	1Lt	Myllylä	HLeLv 34	-	20	20	22
Author's addition							
19	1Lt	Tervo †	HLeLv 34	-	21½		21½
20	WO	Pyötsiä	HLeLv 24	7½	14		21½

† killed in action

When the headquarters published this list, it arrived at those figures by adding to every pilot all confirmed, all unconfirmed and all victories claimed into the regiment's account. In addition those figures contained observation balloons. With the same principles the authors have compiled their earlier victory statistics, though excluding observation balloons.

According to Anglo-Saxon practice, taking into account only witnessed claims, the leading ace is Warrant Officer Ilmari Juutilainen, with 79 victories witnessed by outsiders. Second is Captain Hans

Finnish armed forces C-in-C Marshal Carl Gustaf Emil Mannerheim (left) handing over the highest military decoration, the Mannerheim Cross, at Mikkeli in September 1942. The blue tunic men are LeLv 28 CO Maj Auvo Maunula and 3/LeLv 24 leader Capt Jorma Karhunen (right). Between them is Cadet Yrjö Keinonen, a future general and Finnish Defence Force commander. (SA-kuva)

Fighter pilot's war was not all about shooting down enemy aircraft, but also a good catch. LeLv 24 CO Maj Jorma Karhunen shows his 11 kg pike at Suulajärvi on 5 May 1944. (SA-kuva)

Wind with 64½ victories, third is Major Eino Luukkanen with 43 victories and fourth Captain Olli Puhakka with 41 victories.

To make these figures more comparable, especially with British or American statistics, we have listed for each airman his confirmed, probable and damaged claims: 5 + 0 + 2. In this connection a confirmed victory means one with outside witnesses, a probable one claimed by the pilot as destroyed but without witnesses and a damaged one hitting the target but not seeing it destroyed.

In the right-hand column "Book", the larger figure indicates that a damaged case has been upgraded to confirmed as verified by the Russian loss records. The same applies also to observation balloons and aircraft destroyed on the ground or water, which the Air Force HQ included it its figures. In the re-calculation it turned out that due to the lack of first names, First Lieutenant Lauri Nissinen was credited with two victories of First Lieutenant Aarne Nissinen in his account and Staff Sergeant Urho Lehtovaara gained two victories of Staff Sergeant Martti Lehtovaara.

The confirmation process of the air force headquarters was very strict in the Winter War. Only those claims that were witnessed by outsiders were confirmed. During the first six months of the Continuation War the situation remained much the same, although the headquarters began to confirm claims without neutral witnesses. It became a matter of some sort of trust. By autumn 1942 the air force commander paid attention for the first time to the steeply ascending number of unconfirmed victories. The confirmation process remained the same however. The problem was solved in summer 1943 so that all claims without witnesses were confirmed into the regiment's account and not into the pilot's personal score. This practise continued throughout the war.

Re-Valuation

The air victories presented by the authors earlier were based on the total figures supplied by the air force headquarters. The details were collected from the regiment and squadron statistics. The authors have been aware for years that these statistics were correct in magnitude, but the headquarters produced tens of different types of errors in the confirmation process. Just to mention a few, disregarding the witnesses, confirming to another pilot, confirming in different proportions that in the reports, miss-spelling in names and aircraft types and in broader sense the inconsistent use of the so-called trust factor.

We have gone through again all combat reports and they cover those almost 2,000 cases mentioned earlier. The reports can be found in the military archives assembled per annum and also as copies in the respective archives of the units. With a high level of certainty we can claim that all reports have been found.

As a basis for the entry to this book has been that the shooter had claimed in his report as having destroyed the target. All damaged enemy aircraft have been added to the lists of these claimers. After the opening of the Russian archives just over twenty years ago, many "damaged" cases could be verified from their loss lists as destroyed. In this book we have taken this fact into account when calculating aerial victories.

3/LLv 24 pilots in front of BW-365 at Rantasalmi on 8 August 1941. From left Sgt Nils Katajainen, Cpl Paavo Mellin, Sgt Jouko Huotari, flight leader Capt Jorma Karhunen, deputy leader 1Lt Pekka Kokko, WO Ilmari Juutilainen, 1Lt Georg Strömberg and 2Lt Kim Lindberg. All but Strömberg became aces. (SA-kuva)

In this connection we can state that as late as 1978 the United States produced an official survey called "USAF Historical Study No. 85". The reason was continuous inquiries of the precise aerial victory numbers. The researchers of the study went through all combat reports and verified claims, also using the loss lists of the opposing party. The study also out ruled all previous figures and it was declared as final. The listings presented in this book are neither official nor final, but the spirit in the compilation is the same as described above.

Combat Report

The Finnish combat report consisted of three parts: the title area, the report itself and the filling instructions.

The unit's digit code and flight were to be noted in the title area. Likewise there was a space where the claimer's rank and name were placed, plus the number and type of destroyed and/or damaged aircraft. This space was reserved for the headquarters.

The report itself contained seven paragraphs:
1. Date and time
2. Combat area and altitude
3. Whether destroyed or damaged (type and number)
4. Crash location (or when, where and in which condition last seen)
5. Short description of the combat
6. Witnesses to the incident
7. Notes (tactical or other observations, own damage etc.)

The pilot's aircraft (serial), rank and name were to be shown with the signature.

The reverse side had the filling instructions of the form. It had four points:
1. As certainly destroyed can be marked the following:
   – Crashed (location or area to be stated)
   – Broken in the air
   – Completely on fire
   – Abandoned aircraft (seat number known)
   – Forced down in own territory, without a chance to take off
2. As damaged are considered aircraft, with the engine(s) smoking heavily or on fire (which was often extinguished later) and the aircraft losing altitude when last seen or when from other behaviour it can be deduced that they have lost considerably their airworthiness, but do not fulfil the requirements of point 1.
3. Aircraft, which when last seen continue their flight without losing their airworthiness, shall not be marked as damaged in spite of the signs in point 2.
4. The forms are to be sent daily to the regiment headquarters in one copy. The squadrons are advised to produce a duplicate or take a copy.

The combat report was to be filed on all missions in which a fire-fight took place with the enemy, no matter from which side. The forms were usually filled as teamwork of all combat participants, especially

3/HLeLv 24 has landed to Lappeenranta from a mission on 30 June 1944. Identifiable aces are MSgt Nils Katajainen (on the wing without cap), SSgt Emil Vesa (closest to camera) and MSgt Jouko Huotari (at right). (SA-kuva)

to obtain a witness for a personal air victory. This was due to the heavy competition to be top of the scoring list. The squadron commander then presented the combat reports to his superior, the regiment commander, who normally supported his subordinate. The air force headquarters then dealt with the claims with the methods described earlier.

Many pilots have told the authors about cases where no combat reports were made. There were two main reasons. Firstly the pilot did not see the destruction of his target and he considered the filling of the forms as waste of time. Secondly the shoot down occurred in an area with restricted access and the case was concealed to avoid a punishment. From July 1943 onwards such areas were the Estonian coast and heavily defended enemy-held islands in the Gulf of Finland. The main reason was to preserve the limited number of aircraft.

Many pilots have also claimed during the years that they have shot down so-and-so many aircraft. These persons have forgotten that all combat reports they have filed can be found in the archives. When matching ones cannot be found, the reason has been either of the above mentioned ones. An often-heard excuse has also been a quick transfer to another unit or getting wounded on that mission and posted to the military hospital. Based on the archive records this is hard to believe, because the transferred, wounded and even captured or killed pilots have made the report from their last encounter, or more rightfully on their behalf the pilot's friends on the same mission.

In the researcher's point of view the matter is very simple. A filled combat report is the only document which can be taken into account.

Concerned four aces of HLeLv 34 at Taipalsaari in the middle of the major Soviet offensive of summer 1944. From left squadron CO Maj Eino Luukkanen, WO Ilmari Juutilainen, WO Mauno Fräntilä and 1/HLeLv 34 leader 1Lt Väinö Pokela.

Air Victories

Claiming a higher number of aerial victories than in reality occurred became a sensitive subject as soon as the enemy's loss reports became available. Earlier, pilot's own claims were believed to be true.

All began about fifty years ago, when the British Public Record Office published the highly detailed lists of losses of the German *Luftwaffe* in the Second World War. Researchers both in England and United States observed immediately that their claims were noticeably higher that the respective German losses, starting immediately with difference of 20–30 per cent. The over-claiming was especially heavy in the strategic bomber command. The Americans soon noticed themselves that such figures cannot be valid. When the claims were divided by ten, the figures came closer to the truth. The reason for over-claiming was simple. The destruction of one and the same aircraft was witnessed independently by several persons in good faith and so the victory was confirmed to several gunners. As far as fighter pilots are concerned the phenomenon is the same, especially during large combats.

Accurate statistics of such over-claiming do not exist. However, the problem concerns the air forces of all countries and the scale of the error is dependent on the nationality, culture and organization. Also intentional over-claiming cannot be excluded. The Americans seem to have the least over-claiming while the Russians and Japanese have the most.

On behalf of Finland this over-claiming began from zero, moving upwards along the time and operations area. The Russian loss listings are in many places very accurate and reliable, e.g. in the Winter War and summer 1944 offensive. On the other hand Russian researchers have informed the authors that there are gaps in some periods and fronts. Due to the last mentioned reason we cannot point out in this connection what kind of over-claiming when and where took place. But we have made an observation that, especially in the fierce air battles fought over the Gulf of Finland in 1942–43, there seems to be a bigger difference than normally.

When researching air warfare one could assume at first sight that one's victory is another's loss, i.e. a shot or forced down aircraft always matches the loss of the other side. A phenomenon of this kind could thus be easily verified from the enemy documents. In practise the matter is not so simple, because the aircraft claimed shot down was not always considered as lost by the opponent, but often suitable for repair. Against this background it is obvious that the claim and loss listings never meet accurately.

In the Finnish Air Force, with a constant lack of planes, such badly damaged aircraft (and considered destroyed in the enemy records) have been repaired into flying condition, which a major power would without hesitation strike off.

In the Red air forces an aircraft was not always considered as destroyed though it came down on its own side badly damaged ("pilot making a controlled forced landing"). Not until the technical in-

Fokker D.XXI coded FR-95 of Lentolaivue 32 banks for the photographer over Siikakangas in June 1940.

spection, sometimes after a long time, it was found out that the plane needed a factory repair, or it was found to be in an unrepairable condition or was dismantled for spares.

And finally. The air force headquarters has published the official air victory scores and they remain so until further notice. Even internationally the official scores are according their title, right or wrong.

Aerial Victory List

In connection with each pilot is his aerial victory list. It has eight columns, which are:
- Date
- Time
- Location
- Unit
- Own aircraft
- Victories
- Enemy unit
- Victory type

In this book the types of claimed aircraft have been in some cases amended to correspond with the Russian loss listings. From here we enter directly into another international problem. The recognition of enemy aircraft has always proved to be difficult in all air forces.

The arrival of Hurricanes to the Finnish fronts in late 1941 caused confusion, the reports identified them as I-18s and MiG-3s. Also the appearance of Yak-1s and Yak-7s in summer 1942 were reported as Hurricanes or even Spitfires. We know now from the Russian orders of battle that the latter type never served on the Finnish front. Also the myth of Mustangs lives on persistently. These were never operating in this air space. All identified Mustang claims have been Yak-9s. No wonder, differing from all other Yakovlevs, the "nine" had blunt wing tips like the Mustang.

The victory type column uses the same method as the Air Force Headquarters, but we have corrected all found mistakes and omissions. The key for the letter is the following:

T = confirmed claim, witnessed by outsiders

E = confirmed by the HQ, without witnesses

R = claim confirmed into regiment's account, no witnesses

V = recorded as damaged

If any line has a statement of a Russian unit, this means a loss that they have admitted. The lack of this statement does not indicate the lack of a victory, just that a corresponding enemy case was neither found nor matched.

Finnish style	UK/US style
T	Confirmed
E and R	Probable
V	Damaged

Aces of *Suomen Ilmavoimat* 1939–1945

AALTONEN, Lasse Erik

13½ victories (13½ + 0 + 5) on appr. 300 missions
Planes*: FR-105, FA-4, FA-33 and MT-458

*Planes assigned to the pilot.

Lasse Aaltonen was born at Pori on 03.09.17. He received flying training in 1938 at RAOK 4 and AOK 7. Aaltonen was posted on 01.01.39 as Corporal to LLv 26. He was promoted to Sergeant on 20.05.39, Staff Sergeant on 23.03.40 and Master Sergeant on 01.03.41. Aaltonen was transferred on 16.04.43 to LeLv 34, where he was promoted to Warrant Officer on 26.10.44. After the war Aaltonen served with LeR 1 and resigned on 10.07.56. In civilian life he was an airline pilot. Aaltonen died on 05.08.88. He was awarded with VM2, VM 1, VR 4, VR 4tlk and EK 2.

Fokker D.XXI serial FR-105 of 5/LLv 24 seen at Joroinen in spring 1940. It was flown by LLv 26 pilot Sgt Aaltonen during the latter half of the Winter War, which included a claim of an SB bomber shot down on 17 January 1940.

Right: *Aaltonen was a member of 2/LLv 26 at the beginning of the Continuation War, standing here in front of a FIAT G. 50 coded FA-31 at Joroinen in late June 1941.*

Aaltonen claimed his two FIAT kills in the Winter War in FA-4, which he continued to fly in the next war with 2/LLv 26. It is seen here at Lunkula in August 1941.

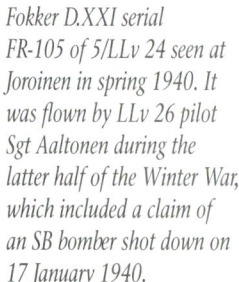

Bf 109G-6, MT-458 of 3/HLeLv 34 flown by MSgt Lasse Aaltonen in July 1944, based at Taipalsaari. (Karolina Hołda)

Mersu *MT-458 was transferred to 3/HLeLv 24 on 6 August 1944, seen here after transfer at Lappeenranta. Before this MT-458 was the assigned fighter of 3/HLeLv 34 pilot Aaltonen for six hectic weeks.*

Date	Time	Area	Unit	Aircraft	Enemy aircraft	Enemy Unit	Victory type
19.12.39	10.50	Muolaanjärvi	26	FR-114	SB	13 SBAP	T
17.01.40	14.20	Muolaa	26	FR-105	SB		T
29.02.40	09.20	Outside Kotka	26	FA-4	DB-3		T
02.03.40		Lahti	26	FA-4	I-153		T
25.06.41	11.55–12.05	Tuusmäki – Haukivesi	26	FA-33	1½ x SB		T
27.10.43	11.30–12.25	Suursaari	34	MT-201	Yak-7B	12 KOAE, KBF	V
27.10.43	11.30–12.25	Suursaari	34	MT-201	La-5		V
08.05.44	07.20–07.50	Haapasaari	34	MT-423	Yak-9	21 KIAP, KBF	T
08.05.44	12.50–13.40	Suursaari	34	MT-423	2 x Yak-9		V
28.06.44	19.55–21.05	Heinjoki	34	MT-458	Yak-9	14 GIAP	T
28.06.44	19.55–21.05	Heinjoki	34	MT-458	Yak-9	14 GIAP	R
29.06.44	07.25–08.25	Tali	34	MT-458	Yak-9	15 ORAK, KBF	T
30.06.44	10.55–12.05	Viipuri	34	MT-458	Yak-9		T
01.07.44	19.00–20.10	Koivisto	34	MT-458	Yak-9		V
03.07.44	06.30–07.40	Tali	34	MT-458	Yak-9	26 GIAP	R
05.07.44	09.45–10.40	Vatnuori	34	MT-458	LaGG-3		T
09.07.44	06.45–07.50	Äyräpää	34	MT-469	Il-2		V

Ahokas, Leo

12 victories (11 + 1 + 1) on 189 missions
Planes: BW-351 and MT-437

Leo Ahokas was born at Jaakkima on 25.04.15. He received flying training in 1936–37 at RAOK 2. On 09.10.39 Ahokas was posted T-LentoR 2 and was transferred on 24.02.40 as Corporal to LLv 22. He was demobilized on 25.07.40. Ahokas returned on 03.06.41 to serve with LLv 32, where he was promoted Sergeant on 03.08.41. He was posted on 11.08.41 to LLv 24, promoted on 14.09.42 to Staff Sergeant and demobilized on 10.11.44. Ahokas acted in civilian life as a chauffeur. He died on 25.10.88. Ahokas was awarded with VM 2, VM 1, VR 4 and VR 4 tlk.

SSgt Leo Ahokas of 3/HLeLv 24 sitting on the threshold of a thick cardboard hut at Lappeenranta on 10 July 1944. The pilots were accommodated in these huts in most air bases. (SA-kuva)

(Above right and below) Brewster Model 239 serial BW-355 of 3/LeLv 24 parked at Römpötti airfield in October 1942. It was regularly flown by Ahokas, who gained three air victories flying it. The machine bears the donor's inscription NOKA behind the squadron's lynx emblem.

Bf 109 G-6, MT-437 of 3/HLeLv 24 flown by SSgt Leo Ahokas in June 1944, based at Lappeenranta. (Karolina Hołda)

Ahokas sitting in the cockpit of a later model Bf 109G-6 with Erla-haube, possibly his MT-480 in mid-July 1944. (SA-kuva)

MT-437 of 3/HLeLv 24 was the assigned fighter of Ahokas from 19 June 1944. Here Sgt Kosti Keskinummi bellied it at Nuijamaa nine days later, after taking hits in combat. The plane's rudder has the squadron's new lynx head emblem. (SA-kuva)

Date	Time	Area	Unit	Aircraft	Enemy aircraft	Enemy Unit	Victory type
26.02.42	08.50–11.00	Juka stop	24	BW-364	MiG-3		E
16.08.42	17.45–18.00	Seiskari – Karavaldai	24	BW-351	I-16		T
22.11.42	13.20–15.05	Kreivinlahti	24	BW-351	Il-4		T
04.05.43	10.40–12.20	Seiskari 10 km S	24	BW-351	I-153		T
20.05.43	13.20–14.20	Kreivinlahti 10 km NW	24	BW-355	La-5	4 GIAP, KBF	T
20.08.43	16.30–18.00	Tolli lighthouse	24	BW-364	LaGG-3		V
23.09.43	12.00–13.45	Yhinmäki	24	BW-355	LaGG-3		T
23.09.43	15.30–16.30	Seiskari	24	BW-355	LaGG-3		T
29.05.44	19.20–20.35	Kronstadt	24	MT-201	La-5		T
20.06.44	02.45–03.45	Someri	24	MT-437	La-5	159 IAP	T
22.06.44	18.20–19.00	Hevossaari	24	MT-439	Pe-2		T
30.06.44	18.50–19.50	Ihantalanjärvi	24	MT-443	Il-2		T
10.07.44	18.55–19.40	Äyräpää	24	MT-480	La-5		T

ALAKOSKI, Klaus Jalmari

28 victories (22 + 6 + 3) on 239 missions
Planes: FA-15, MT-214 and MT-445

Klaus Alakoski was born at Helsinki on 17.08.21. He received flying training in 1940–41 at SOK 3. Alakoski was posted on 06.07.41 as Private to LLv 26, where promoted to Corporal a week later. He was promoted to Sergeant on 15.12.41. On 14.06.42 Alakoski was posted to LeLv 30 and back to LeLv 26 on 07.11.42. He was transferred on 16.04.43 to LeLv 34 and promoted to Staff Sergeant on 20.07.43. Alakoski was demobilized on 11.11.44. In civilian life he was a laboratory manager. He died on 24.09.88. Alakoski was awarded VM 2, VM 1, VR 4 and VR 4 tlk.

FIAT G.50, FA-15 of 3/LeLv 26 flown by Sgt Klaus Alakoski in April 1943, based at Kilpasilta. (Karolina Hołda)

Alakoski and his mechanics of 3/LeLv 26 in front of FIAT G.50 serial FA-15 at Kilpasilta shortly before Alakoski was transferred to LeLv 34, which took place on 16 April 1943.

Sgt Alakoski taxies FA-17 of LLv 26 at Helsinki Malmi in March 1942, with mechanics on the wings pointing the way.

Bf 109G-6, MT-445 of 3/HLeLv 34 flown by SSgt Klaus Alakoski in June 1944, based at Taipalsaari. (Karolina Hołda)

Mersu MT-445 of 3/HLeLv 34 was the assigned plane of Alakoski from 20 June 1944. It is seen here after the hostilities in the latter half of September 1944 at Selänpää.

Date	Time	Area	Unit	Aircraft	Enemy aircraft	Enemy Unit	Victory type
13.08.41	13.00–14.10	Tuulos-Aunus	26	FA-28	I-153		T
16.05.43	05.10–06.30	Lavansaari	34	MT-209	I-153		E
17.05.43	19.10–20.15	Peninsaari – Lavansaari	34	MT-213	I-153		E
18.05.43	19.30–20.40	Peninsaari	34	MT-213	I-153	71 KIAP, KBF	V
21.05.43	18.40–19.55	Lavansaari	34	MT-220	2 x LaGG-3	13 KIAP, KBF	E
04.07.43	20.35–21.30	Tytärsaari	34	MT-223	Boston		V
26.07.43	14.40–15.40	Suursaari	34	MT-223	La-5		T
07.08.43	19.30–20.50	Vaindlo lighthouse	34	MT-209	Boston		V
04.09.43	13.05–13.55	Seiskari	34	MT-229	LaGG-3		T
07.09.43	13.05–14.05	Tytärsaari 20 km SW	34	MT-229	2 x Yak-7	13 KIAP, KBF	T
08.09.43	12.05–12.50	Tytärsaari 20 km NW	34	MT-209	Il-2	35 ShAP, KBF	T
17.05.44	10.30–11.10	Hamina	34	MT-425	Pe-2	12 GBPAB, KBF	T
17.05.44	10.30–11.10	Hamina	34	MT-425	Yak-9		T
19.05.44	04.35–05.00	Ristisaari	34	MT-416	Pe-2	12 GPBAP, KBF	T
14.06.44	13.35–14.40	Kuuterselkä	34	MT-415	Il-2	566 ShAP	T
19.06.44	16.00–17.15	Leipäsuo	34	MT-410	Il-2	872 ShAP	T
19.06.44	16.00–17.15	Perkjärvi	34	MT-410	Airacobra	196 IAP	T
20.06.44	09.00–10.00	Perkjärvi	34	MT-427	Airacobra		T
22.06.44	18.25–19.40	Viipuri	34	MT-406	Airacobra		T
28.06.44	19.55–21.05	Viipuri	34	MT-445	2 x Il-2		R
28.06.44	19.55–21.05	Kämäränjärvi	34	MT-445	Yak-9		R
29.06.44	07.25–08.25	Noskuanselkä	34	MT-445	2 x Pe-2		T
30.06.44	20.05–21.10	Tammisuo	34	MT-445	Warhawk	191 IAP	T
01.07.44	18.50–20.00	Makslahti	34	MT-445	Yak-9		V
03.07.44	06.10–07.20	Tali – Juustila	34	MT-445	2 x Il-4	815 BAP	T
09.07.44	19.50–20.05	Äyräpää	34	MT-433	Il-2		R

ALAPURO, Veikko Sakari

5 victories (4 + 1 + 1) on 211 missions
Planes: CUw-555 and CU-584

Sakari Alapuro was born at Ruokolahti on 13.04.20. He received flying training in 1939–41 at SOK 1 and was accepted on 04.06.41 to KadK. Alapuro was posted on 20.06.41 as 2nd Lieutenant to LLv 32. On 01.07.42 he became a cadet at MaaSK, and was promoted on 01.12.42 to 1st Lieutenant. Alapuro was posted on 22.06.43 to LeLv 32 and became on 27.06.44 the 3rd Flight leader. After the war he served with LeR 1, 3. Lsto and IlmavE, resigning on 17.04.58 as Lieutenant Colonel. Alapuro died on 17.12.94. He was awarded with VR 4, VR 3 and VR 3 tlk.

Curtiss Hawk 75A-2, CU-574 flown by 3/HLeLv 32 leader 1Lt Sakari Alapuro in July 1944, based at Uomaa. (Karolina Hołda)

Five young second lieutenants of LLv 32 in front of CUw-553 at Lappeenranta in August 1941. From the left: Jaakko Hillo, Juho Nyholm, Kyösti Karhila, Reino Lampelto and Sakari Alapuro. Though CU-553 was the official designation, CUw-553 was painted on the plane.

Fokker FR-95 forced landed into the splinter box of HLeLv 32 plane CU-574 at Nurmoila on 26 February 1944. Alapuro claimed one observation balloon shot down in CU-574, on 1 July 1944.

Date	Time	Area	Unit	Aircraft	Enemy aircraft	Enemy Unit	Victory type
03.09.41	10.55–11.10	Suuri-Harvasuo	32	CU-553	I-153		E
12.09.41	16.00–16.05	Kasilovo	32	CU-558	R-Z		T
15.09.41	15.45–17.25	Rukonjärvi	32	CU-570	Il-4	1 AP, KBF	E
28.03.42	17.40–18.00	Lavansaari – Suursaari	32	CU-552	I-153		V
15.06.42	13.15–14.25	Mergino	32	CU-552	Yak-1		T
28.06.44	18.20–20.00	Vitele	32	CU-584	LaGG-3	260 SAD	T

15 victories (13 + 2 + 2) on appr. 300 missions
Planes: BW-383

Martti Alho was born at Padasjoki on 20.06.18. He received flying training in 1938–39 at RAOK 4 and AOK 8. He was posted on 01.03.39 as a Sergeant to LLv 24. Alho was promoted to Staff Sergeant on 22.03.40, Master Sergeant on 05.10.41 and Warrant Officer on 16.05.43. Alho was killed in a flying accident on 05.06.43, when his BW-392 stalled in a steep bank to the ground at Tampere. Alho was awarded with VM 2, VM 1, VR 4, VR 4 tlk and VR 3.

From the left mechanic SSgt Lauri Kitinoja, 4/LLv 24 leader Capt Per Sovelius and SSgt Martti Alho at Immola on 25 August 1941, in front of Alho's Brewster BW-383. (SA-kuva)

This time 1Lt Urho Sarjamo of 4/LLv 24 is attaching his parachute and is about to board Alho's BW-383 at Rantasalmi in early August 1941.

Date	Time	Area	Unit	Aircraft	Enemy aircraft	Enemy Unit	Victory type
19.12.39	15.20	Oravaniemi	24	FR-117	½ DB-3	6 DBAP	T
02.02.40	11.15	Virolahti	24	FR-81	½ DB-3	6 DBAP	T
14.02.40	15.15	Nuijamaa	24	FR-89	½ SB	48 SBAP	T
09.07.41	05.20–05.40	Lahdenpohja	24	BW-383	I-153		E
17.09.41	16.20–16.40	Pyhäjärvi – Prääsä	24	BW-383	MiG-3	179 IAP	E
07.10.41	09.30–11.15	Suopohja	24	BW-383	I-16	155 IAP	T
07.10.41	09.30–11.15	Suopohja	24	BW-383	I-16		V
26.02.42	08.45–11.00	Kumsjärvi	24	BW-383	MiG-3		T
26.02.42	08.45–11.00	Kärkijärvi	24	BW-383	MiG-3		E
12.08.42	17.10–19.20	Tolli lighthouse	24	BW-364	I-16		V
26.10.42	12.15–14.00	Kreivinlahti	24	BW-383	I-16		T
22.11.42	12.15–14.15	Sepeleva lighthouse	24	BW-383	Yak-7	21 IAP, KBF	T
23.11.42	10.50–12.30	Seiskari	24	BW-383	Pe-2	73 AP, KBF	T
10.03.43	15.15–16.25	Lavansaari	24	BW-383	LaGG-3		T
18.04.43	17.00–18.20	Kreivilä 10 km W	24	BW-383	LaGG-3		T
18.04.43	17.00–18.20	Yhinmäki	24	BW-383	La-5		T
21.04.43	08.00–09.30	Ystinski	24	BW-383	Yak-1		T
02.05.43	10.00–11.30	Kronstadt	24	BW-383	½ LaGG-3		T
04.05.43	10.40–11.40	Seiskari	24	BW-383	I-153		T

Brewster 239, BW-383 of 4/LLv 24 flown by SSgt Martti Alho in August 1941, based at Rantasalmi. (Andrzej M. Olejniczak)

BW-383 of 4/LLv 24 taxied into a hole at Rantasalmi on 9 August 1941. The minor damage was repaired in one day.

6 victories (5 + 1 + 0) on appr. 80 missions
Planes: BW-370

Onni Avikainen was born at Valkeala on 05.08.18. He received flying training in 1939–41 at RAOK 6 and was transferred on 14.10.41 as Corporal to LLv 24. On 16.02.42 Avikainen was promoted to Sergeant. He was transferred on 26.09.42 to LeLv 6 and on 16.11.42 further to LeLv 30. On 05.04.43 he returned to LeLv 24, where he was promoted to Staff Sergeant on 21.09.43. Avikainen was grounded due to medical conditions on 10.05.44 and he was demobilized on 10.11.44. Avikainen died on 06.02.61. He was awarded with VM 2, VM 1 and VR 4.

Brewster 239, BW-370 of 2/LeLv 24 flown by Sgt Onni Avikainen in June 1943, based at Suulajärvi. (Karolina Hołda)

Sgt Erkki Siponmaa ditched 2/LeLv 24 Brewster BW-370 into Lake Immola on 1 May 1943. Designed as a naval fighter it stayed afloat and was towed ashore. Avikainen flew this plane on many occasions and claimed two kills in it.

Date	Time	Area	Unit	Aircraft	Enemy aircraft	Enemy Unit	Victory type
09.03.42	14.20–14.50	Uikujärvi	24	BW-386	MiG-3		T
14.03.42	11.00–11.15	Siikasaari	24	BW-366	MiG-3		E
18.04.43	17.00–18.20	Karavaldai	24	BW-384	LaGG-3		T
16.05.43	18.30–19.30	Peninsaari – Lavansaari	24	BW-383	Yak-7	13 KIAP, KBF	T
05.06.43	14.30–15.50	Tolli lighthouse	24	BW-370	Yak-1		T
05.06.43	14.30–15.50	Tolli lighthouse	24	BW-370	Pe-2		T

BERG, Paavo David

10½ victories (10½ + 0 + 1) on appr. 100 missions
Planes: GL-279, FR-116, CUw-553 and CU-570

Pate Berg was born at Lahti on 23.11.11. He received flying training in 1930–31 and was accepted to KadK on 02.06.31. Berg was transferred on 31.01.38 as 1ˢᵗ Lieutenant to LLv 26. On 27.03.40 he was posted to LLv 32 and was promoted on 30.04.40 to Captain. On 18.06.41 Berg was appointed to lead the 1ˢᵗ Flight. Berg was killed on 01.11.41, when I-16s (SrLt V.F. Golubev and SrLt G.D. Tsokolayev) of 13 AP, KBF shot down his CU-570 at Hanko. He was awarded with VR 4.

Fokker D.XXI, FR-116 of 1/LLv 32 flown by the flight leader Capt Paavo Berg in July 1941, based at Utti.
(Andrzej M. Olejniczak)

Date	Time	Area	Unit	Aircraft	Enemy aircraft	Enemy Unit	Victory type
02.02.40	10.40	Bromarv	26	GL-263	I-15bis	38 IAP	T
18.02.40	11.30	Kouvola	26	GL-279	2 x SB		T
18.02.40	11.30	Utti	26	GL-279	SB		V
19.02.40	15.25	Sippola	26	GL-279	I-153		T
20.02.40	10.50	Kouvola	26	GL-280	SB	48 SBAP	T
17.07.41	11.00–12.00	Kilpeenjoki	32	CU-553	I-153	10 KOAE, KBF	T
03.09.41	10.55–11.10	Rajajoki	32	CU-551	I-153	7 IAP	T
19.09.41	13.20–13.30	Ohalatva	32	CU-563	3 x MiG-3	7 IAP	T
01.11.41	14.00–14.05	Tvärminne	32	CU-570	½ I-16		T

1/LLv 32 leader Capt Berg collided his FR-116 with the car of some war correspondents who, unauthorized, drove too far onto the Utti airfield on 9 July 1941. (SA-kuva)

Above: Capt Berg points out patched bullet holes of CUw-553 to his mechanic SSgt Lauri Manelius at Lappeenranta on 23 September 1941. C/n 13663 on the fin tells us this a Hawk 75A-6. (SA-kuva)

1/LLv 32 leader Berg with his dogs Hilu and Heku at Lappeenranta in September 1941. The planes from right are CUw-562 and 563. The small w stood for Twin Wasp engine, painted only on the planes.

No LLv 32 Hawks were assigned to any pilot, thus CUw-551 was one of many flown by Berg. It is seen here at Lappeenranta in September 1941. Nine different pilots claimed a total of 10½ victories flying it.

CUw-570 was preferred by Berg, without the tactical tail number. It is seen here at Nummela shortly before Berg was shot down and killed in it on 1 November 1941.

BREMER, Aulis Nathanael

7½ victories (5½ + 2 + 3) on appr. 250 missions
Planes: FR-160, CUw-556 and CU-559

Aulis Bremer was born at Kuopio on 07.05.11. He was accepted on 06.06.32 to KadK and was posted on 16.05.35 as 2nd Lieutenant to LLv 12. Bremer was appointed on 10.10.39 as 1st Lieutenant 3/LLv 12 leader. On 23.02.40 Bremer was transferred to lead 1/LLv 22. On 29.03.40 He was transferred to LLv 32 and appointed on 18.06.41 to head 3/LLv 32 and on 04.08.41 came the promotion to Captain. On 01.11.41 he took over the 2nd Flight. Bremer was posted on 28.09.43 to T-LeLv 35 and further on 04.07.44 to lead 1/HLeLv 26. After the war he served with LeR 2, LeR 1, 3. Lsto and HämLsto resigning on 05.05.59 as Lieutenant Colonel. Bremer died on 30.04.87. He was awarded with VR 4, VR 3, VR 3 tlk and VR 2.

Brewster BW-352 of 3/LLv 22 leader 1Lt Aulis Bremer as seen at Hollola in March 1940. The flight was disbanded by the end of the month and the Brewsters given to LLv 24.

Fokker D.XXI, FR-160 of 3/LLv 32 flown by the flight leader 1Lt Aulis Bremer in June 1941, based at Vesivehmaa. (Karolina Hołda)

FR-160 of the re-established 3/LLv 32 seen at Vesivehmaa on 29 June 1941. It was flown by the flight leader Bremer.

Fokker D.XXI, FR-113 of 2/LLv 32 flown by 1Lt Aulis Bremer in June 1941, based at Hyvinkää. (Karolina Hołda)

FR-113 of 2/LLv 32, lightly camouflaged, at Hyvinkää on 25 June 1941. It was piloted by the flight deputy leader 1Lt Bremer.

Date	Time	Area	Unit	Aircraft	Enemy aircraft	Enemy Unit	Victory type
21.08.41	14.50–15.05	Pölläkkälä	32	CU-563	I-153		T
23.08.41	09.45–10.30	Muolaanjärvi	32	CU-563	2 x I-153		T
27.03.42	15.55–17.55	Suursaari	32	CU-556	½ I-153	11 AP, KBF	T
07.04.42	15.20–16.00	Lumisuo	32	CU-556	I-16		T
07.04.42	15.20–16.00	Lumisuo	32	CU-556	I-16		E
06.07.42	20.30–21.35	Ladoga	32	CU-552	Pe-2		V
07.07.42	15.40–17.20	Lotinanpelto	32	CU-552	LaGG-3		T
05.09.42	11.35–13.15	Mergino	32	CU-564	LaGG-3		E
05.09.42	11.35–13.15	Mergino	32	CU-564	LaGG-3		V
23.10.42	14.05–15.10	Mantsinsaari	32	CU-559	Pe-2		V

Curtiss Hawk 75A-2, CUw-556 flown by 2/LeLv 32 leader Capt Aulis Bremer in June 1942, based at Nurmoila. (Karolina Hołda)

CUw-556 of LeLv 32 in front of the service hangar at Nurmoila in June 1942. Though not seen here, the wing swastikas were painted in error as mirror images.

From the left: 2/LeLv 32 leader Capt Aulis Bremer and WO Eino Koskinen at Nurmoila in front of a Curtiss Hawk on 9 July 1942, two days after a victorious combat. (SA-kuva)

6 victories (3 + 3 + 1) on 145 missions
Planes: IT-15 and MT-416

Matti Durchman was born at Nurmes on 06.08.19. He received flying training in 1940–42 at SOK 2 and was posted on 17.04.42 as Sergeant to LLv 6. On 16.11.42 Durchman was transferred to LeLv 30 and on 26.05.44 further to HLeLv 34. He was promoted on 01.09.44 to Staff Sergeant and demobilized on 11.11.44. Durchman acted in civilian life as a special dental technician. He died on 05.02.88. Durchman was awarded with VM 2, VM 1 and EK 2.

Polikarpov I-153, IT-15 of 3/LeLv 6 flown by Sgt Matti Durchman in October 1942, based at Römpötti. (Karolina Hołda)

The four captured I-153s of 3/LeLv 6 in a row at Römpötti on 30 October 1942. Sgt Durchman stands in front of the closest plane, which is his IT-15. The rest are IT-20, 19 and 18. (Finnish Air Force)

Date	Time	Area	Unit	Aircraft	Enemy aircraft	Enemy Unit	Victory type
14.06.44	13.35–14.45	Siesjärvi	34	MT-416	Airacobra		R
22.06.44	17.05–18.25	Säiniö	34	MT-445	La-5		T
23.06.44	12.15–13.25	Hevossaari	34	MT-434	Il-4		T
02.07.44	10.55–12.10	Viipurinlahti	34	MT-416	La-5	191 IAP	R
03.07.44	11.05–12.20	Viipuri	34	MT-416	Il-2		R
03.07.44	11.05–12.20	Viipuri	34	MT-416	Il-2		V
15.07.44	04.05–05.10	Rättijärvi	34	MT-406	La-5		R

EHRNROOTH, Erkki Olavi

5 victories (5 + 0 + 1) on appr. 80 missions
Planes: FA-1 and CUw-557

Olavi Ehrnrooth was born at Oulu on 04.07.07. He was accepted to KadK on 01.10.28 and he served as 1st Lieutenant at LAs 6 and ISK. Ehrnrooth was promoted on 06.12.36 to Captain and on 06.12.39 was transferred to lead KoeL. He was posted on 25.07.40 to command LLv 34 and was promoted on 11.02.41 to Major. On 12.07.41 Ehrnrooth was posted to command LLv 32 and on 19.01.43 further to command LeLv 34. He was killed in a flying accident on 27.03.43, when his PY-25 hit the ground during low-level manoeuvres at Utti. Ehrnrooth was awarded with VR 4, VR 3, VR 2, EK 2 and EK 1.

FIAT G.50, SA-1 of KoeL flown by the flight leader Capt Olavi Ehrnrooth in January 1940, based at Tampere. (Janusz Światłoń)

The first FIAT, SA-1, was tested by KoeL (Test Flight) at Tampere during January and February 1940, and also used for the State Aircraft Factory defence. Ehrnrooth claimed two kills flying it on intercept missions. (State Aircraft Factory)

Date	Time	Area	Unit	Aircraft	Enemy aircraft	Enemy Unit	Victory type
13.01.40		Tuulos	K	FA-1	SB		T
29.01.40		Lempäälä	K	FA-1	DB-3		T
17.02.40		Våtskär	K	FA-1	SB		V
02.03.40	12.45	Nokia	K	FR-101	DB-3		T
16.07.41	14.45–14.50	Nurmi station	32	CU-556	½ I-153		T
21.07.41	10.05–10.10	Kärkijärvi	32	CU-557	I-153	155 IAP	T
28.03.42	05.30–07.30	Suursaari	32	CU-571	½ I-153		T

LLv 32 commander Maj Ehrnrooth in the cockpit of a Curtiss Hawk at Suulajärvi, during the re-capture of the island of Suursaari in the Gulf of Finland.

Ehrnrooth about to take off from Suulajärvi in CUw-563 for a reconnaissance mission to Suursaari on 22 March 1942. A week later his fighters flew 60 missions supporting the invasion of Suursaari.

On 19 January 1943 Ehrnrooth took over the command of the new LeLv 34. In this post his assigned plane was Bf 109G-2 serial MT-201, seen here at Helsinki Malmi in March 1943.

ERKINHEIMO, Niilo Johannes

10¾ victories (8¾ + 2 + 3) on appr. 180 missions
Plane: CUw-558

Niilo Erkinheimo was born at Kortesjärvi on 28.05.20. He received flying training in 1940–41 at SOK 3 and was posted on 02.03.41 as Private to LLv 32. Erkinheimo was promoted to Corporal on 12.07.41, Sergeant on 26.09.41 and Staff Sergeant on 01.02.43. He was transferred on 25.03.43 to LeLv 34. Erkinheimo drowned in a flying accident on 16.11.43, when his MT-223 caught fire mid-air and went in the sea near Tytärsaari. Erkinheimo was awarded with VM 2, VM 1 and VR 4.

Though LLv 32 did not assign planes to pilots, Erkinheimo preferred to fly CUw-558, which he used in making several claims. It is seen here at Lappeenranta in August 1941.

1/LeLv 32 pilot Erkinheimo sitting in the cockpit of a Curtiss Hawk at Nurmoila. He became an ace on 20 July 1942.

Date	Time	Area	Unit	Aircraft	Enemy aircraft	Enemy Unit	Victory type
05.11.41	13.45–15.00	Hanko	32	CU-568	¼ I-16		T
28.03.42	05.30–07.20	Suursaari	32	CU-551	½ I-16	11 AP, KBF	T
03.04.42	18.00–19.10	Seiskari	32	CU-551	I-153		T
07.06.42	09.45–10.55	Aunusjoki 10 km SW	32	CU-558	MiG-3		T
28.06.42	05.50–06.50	Mouth of River Svir	32	CU-558	½ Pe-2		T
04.07.42	18.10–19.00	Svir power plant	32	CU-558	Pe-2		V
05.07.42	16.30–17.00	Mäkriä lake	32	CU-558	Pe-2	119 ORAE	T
05.07.42	16.30–17.00	Mäkriä lake	32	CU-558	Pe-2		V
20.07.42	12.45–14.35	Krestnojärvi	32	CU-558	LaGG-3		T
21.07.42	08.55–09.40	Nulbenskoje lake	32	CU-558	½ LaGG-3		T
25.07.42	21.20–22.10	Aunus – Nurmoila	32	CU-558	MBR	58 AE KBF	T
05.05.43	16.30–17.20	Lavansaari	34	MT-203	Boston	15 ORAP, KBF	T
05.05.43	16.30–17.20	Lavansaari 10 km N	34	MT-203	La-5	4 GIAP, KBF	T
16.05.43	18.10–19.05	Seiskari	34	MT-203	LaGG-3		E
22.05.43	10.25–11.00	Lavansaari	34	MT-230	LaGG-3		E
28.05.43	04.10–04.50	Selänpää – Immola	34	MT-201	DB-3F		V

CUw-558 of LeLv 32 poses for the air force photographer at Nurmoila on 2 August 1942. Ten different pilots scored 17 air victories in it, five of them by Erkinheimo. (Finnish Air Force)

Curtiss Hawk 75A-6, CUw-558 flown by 1/LeLv 32 pilot Sgt Niilo Erkinheimo in August 1942, based at Nurmoila. (Karolina Hołda)

Bf 109G-2 serial MT-203 was assigned to 2/LeLv 34 leader Capt Kullervo Lahtela, but Erkinheimo was the secondary pilot. It is seen here at Utti in April 1943, with the squadron CO Maj Eino Luukkanen standing in front of it.

EVINEN, Veikko Arvid

6 victories (6 + 0 + 0) on appr. 150 missions
Planes: FR-114, MT-203 and CU-581

Veikko Evinen was born at Helsinki on 27.08.18. He received flying training in 1938–39 at IRUK 8 and was accepted to KadK on 27.05.39. Evinen was posted on 27.03.40 as 2nd Lieutenant to LLv 32 and was promoted to 1st Lieutenant on 23.01.41. He was transferred on 28.11.41 to T-LLv 35 and on 28.05.42 back to LeLv 32. Evinen was transferred on 26.03.43 to LeLv 34 and on 20.02.44 back to HLeLv 32 to lead the 3rd Flight. He was promoted to Captain on 24.05.44. He died of his wounds on 25.06.44, after the flak had shot down his CU-581 the previous day at Lake Ladoga near Vitele. Evinen was awarded with VR 4 and VR 3.

Fokker D.XXI, FR-114 of 1/LLv 32 flown by 1Lt Veikko Evinen in July 1941, based at Utti.
(Karolina Hołda)

FR-114 of 1/LLv 32 parked at Utti in early July 1941. It was the assigned plane of the flight deputy leader 1Lt Evinen. Evinen scored the unit's first two victories on 25 June 1941, the opening day of the Continuation War.

Curtiss Hawk 75A-2, CU-581 flown by 3/HLeLv 32 leader Capt Veikko Evinen in May 1944, based at Nurmoila. (Karolina Hołda)

Date	Time	Area	Unit	Aircraft	Enemy aircraft	Enemy Unit	Victory type
25.06.41	08.00–08.10	Malmi - Helsinki	32	FR-116	2 x SB	4 SAD	T
04.07.41	09.10–09.25	Hamina	32	FR-114	½ Pe-2	58 SBAP	T
10.08.41	17.40–17.45	Kirvu	32	CU-553	½ I-16		T
03.09.41	10.55–11.10	Suuri-Harvasuo	32	CU-560	I-153	7 IAP	T
02.05.43	10.00–11.00	Someri 15 km S	34	MT-203	La-5		T
26.07.43	10.15–11.15	Haapasaari – Lavansaari	34	MT-225	Il-2	7 GShAP, KBF	V

CU-581 of LeLv 32 taxies at Nurmoila on 26 October 1943, on this occasion piloted by SSgt Väinö Virtanen. The next fighter is CU-577. Evinen was shot down in CU-581 on 25 June 1944 on the east coast of Lake Ladoga. He died the next day of his wounds. (SA-kuva)

Fräntilä, Mauno Mikael

5½ victories (4½ + 1 + 2) on 380 missions
Planes: FR-76, CU-572 and MT-410

Mauno Fräntilä was born at Kauhava on 30.04.17. He received the flying training in 1936–37 at RAOK 3 and AOK 8. Fräntilä was posted on 01.03.39 as Corporal to LLv 24, where promoted on 06.12.39 to Sergeant and on 12.03.40 to Staff Sergeant. On 31.08.40 he was transferred to LLv 32 and promoted on 26.08.41 to Master Sergeant. On 09.02.43 Fräntilä was posted to LeLv 34 and on 06.03.44 to HLeLv 30. He became a Warrant Officer on 02.06.44 and was transferred on 15.06.44 to HLeLv 34 and on 05.08.44 back to HLeLv 30. After the war Fräntilä served as a tutor in LeSK and resigned on 12.02.56. He became later a farmer. Fräntilä died on 08.07.09. He was awarded with VM 1, VR 4 and VR 4 tlk.

2/LLv 32 pilot SSgt Fräntilä boarding CUw-553 at Lappeenranta in mid-August 1941. The squadron flew top cover for the troops taking back the Karelian Isthmus. (SA-kuva)

Three aces of 3/HLeLv 34 at Taipalsaari in early August 1944. From left WO Oiva Tuominen, WO Mauno Fräntilä and SSgt Klaus Alakoski.

Fräntilä's last assigned plane was Mersu MT-487 of 2/HLeLv 30, seen here at Kymi shortly after the war in September 1944.

Date	Time	Area	Unit	Aircraft	Enemy aircraft	Enemy Unit	Victory type
17.01.40	14.00	Muolaanjärvi	24	FR-90	SB	54 SBAP	T
04.02.40	14.20	Skinnarvik	24	FR-117	SB		V
29.02.40	08.40	Ruokolahti – Valkjärvi	24	FR-76	I-16		V
28.03.42	17.40–18.00	Lavansaari – Suursaari	32	CU-572	I-16	71 IAP, KBF	T
28.03.42	17.40–18.00	Lavansaari – Suursaari	32	CU-572	½ I-153		T
02.06.43	14.15–15.15	Seiskari	34	MT-217	LaGG-3		E
20.06.44	07.10–08.05	Ristiniemi	34	MT-410	Il-2	35 ShAP, KBF	T
30.06.44	10.45–12.00	Tammisuo	34	MT-422	Yak-9	404 IAP	T

2/LeLv 34 pilot Fräntilä sitting on the cowling of MT-225 at Helsinki Malmi in August 1943. The plane was assigned to the flight leader Capt Lahtela.

MT-217 of 1/LeLv 34 pilot 1Lt Väinö Pokela at Utti on 2 June 1943. On this particular day Fräntilä claimed a LaGG-3 shot down flying this plane.

Gerdt, Aimo Emil

6 victories (6 + 0 + 5) on appr. 250 missions
Planes: CUw-568 and CU-503

Aimo Gerdt was born at Heinävesi on 23.11.19. He received flying training in 1939–40 at RAOK 6 and was posted on 30.09.40 as Corporal to LLv 32, where he was promoted on 03.08.41 to Sergeant and on 01.08.42 to Staff Sergeant. On 26.03.43 Gerdt was transferred to LeLv 34 and on 06.03.44 further to HLeLv 30. He was promoted on 21.09.44 to Master Sergeant and demobilized on 14.11.44. Gerdt recruited on 01.06.45 and became an instructor at LeSK, retiring on 23.11.59 as Warrant Officer. Gerdt died on 21.04.75. He was awarded with VM 2, VM 1 and VR 4.

Date	Time	Area	Unit	Aircraft	Enemy aircraft	Enemy Unit	Victory type
22.08.41	19.35–19.45	Yskjärvi	32	CU-565	I-153		T
23.12.41	13.20–15.10	Ohalatva	32	CU-568	⅓ R-5		T
08.01.42	14.10–15.30	Lumisuo	32	CU-566	½ I-15bis		T
28.03.42	17.40–18.00	Lavansaari – Suursaari	32	CU-551	I-153		T
18.06.42	13.20–14.35	Uuksu – Valamo	32	CU-568	Pe-2		V
23.06.42	01.05–01.45	Lotinanpelto	32	CU-552	U-2	716 AP	T
24.06.42	13.20–14.10	Ladoga – Pisi	32	CU-503	Pe-2		V
07.11.42	12.15–14.00	Ljugovitsa	32	CU-503	2 x MiG-3		T
27.07.43	05.25–06.35	Lavansaari	34	MT-206	LaGG-3		V
02.08.43	05.20–06.20	Seiskari – Tolli	34	MT-204	2 x La-5		V

Curtiss Hawk 75A-3, CU-503 flown by 3/LeLv 32 pilot Sgt Aimo Gerdt in June 1942, based at Nurmoila. (Karolina Hołda)

3/LeLv 32 pilot Gerdt flew CU-503 on many occasions, claiming two kills. It is seen here at Nurmoila in June 1942. Eight different pilots scored 9¾ air victories in CU-503.

HALONEN, Eero Martti Olavi

17 victories (14 + 3 + 3) on 81 missions
Planes: MT-241 and MT-202

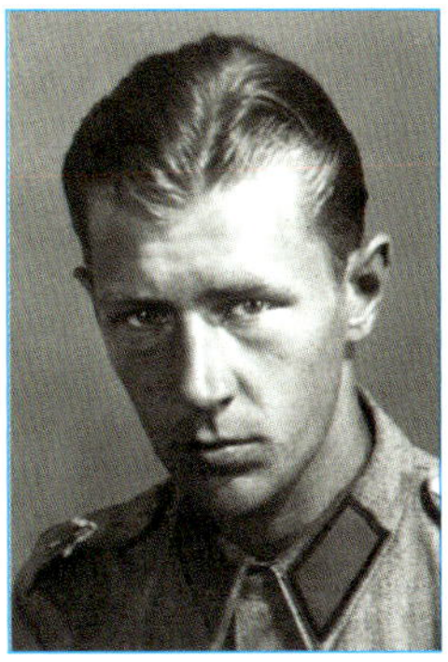

Eero Halonen was born at Kuolemanjärvi on 12.08.19. He received flying training in 1940–42 at AOK 11 and was posted as Corporal on 19.03.43 to LeLv 24. Halonen was promoted on 08.04.43 to Sergeant and on 26.10.44 to Staff Sergeant and was demobilized on 10.11.44. In civilian life Halonen was a merchant in Sweden. He was awarded with VM 2, VM 1 and VR 4.

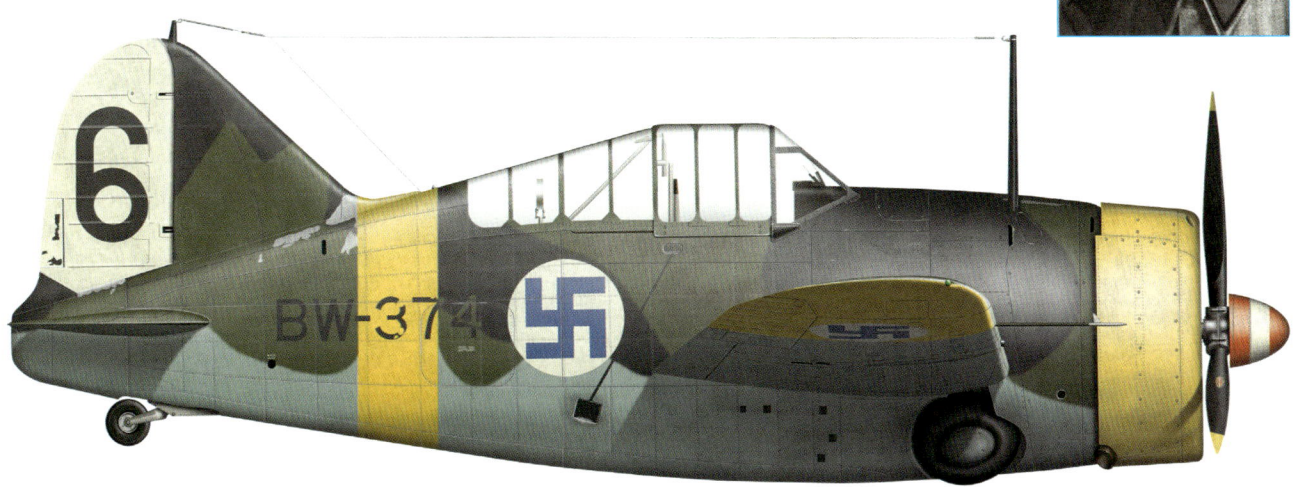

Brewster 239, BW-374 of 2/HLeLv 24 flown by Sgt Eero Halonen in April 1944, based at Suulajärvi. (Andrzej M. Olejniczak))

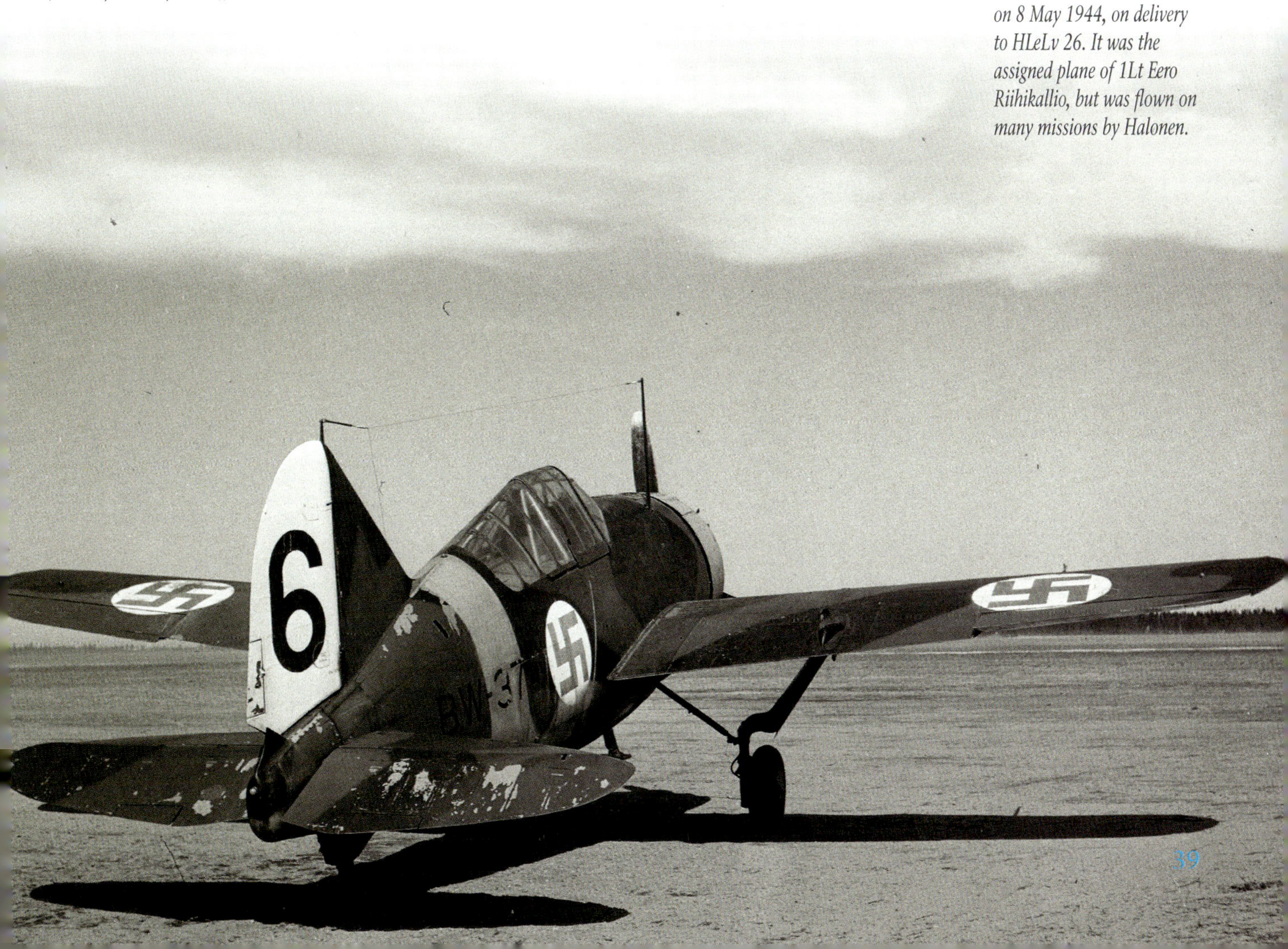

Brewster BW-374 of 2/HLeLv 24 at Suulajärvi on 8 May 1944, on delivery to HLeLv 26. It was the assigned plane of 1Lt Eero Riihikallio, but was flown on many missions by Halonen.

Brewster BW-374 of 2/HLeLv 24 seen from another angle at Suulajärvi on 8 May 1944. (SA-kuva)

From left 1Lt Atte Nyman and Sgt Eero Halonen of 2/HLeLv 24 standing by the wreck of 1Lt Jorma Saarinen's Mersu MT-478 at Antrea on 18 July 1944. It was the last operational loss of HLeLv 24. (SA-kuva)

Date	MT-437	Area	Unit	Aircraft	Enemy aircraft	Enemy Unit	Victory type
31.08.43	15.20–15.50	Saarenpää	24	BW-384	Il-2	7 GShAP, KBF	V
09.06.44	16.50–17.30	Rajajoki – Valkeasaari	24	MT-206	2 x La-5		T
13.06.44	09.00–10.00	Suulajärvi	24	MT-213	La-5		V
14.06.44	08.50–09.50	Siesjärvi	24	MT-227	Airacobra		R
20.06.44	08.30–09.40	Vatnuori – Koivisto	24	MT-241	Yak-1		T
20.06.44	08.30–09.40	Vatnuori – Koivisto	24	MT-241	LaGG-3		R
20.06.44	08.30–09.40	Vatnuori – Koivisto	24	MT-241	Yak-9		V
20.06.44	19.15–20.15	Muolaanjärvi	24	MT-241	La-5		T
20.06.44	19.15–20.15	Muolaanjärvi	24	MT-241	Pe-2		T
21.06.44	14.50–16.10	Pero station	24	MT-241	Airacobra		T
23.06.44	07.30–08.05	Pero station	24	MT-213	La-5		T
26.06.44	10.30–11.05	Tali – Tammisuo	24	MT-449	2 x Il-2		T
29.06.44	16.20–17.30	Sorvali	24	MT-463	Yak-9	401 IAP	T
01.07.44	04.10–05.20	Teikarsaari	24	MT-450	La-5		R
05.07.44	12.20–13.25	Tiurinsaari	24	MT-202	LaGG-3		T
09.07.44	14.00–15.20	Äyräpää	24	MT-202	Yak-9		T
15.07.44	08.50–10.10	Äyräpää	24	MT-202	Il-2		T
15.07.44	08.50–10.10	Äyräpää	24	MT-202	La-5		V

6 victories (4 + 2 + 3) on appr. 80 missions
Planes: MS-305 and MSv-617

Lars Hattinen was born at Kuorevesi on 16.05.23. He received flying training in 1940–42 at AOK 11 and was posted 03.04.43 as Corporal to LeLv 28. Hattinen was promoted on 20.08.43 to Sergeant on and on 13.07.44 to Staff Sergeant. He was demobilized on 31.12.44. In civilian life Hattinen became an airline captain. He was killed on 03.01.61 in Koivulahti flying accident. Hattinen was awarded with VM 1.

Mörkö-Morane MSv-631 of 1/HLeLv 28 flown by SSgt Lars Hattinen in July 1944, based at Värtsilä. (Karolina Hołda)

Date	Time	Area	Unit	Aircraft	Enemy aircraft	Enemy Unit	Victory type
23.06.44	12.20–13.05	Latva	28	MS-318	Pe-2		V
25.06.44	12.50–13.35	Äänislinna	28	MS-308	Boston		T
26.06.44	19.30–20.20	Aitalampi	28	MS-639	Yak-9	197 IAP	T
08.07.44	11.45–12.45	Suojärvi	28	MS-305	Airacobra		V
12.07.44	18.30–19.55	Kitelä	28	MS-305	La-5		T
16.07.44	18.00–19.05	Urusjärvi	28	MSv-631	La-5	760 IAP	R
27.07.44	21.00–22.45	Tolvajärvi	28	MS-644	Airacobra		V
30.07.44	09.05–10.15	Tolvajärvi	28	MSv-617	Airacobra	773 IAP	T
30.07.44	09.05–10.15	Tolvajävi	28	MSv-617	Airacobra		R

This MSv-631 was the first Mörkö-Morane to arrive with HLeLv 28 and was assigned to 2Lt Kalle Kyllönen. It was flown by Hattinen to claim an La-5 as the first kill of the type on 16 July 1944. Hattinen achieved all his six victories in summer 1944 battles.

HILLO, Jaakko Juho

8⅓ victories (7⅓ + 1 + 0) on 220 missions
Planes: CUw-564 and CU-580

Jaakko Hillo was born at Helsinki on 09.07.21. He received flying training in 1939–40 at SOK 1 and was posted on 27.03.41 to LLv 32, where promoted to 2nd Lieutenant five days later. On 04.06.41 Hillo was accepted to KadK, but in the Continuation War mobilization was transferred on 20.06.41 to LLv 32. On 01.07.42 he continued as cadet at MaaSK and was promoted to 1st Lieutenant on 01.12.42. He was posted on 04.06.43 to LeLv 32. After the war Hillo served with LeR 1 and LeV as a test pilot until resigning on 10.05.47. In civilian life he became an airline captain and held the rank of Captain. Hillo died on 22.02.19. He was awarded with VR 4, VR 4 tlk and VR 3.

1/LLv 32 pilot 2Lt Hillo scored his first air victory in this Curtiss CUw-565. It is seen here at Immola, fresh from an overhaul and departing for Suulajärvi on 19 March 1942.

1/LLv 32 pilots and future aces at Lappeenranta on 23 September 1941. From left Sgt Mauno Kirjonen, flight leader Capt Paavo Berg, Sgt Yrjö Pallasvuo and 2Lt Jaakko Hillo. (SA-kuva)

Date	Time	Area	Unit	Aircraft	Enemy aircraft	Enemy Unit	Victory type
18.08.41	14.10–14.20	Sintola	32	CU-565	I-153	7 IAP	T
22.08.41	19.35–19.45	Yskjärvi	32	CU-564	I-153		T
03.09.41	10.55–11.10	Suuri-Harvasuo	32	CU-558	I-153	7 IAP	E
28.03.42	05.30–07.30	Suursaari	32	CU-553	I-16		E
07.06.42	09.45–10.55	Aunusjoki 10 km SW	32	CU-503	MiG-3	524 IAP	T
28.05.44	09.50–10.20	Sotkusha	32	CU-584	La-5		T
12.06.44	15.00–17.00	Tenenitsi	32	CU-505	U-2		T
15.06.44	12.45–13.15	Svir power plant	32	CU-584	⅓ Il-2	957 ShAP	V
22.06.44	13.50–15.40	Kuittinen	32	CU-580	½ Il-2	257 ShAD	T
01.07.44	20.20–21.15	Vieljärvi	32	CU-578	½ Il-2	839 ShAP	T

Curtiss Hawk 75A-2, CU-580 flown by 1/LeLv 32 pilot 1Lt Jaakko Hillo in October 1943, based at Nurmoila. (Karolina Hołda)

Hillo piloting LeLv 32 plane CU-580 over the River Svir on 16 October 1943, leading a top cover division to a Fokker C.X on a reconnaissance and photography mission. (SA-kuva)

Huhanantti, Tatu Mauri

7 victories (7 + 0 + 2) on appr. 100 missions
Planes: FR-94

Tatu Huhanantti was born at Oulu on 23.06.14. He received flying training in 1933–34 at IRUK 3 and was accepted to KadK on 06.06.34 and was posted on 16.05.38 as 2nd Lieutenant to LLv 24. On 20.02.37 followed a tour at IlmK and return to LLv 24 on 01.06.38 as 1st Lieutenant. Huhanantti was killed on 29.02.40, when a I-16 of 68 IAP shot down his FR-94 at Ruokolahti. Huhanantti was awarded with VR 4.

Fokker D.XXI serial FR-76 of LLv 24 in front of a Utti hangar on 16 December 1937. It was armed with two 20 mm Oerlikon cannons and was also thus equipped two years later, when 3/LLv 24 pilot Huhanantti claimed air victories with it. (Finnish Air Force)

Date	Time	Area	Unit	Aircraft	Enemy aircraft	Enemy Unit	Victory type
20.12.39	14.10–15.00	Lempaala	24	FR-76	SB	24 SBAP	T
23.12.39	10.15	Leipäsuo	24	FR-76	½ R-5	16 KAO	T
20.01.40	15.00	Hämeenlinna – Riihimäki	24	FR-91	2 x SB	35 SBAP	T
02.02.40	13.15	Virtasalmi	24	FR-94	SB	24 SBAP	T
02.02.40	13.15	Pieksämäki	24	FR-94	SB		V
13.02.40		Summa	24	FR-94	I-15bis		V
15.02.40	11.00	Räisälä	24	FR-94	SB	10 SBAP	E
21.02.40	09.30	Antrea	24	FR-94	I-15bis		T
21.02.40	12.00	Simola	24	FR-94	½ SB	54 SBAP	T

17½ victories (16½ + 1 + 2) on 291 missions
Planes: BW-353 and MT-440

Jussi Huotari was born at Nuijamaa on 23.11.18. He received flying training in 1939–40 at RAOK 6 and was posted on 01.11.40 as Corporal to LLv 24. Huotari was promoted to Sergeant on 23.07.41, Staff Sergeant on 12.04.42 and Master Sergeant on 22.10.42. After the war he served with LeR 3/3. Lsto/KarLsto, resigning on 15.11.58 as a Warrant Officer. Thereafter Huotari worked at the railways as an engine warmer. He died on 13.08.04. Huotari was awarded with VM 2, VM 1, VR 4, VR 4 tlk and VR 3.

Brewster 239, BW-353 of 3/LLv 24 flown by Sgt Jouko Huotari in September 1941, based at Mantsi. (Andrzej M. Olejniczak)

BW-353 of 3/LLv 24 was assigned to Huotari on 30 June 1941. It was equipped with retractable skis on 30 December 1941 and is seen here on Kontupohja harbour ice.

Huotari's BW-353 parked on the dry beach of Lake Ladoga at Mantsi in late September 1941. He served with 3/LLv 24 throughout the Continuation War.

3/HLeLv 24 pilots at Lappeenranta on 10 July 1944. From left SSgt Emil Vesa, squadron CO Maj Jorma Karhunen, SSgt Leo Ahokas and MSgt Jouko Huotari. (SA-kuva)

Huotari entering the cockpit of his Mersu MT-240 at Suulajärvi on 8 May 1944, about to take off for an intercept mission over the Karelian Isthmus. (SA-kuva)

Date	Time	Area	Unit	Aircraft	Enemy aircraft	Enemy Unit	Victory type
09.07.41	05.10–05.20	Huuhanmäki	24	BW-355	I-153		V
19.07.41	13.05–13.10	Käkisalmi	24	BW-353	I-153		T
12.08.41	13.00–13.30	Antrea – Kirvu	24	BW-353	I-153	65 ShAP	T
23.09.41	14.00–14.15	Petäjäselkä	24	BW-353	I-16	155 IAP	T
26.02.42	08.45–11.00	Liistepohja	24	BW-387	MiG-3		V
09.03.42	14.20–14.50	Uikujärvi	24	BW-380	MiG-3		T
28.03.42	07.45–08.15	Lavansaari	24	BW-353	2 x I-153	11 AP, KBF	T
14.08.42	11.00–11.10	Krasnaja Gorka	24	BW-353	Hurricane		T
16.08.42	17.45–18.00	Seiskari – Karavaldai	24	BW-353	2 x I-16		T
28.09.43	17.15–18.30	Sepeleva 10 km NW	24	BW-355	½ LaGG-3		T
04.06.44	09.20–10.00	Revonnenä	24	MT-247	La-5		T
06.06.44	08.15–09.10	Varsolovo	24	MT-237	Airacobra		T
21.06.44	15.00–16.20	Viipuri	24	MT-440	Airacobra		T
28.06.44	10.25–11.15	Lyykylänjärvi	24	MT-440	Yak-9		R
28.06.44	14.00–14.55	Mannikkala	24	MT-440	Il-2		T
30.06.44	10.50–12.00	Tammisuo	24	MT-440	Il-4	113 BAD	T
03.07.44	11.50–12.40	Kärstilä	24	MT-440	La-5	10 GIAP	T
09.07.44	18.55–19.50	Äyräpää	24	MT-441	La-5		T

HYRKKI, Tuomo Uuno Martti

5 victories (5 + 0 + 2) on appr. 150 missions
Planes: MS-301 and MS-657

Tuomo Hyrkki was born at Virrat on 10.10.12. He received flying training in 1935–36 at IRUK 3 and was accepted to KadK on 03.06.36. Hyrkki was posted on 16.05.38 as 2nd Lieutenant to LLv 10. On 15.12.39 he was transferred to LLv 28. Promotion to 1st Lieutenant followed on 31.01.40 and to Captain on 22.05.42. Hyrkki was appointed to lead 3/LeLv 28 on 26.11.42 and became acting commander of HLeLv 28 on 17.05.44. He was transferred on 13.06.44 as a liaison officer to II AKE. After the war he served with LeR 1 and IlmavE. Hyrkki was killed in a flying accident on 11.12.45, when his FK-99 hit a hill in a snow fall at Pirkkala. Hyrkki was awarded with VR 4 and VR 3.

Morane MS 319 in early 1941 at Naarajärvi, assigned to Hyrkki who continued to fly it with 1/LLv 28 until it was damaged in an air raid on 5 September 1941.

Morane MS-607 of 1/LLv 28 was then assigned to Hyrkki. It is seen here taking off at Viitana on 17 March 1942 with 1Lt Olli Puhakka at the controls. (SA-kuva)

Date	Time	Area	Unit	Aircraft	Enemy aircraft	Enemy Unit	Victory type
17.02.40	11.30	Utö	28	MS-301	DB-3		V
17.02.40	14.45	Korppoo	28	MS-301	DB-3	53 DBAP	T
02.09.41	13.45–15.20	Säämäjärvi	28	MS-607	I-16		T
04.12.41	13.00–14.45	Liistepohja	28	MS-618	I-153	197 IAP	T
08.06.43	11.50–13.05	Sumeri	28	MS-657	2 x I-16	197 IAP	R
07.03.44	11.05–11.20	Maaselkä	28	MS-657	Tomahawk		V

IKONEN, Heikki Sakari

6¾ victories (4¾ + 2 + 1) on 204 missions
Planes: FR-102 and BW-386

Sakke Ikonen was born at Mäntyharju on 17.10.16. He received flying training in 1937–38 at RAOK 3 and AOK 6 and was posted on 28.02.39 as Sergeant to LLv 24. He was promoted to Staff Sergeant on 31.12.39 and Master Sergeant on 25.01.40. Transfer as an instructor to LeSK came on 01.02.43 and further to T-LeLv 35 on 02.11.43. After the war Ikonen served with LeR 1 and 2. Lsto resigning on 15.05.55 as a Warrant Officer. In civilian life he was an office manager. Ikonen died on 18.03.12. He was awarded with VM 2, VR 4 and VR 4 tlk.

Fokker D.XXI serial FR-95 photographed at Siikakangas in June 1940 belonging to LLv 32. Though it was not Ikonen's assigned fighter in the Winter War, he flew it with 4/LLv 24 on several occasions.

Date	Time	Area	Unit	Aircraft	Enemy aircraft	Enemy Unit	Victory type
19.12.39	10.30	Kilpola lake	24	FR-102	SB	13 SBAP	T
25.12.39	12.00	Enso	24	FR-102	½ DB-3	6 DBAP	T
01.02.40	15.15–15.30	Viipuri	24	FR-102	I-16		E
09.02.40	13.30	Summa	24	FR-102	¼ R-5		T
30.06.41	09.15–10.45	Outside Porvoo	24	BW-386	MBR-2	58 OAE	T
17.09.41	16.20–16.40	Pyhäjärvi – Prääsä	24	BW-386	MiG-3		E
26.10.41	15.15	Monastirskaja	24	BW-386	MiG-3		V
12.08.42	17.20–19.30	Tolli lighthouse	24	BW-370	I-16		T
22.11.42	12.15–14.15	Sepeleva lighthouse	24	BW-386	Spitfire		T

BW-386 taking off at Immola on 2 September 1941. It was assigned to Ikonen for 18 months, suffering no damage during this time.

Ikonen sitting on the cockpit sill of his 4/LLv 24 Brewster BW-386 at Ranta-salmi in August 1941.

BW-370 was the assigned plane of the acting flight leader 1Lt Aulis Lumme of 4/LLv 24. Ikonen claimed one Polikarpov I-16 destroyed with it on 12 August 1942, when this photo was taken at Römpötti.

INEHMO, Martti Olavi Kalervo

8 victories (3 + 5 + 1) on 87 missions
Planes: MS-602 and MS-318

Martti Inehmo was born at Rauma on 12.08.14. He received flying training in 1933–34 at IRUK 3. In the Winter War mobilization Inehmo was posted on 16.10.39 as 2nd Lieutenant to T-LentoR 2. He was transferred on 15.02.40 to LLv 28 and demobilized on 26.06.40. In the Continuation War mobilization Inehmo was posted on 18.06.41 to LLv 28. Inehmo posted missing in action on 26.12.41, when his MS-618 did not return from Liistepohja direction. Inehmo was awarded with VR 4 and VR 4 tlk.

Inehmo was assigned this 2/LLv 28 Morane serial MS-602 on 18 June 1941 and flew it for seven weeks. It is seen here undergoing gun calibration at Joensuu in early July 1941.

Date	Time	Area	Unit	Aircraft	Enemy aircraft	Enemy Unit	Victory type
11.03.40	15.15	Loviisa	28	MS-304	DB-3	7 DBAP	T
09.07.41	14.40–15.10	Konnitsa – Ladoga	28	MS-602	MiG-3		E
23.07.41	19.15–19.30	Nuosjärvi	28	MS-602	SB		V
09.09.41	10.00–10.45	Svir	28	MS-613	2 x I-16		T
09.09.41	10.00–10.45	Svir	28	MS-613	I-153		E
09.10.41	11.15–11.35	Suopohja – Mäkräjärvi	28	MS-327	3 x MiG-3		E

Inehmo's next assigned plane in 2/LLv 28 was MS318, which is running here at Viitana in November 1941. The leading edge of the fin carries all his eight victory bars. The serial of the Moranes (MS301–MS330) arrving in 1940 had no dash printed on the plane.

25½ victories (23½ + 2 + 2) on 247 missions
Planes: BW-377, MT-206, MT-450 and MT-463

Tappi Järvi was born at Hämeenlinna on 24.01.19. He received flying training in 1940–41 at SOK 2 and in Continuation War mobilization on 18.06.41 was posted as Private to LLv 30. He became a Corporal on 03.07.41 and was transferred on 11.08.41 to LLv 24. Järvi was promoted to Sergeant on 30.10.41, Staff Sergeant on 16.11.42 and Master Sergeant on 16.07.44. He was demobilized on 10.11.44. In civilian life Järvi was a textile merchant and held the rank of Warrant Officer. He died on 20.08.82. Järvi was awarded VM2, VM 1, VR 4, VR 4 tlk and VR 3.

4/LeLv 24 Brewster BW-377 was assigned to Sgt Järvi. It is seen here taxiing at Römpötti on 4 October 1942, but this time with 1Lt Hans Wind at the controls. (SA-kuva)

BW-377 of 4/LeLv 24 in the blast pen at Hirvas in June 1942. During ten months Järvi scored 7½ air victories flying it.

Bf 109G-6/R6, MT-450 of 2/HLeLv 24 flown by SSgt Tapio Järvi in June 1944, based at Lappeenranta. (Karolina Hołda)

WNr 165278 arrived at 2/HLeLv 24 at Lappeenranta on 22 June 1944. It was assigned to Järvi who flew it on a few missions before the German markings were painted over and this became MT-450.

2/HLeLv 24 pilots check their maps at Lappeenranta on 30 June 1944. From left deputy flight leader 1Lt Aulis Lumme, Sgt Kosti Kaloinen, 2Lt Johannes Brotherus and SSgt Tapio Järvi. (SA-kuva)

Järvi closing the canopy of his 2/HLeLv 24 Mersu MT-246 prior to an escort mission on 12 May 1944 from Suulajärvi. In addition to Puro and Saarinen he became one of the top scorers of his flight. (SA-kuva)

Date	Time	Area	Unit	Aircraft	Enemy aircraft	Enemy Unit	Victory type
07.10.41	09.30–11.15	Suopohja	24	BW-393	I-153		T
26.10.41	15.15	Monastirskaja	24	BW-383	MiG-3		E
06.08.42		Seiskari	24	BW-377	I-16		T
22.11.42	09.15–09.40	Kreivinlahti	24	BW-377	Spitfire		T
10.03.43	15.15–16.25	Saeiskari	24	BW-377	LaGG-3		T
18.04.43	17.00–18.20	Seiskari – Yhinmäki	24	BW-377	2 x LaGG-3		T
21.04.43	08.00–09.00	Seiskari 25 km SW	24	BW-377	Yak-1		T
02.05.43	10.00–11.30	Kronstadt	24	BW-377	½ LaGG-3		T
05.06.43	14.30–15.50	Kronstadt	24	BW-377	Yak-1		T
05.06.43	14.30–15,50	Kronstadt	24	BW-377	Il-2		V
07.06.44	11.10–11.55	Siestarjärvi	24	MT-206	Airacobra	196 IAP	T
10.06.44	06.25–07.40	Retunkylä	24	MT-206	Pe-2	140 BAP	T
10.06.44	09.55–11.00	Kivennapa	24	MT-206	2 x Pe 2		T
13.06.44	09.00–10.00	Halolanjärvi	24	MT-206	Pe-2		T
23.06.44	07.30–08.10	Pero - Mäkärlahti	24	MT-450	2 x Il-2	872 ShAP	T
26.06.44	10.30–11.10	Lyykylänjärvi	24	MT-450	2 x Il-2		T
28.06.44	09.00–09.40	Tali	24	MT-450	Il-2		T
28.06.44	11.00–11.45	Lyykylänjärvi	24	MT-450	2 x Il-2		T
29.06.44	18.30–19.35	Kärstilänjärvi	24	MT-452	Il-2	15 GShAP	T
02.07.44	19.30–20.45	Suomenvesi	24	MT-463	Il-2	703 ShAP	T
03.07.44	19.40–20.10	Noskuanselkä	24	MT-463	Pe-2	34 GBAP	T
05.07.44	09.45–10.55	Vatnuori	24	MT-470	Yak-9		R
05.07.44	09.45–10.55	Vatnuori	24	MT-470	Yak-9		V

Joensuu, Antti Ilmari

5 victories (5 + 0 + 1) on appr. 100 missions
Planes: GL-256 and FA-35

Ilmari Joensuu was born at Pori on 19.03.16. He received flying training in 1938–38 at RAOK 5 and was posted on 16.11.39 as Corporal to LLv 26. Joensuu was promoted to Sergeant on 23.03.40 and demobilized on 16.05.40. At the Continuation War mobilization Joensuu was posted on 19.06.41 to LLv 26. On 28.02.42 he started the RUK and was accepted as a cadet to MaaSK on 01.07.42. Joensuu was posted on 27.06.43 as 1st Lieutenant to LeLv 26 and was transferred on 27.06.44 to HLeLv 34. He resigned on 22.01.45 and acted in civilian life as an airline captain. Joensuu was killed on 08.11.63 in Maarianhamina flying accident. He was awarded with VM 1, EK 2 and VR 4.

Joensuu standing on front of 2/LLv 26 FIAT FA-31 at Joro-inen in late June 1941. He was a very tall pilot at 195 cm, giving him the nick-name "Long-Jim".

In the Winter War Joensuu flew this 2/LLv 26 Gladiator GL-256, claiming four air victories. It is serviced here at Konnunsuo with the next owner, 2/LLv 12, on 2 March 1940. (Finnish Air Force)

Date	Time	Area	Unit	Aircraft	Enemy aircraft	Enemy Unit	Victory type
12.02.40	11.45	Jänisjärvi	26	GL-256	SB	18 SBAP	T
13.02.40	14.40	Havuvaara	26	GL-256	I-15bis	49 IAP	T
20.02.40		Kurkijoki	26	GL-256	SB	54 SBAP	T
25.02.40	16.15	Muolaanjärvi	26	GL-256	R-5	4 LBAP	T
13.08.41	13.00–14.10	Tuulos-Aunus	26	FA-35	I-153	195 IAP	T
13.08.41	13.00–14.10	Tuulos-Aunus	26	FA-35	I-153		V

FIAT FA-35 of 2/LLv 26 was assigned to Sgt Aaltonen at the beginning of the Continuation War, but it made Joensuu an ace on 13 August 1941. It is here at Rantasalmi in late June 1941.

JUTILA, Lauri Olavi

7½ victories (7½ + 0 + 0) on appr. 150 missions
Planes: HC452, CUw-566 and MT-230

Lauri Jutila was born at Kokemäki on 14.12.13. He received flying training in 1938–39 at RAOK 5 and AOK 9 and was posted on 01.09.39 as Corporal to LeR 4. Jutila was demobilized on 30.05.40 and at the Continuation War mobilization was posted on 18.6.41 as Sergeant to LLv 30. He was transferred on 01.07.41 to LLv 32, where he was promoted to Staff Sergeant on 29.10.41 and Master Sergeant on 20.03.43. Jutila was transferred on 25.03.43 to LeLv 34. He was killed on 17.06.43, when an Il-4 (SrLt J.V. Belov) of 1 GMTAP, KBF shot down his MT-214 at Jääski. Jutila was awarded with VM 2, VM 1 and VR 4.

Curtiss Hawk 75A-1, CUw-566 flown by 1/LLv 32 pilot Sgt Lauri Jutila in September 1941, based at Lappeenranta. (Janusz Światłoń)

Hawker Hurricane I, HC452 flown by 1/LeLv 32 pilot SSgt Lauri Jutila in June 1942, based at Suulajärvi. (Karolina Hołda)

Jutila taxies HC452 of LeLv 32 at Vesivehmaa on 23 June 1942, on the way to Helsinki Malmi delivering the plane to LeLv 26.

1/LLv 32 pilot Jutila in front of Hurricane HC-451 at Suulajärvi in January 1942. Jutila was the only Hurricane pilot who later became an ace flying Curtiss Hawks and Messerschmitts.

Mechanic Paavo Korpinen stands in front of 2/LeLv 34 Mersu MT-214 at Utti shortly before 17 June 1943, when Jutila was shot down and killed in this plane by the return fire of his victim.

Date	Time	Area	Unit	Aircraft	Enemy aircraft	Enemy Unit	Victory type
28.03.42	17.40–18.00	Lavansaari – Suursaari	32	CU-566	2 x I-153		T
24.06.42	13.20–14.10	Ladoga – Pisi	32	CU-551	LaGG-3	524 IAP	T
30.06.42	11.30–12.05	Lotinanpelto	32	CU-503	LaGG-3		T
04.05.43	10.20–11.40	Lavansaari	34	MT-209	La-5	4 GIAP, KBF	T
04.05.43	10.20–11.40	Suursaari – Tytärsaari	34	MT-209	La-5	4 GIAP, KBF	T
23.05.43	15.30–16.20	Tytärsaari – Lavansaari	34	MT-230	½ I-16	71 KIAP, KBF	T
17.06.43	02.45–03.00	Jääski	34	MT-214	Il-4	1 GMTAP, KBF	T

JUUTILAINEN, Eino Ilmari

94 victories (79 +15 +10) on 437 missions
Planes: FR-106, BW-364, MT-222, MT-426 and MT-457

Ilmari Juutilainen was born at Lieksa on 21.02.14. He received flying training at SIPL and Karhumäki civil pilot courses and military pilot courses in 1935–36 at RAOK 2 and AOK 5. Juutilainen was posted on 16.02.37 as Corporal to LLv 12. He was transferred on 03.03.39 as Sergeant to LLv 24. On 31.12.39 Juutilainen was promoted to Staff Sergeant, 25 days later to Master Sergeant and on 11.03.41 to Warrant Officer. On 08.02.43 Juutilainen was transferred to LeLv 34. After the war he served with LeR 3 and on 16.05.47 resigned, but enlisted again on 08.09.48 for ten months to PLeLv 41 taking twin-engine courses. In civil life he made his living as a small plane pilot. Juutilainen died on 21.02.99. He was awarded with VM 1, VR 4, VR 4 tlk, VR 3, EK 2, VR 3 tlk and EK plus MHR twice, on 26.04.42 and on 28.06.44.

Fokker D.XXI serial FR-106 of LLv 32 on a visit to Utti in summer 1940. During the Winter War it was assigned to Juutilainen flying with 3/LLv 24. The rudders of several Fokkers were changed due to damage caused by the spring thaw take-offs and landings.

Juutilainen taxies his BW-364 of 3/LLv 24 on the ice of Kontupohja harbour. The last kill marking on the fin was added on 28 January 1942.

Juutilainen about to take-off in his BW-364 from Äänislinna to Kontupohja on 18 December 1941. There are twelve victory markings on the fin.

Brewster 239, BW-364 of 3/LLv 24 flown by WO Ilmari Juutilainen in February 1942, based at Kontupohja. (Karolina Hołda)

BW-364 of 3/LLv 24 ready for a mission at Kontupohja harbour in mid-February 1942. There are 16 black and white victory markings on the fin, now bars.

Juutilainen stands in front of his BW-364 at Hirvas wearing the Mannerheim Cross, the highest military decoration in Finland. He received this on 26 April 1942 after gaining 20 air victories in the Continuation War.

Juutilainen revving up his BW-364 at Hirvas in May 1942. Before flying to Römpötti on 1 August 1942 he added two more victories.

Brewster 239, BW-364 of 3/LeLv 24 flown by WO Ilmari Juutilainen in November 1942, based at Suulajärvi. (Karolina Hołda)

Date	Time	Area	Unit	Aircraft	Enemy aircraft	Enemy Unit	Victory type
19.12.39	11.55	Rautu	24	FR-108	SB	44 SBAP	T
31.12.39	10.30	Uomaa	24	FR-106	I-16		T
29.02.40	08.40	Ruokolahti – Valkjärvi	24	FR-96	I-16		V
09.07.41	05.10–05.20	Lahdenpohja – Ladoga	24	BW-364	2 x I-153		T
21.07.41	09.30–11.15	Käkisalmi	24	BW-364	I-153	7 IAP	T
01.08.41	16.00–16.10	Rautjärvi	24	BW-364	2 x MiG-3		T
12.08.41	13.00–13.30	Antrea – Kirvu	24	BW-364	2 x I-153	65 ShAP	T
12.08.41	13.00–13.30	Antrea – Kirvu	24	BW-364	I-153		E
26.09.41	10.15	Derevjannoje	24	BW-364	I-153	65 ShAP	T
26.09.41	11.30	Derevjannoje	24	BW-364	I-16		T
26.09.41	11.30	Derevjannoje	24	BW-364	I-15bis		T
28.09.41	09.45–10.00	Kontupohja	24	BW-364	MiG-3		E
14.12.41	13.05–13.15	Uikujärvi	24	BW-364	MiG-3		T
28.01.42	10.35–10.40	Huhareva	24	BW-364	Hurricane		T
06.02.42	08.05–10.20	Romantsi – Kärkijärvi	24	BW-364	2 x SB		T
09.03.42	14.20–14.50	Uikujärvi	24	BW-364	MiG-3		T
14.03.42	11.00–11.15	Kärkijärvi	24	BW-364	MiG-3		T
14.03.42	11.00–11.15	Kärkijärvi	24	BW-364	MiG-3		V
28.03.42	07.45–08.15	Lavansaari	24	BW-364	2 x I-153	11 AP, KBF	T

Bf 109G-2, MT-222 of 1/LeLv 34 flown by WO Ilmari Juutilainen in May 1943, based at Helsinki Malmi. (Karolina Hołda)

Mersu *MT-222* of 1/LeLv 34 was assigned to Juutilainen on 10 May 1943 for one year. It is seen here in late May at Helsinki Malmi. Behind is the air force HQ VIP transport Douglas DC-2.

Date	Time	Area	Unit	Aircraft	Enemy aircraft	Enemy Unit	Victory type
25.06.42	13.15–13.45	Seesjärvi	24	BW-364	Hurricane		E
25.06.42	13.15–13.45	Seesjärvi	24	BW-364	Hurricane		V
25.06.42	13.15–13.45	Sekehe	24	BW-364	Hurricane	152 IAP	T
13.08.42	11.20	Tolli lighthouse	24	BW-364	Pe-2		T
16.08.42	17.45–18.00	Seiskari – Karavaldai	24	BW-364	2 x I-16		T
18.08.42	20.00–21.20	Kreivinlahti – Kronstadt	24	BW-364	3 x I-16		T
20.09.42	13.00–14.40	Peninsaari	24	BW-364	Spitfire ?		T
20.09.42	13.00–14.40	Peninsaari	24	BW-364	MiG-1		E
20.10.42	09.15–10.15	Seiskari	24	BW-364	Il-4		E
26.10.42	11.00–12.00	Patterilahti	24	BW-355	Hurricane		T
22.11.42	13.20–15.55	Kreivinlahti	24	BW-364	Spitfire		T
23.11.42	11.45–13.15	Oranienbaum	24	BW-364	Tomahawk		E
24.03.43	14.00–15.00	Vaindlo lighthouse	34	MT-212	Pe-2		T
05.06.43	15.15–16.25	Oranienbaum	34	MT-222	LaGG-3		T
05.06.43	15.15–16.25	Oranienbaum	34	MT-222	Tomahawk		T
05.06.43	15.15–16.25	Karavaldai lake	34	MT-222	I-16		T
24.06.43	09.35–10.25	Lavansaari	34	MT-205	I-153		T
10.07.43	15.35–16.50	Seiskari	34	MT-217	2 x I-153		T
10.07.43	15.35–16.50	Seiskari	34	MT-217	Lightning ?		T
19.07.43	15.20–15.50	Seiskari 20 km NW	34	MT-222	LaGG-3		E
19.07.43	15.20–15.50	Seiskari 20 km NW	34	MT-222	LaGG-3		V
26.07.43	15.10–16.10	Seiskari	34	MT-217	Boston		V
10.08.43	15.55–16.45	Seiskari	34	MT-222	LaGG-3		T
31.08.43	15.35–16.25	Saarenpää 10 km S	34	MT-207	La-5		T
11.09.43	14.10–14.55	Sepeleva lighthouse	34	MT-222	LaGG-3		T
11.09.43	14.10–14.55	Sepeleva lighthouse	34	MT-222	La-5		V
23.09.43	13.10–13.50	Sepeleva – Kreivinlahti	34	MT-222	Yak-1		T
23.09.43	13.10–13.50	Sepeleva – Kreivinlahti	34	MT-222	La-5		T
23.09.43	13.10–13.50	Sepeleva – Kreivinlahti	34	MT-222	La-5		V
27.10.43	14.40–15.35	Kronstadt – Tolli	34	MT-222	2 x La-5		T
27.10.43	14.40–15.35	Kronstadt – Tolli	34	MT-222	La-5		V
04.11.43	14.10–15.15	Sepelava lighthouse	34	MT-222	Yak-7B	13 KIAP, KBF	T
04.11.43	14.10–15.15	Sepelava lighthouse	34	MT-222	Yak-7B	13 KIAP, KBF	V
06.03.44	17.10–18.05	Sepeleva lighthouse	34	MT-222	2 x Pe-2	12 GPBAP, KBF	T
06.03.44	17.10–18.05	Seiskari	34	MT-222	Yak-9		T
25.03.44	15.00–15.45	Oranienbaum	34	MT-243	La-5		T
08.05.44	07.20–07.50	Luppi – Suursaari	34	MT-418	Yak-9	21 KIAP, KBF	T
09.06.44	12.05–12.50	Vaskisavotta–Valkeasaari	34	MT-426	2 x La-5		T
10.06.44	06.20–07.30	Vaskisavotta	34	MT-426	Tu-2	132 BAP	R
10.06.44	06.20–07.30	Valkeasaari	34	MT-426	2 x La-5		R
10.06.44	15.45–16.15	Vammelsuu	34	MT-424	Il-2		R
20.06.44	07.10–08.00	Ristiniemi – Tiurinsaari	34	MT-426	Yak-9	4 GIAP	T
20.06.44	07.10–08.00	Ristiniemi – Tiurinsaari	34	MT-426	2 x Il-4		T
20.06.44	07.10–08.00	Ristiniemi – Tiurinsaari	34	MT-426	Yak-9	4 GIAP	R
20.06.44	17.50–19.00	Viipuri	34	MT-423	Airacobra		T
22.06.44	08.50–10.00	Heinjoki	34	MT-426	La-5		T
26.06.44	12.55–14.00	Juustila – Tali	34	MT-422	Airacobra		T
26.06.44	12.55–14.00	Juustila – Tali	34	MT-422	Il-2		T
26.06.44	12.55–14.00	Juustila – Tali	34	MT-422	Yak-9	159 IAP	R
28.06.44	15.00–16.10	Tammisuo	34	MT-459	Yak-9		T
29.06.44	18.00–17.10	Hovinmaa	34	MT-422	La-5		V
30.06.44	10.50–12.15	Juustila – Säiniö	34	MT-457	Yak-9	14 GIAP	T
30.06.44	10.50–12.15	Juustila – Säiniö	34	MT-457	Yak-9	14 GIAP	R
30.06.44	10.50–12.15	Juustila – Säiniö	34	MT-457	Airacobra	403 IAP	T
30.06.44	10.50–12.15	Juustila – Säiniö	34	MT-457	Airacobra	196 IAP	R

Date	Time	Area	Unit	Aircraft	Enemy aircraft	Enemy Unit	Victory type
30.06.44	10.50–12.15	Juustila – Säiniö	34	MT-457	La-5		R
30.06.44	10.50–12.15	Juustila – Säiniö	34	MT-457	Il-2	872 ShAP	R
01.07.44	10.10–11.25	Vatnuori	34	MT-457	Yak-9	404 IAP	T
01.07.44	14.10–15.30	Juustila	34	MT-457	2 x Il-2	448 ShAP	T
02.07.44	06.15–07.20	Säiniö	34	MT-457	Airacobra	196 IAP	T
03.07.44	11.05–12.20	Ihantala	34	MT-457	Yak-9	29 GIAP	R
04.07.44	04.50–06.10	Pölläkkälä	34	MT-457	Il-2		R
04.07.44	04.50–06.10	Pölläkkälä	34	MT-457	Airacobra	196 IAP	R
05.07.44	09.40–10.55	Tuppura – Vatnuori	34	MT-457	2 x Yak-9	13 KIAP, KBF	T
05.07.44	09.40–10.55	Tuppura – Vatnuori	34	MT-457	Yak-9		R
06.07.44	06.25–07.35	Juustila	34	MT-457	Il-4		R
09.07.44	11.40–12.40	Vuoksenranta	34	MT-426	Airacobra		T
15.07.44	10.00–11.15	Paakkola	34	MT-457	Yak-9		R
03.09.44	10.45–11.40	Valkjärvi	34	MT-498	Li-2		V

Bf 109G-6, MT-426 of 1/HLeLv 34 flown by WO Ilmari Juutilainen in June 1944, based at Lappeenranta. (Karolina Hołda)

Mersu MT-426 of 1/HLeLv 34 in front of the Taipalsaari servicing hangar in July 1944. It was assigned to Juutilainen and he flew it successfully during the previous month. By now Juutilainen had already received his last Mersu, MT-457.

KALIMA, Martti Tauno Johannes

10½ victories (10½ + 0 + 0) on 285 missions
Planes: FR-148, MS-326, MS-642 and MS-622

Martti Kalima was born at Helsinki on 27.09.16. He received flying training in 1938–39 at IRUK 8 and was accepted to KadK on 30.05.39. Kalima was posted on 23.06.41 as 1st Lieutenant to LLv 30. He was transferred on 21.09.41 to LLv 110 and back to LLv 30 on 01.11.41. Kalima was posted on 01.08.42 to LeLv 14 and appointed to lead the 2nd Flight on 23.10.43. He was promoted four weeks later to Captain. On 14.06.44 Kalima was posted to Germany leading a detachment of pilots for night fighter training. He was transferred on 14.09.44 to HLeLv 30. After the war he served with HLeLv 33, PLeLv 41 and HävLv 11 commander, resigning as a Major on 06.02.55. Kalima died on 23.06.02. He was awarded with VR 4, VR 3 and VR 3 tlk.

FR-148 of 3/LLv 30 made a successful forced landing into a field at Kuusio on 4 August 1941, on a flight from Utti to Turku. This plane was assigned to Kalima for one year.

From left 1Lts Esko Lehtonen and Aaro Virkkunen of LtueKäär/LLv 14 in front of the tail of Kalima's Fokker FR-148 at Tiiksjärvi on 4 November 1941. The victory markings consist of both air and ground kills. (SA-kuva)

Kalima's plane FR-148 of Lentue Käär in field servicing at Tiiksjärvi on 1 November 1941. This flight was formally 3/LLv 30, but subjected to LLv 14. (SA-kuva)

Fokker D.XXI, FR-148 of LtueKäär/LeLv 14 flown by the flight deputy leader
1Lt Martti Kalima in June 1942, based at Tiiksjärvi. (Karolina Hołda)

Date	Time	Area	Unit	Aircraft	Enemy aircraft	Enemy Unit	Victory type
06.07.41	03.20–03.50	Stortervo	30	FR-148	SB	117 RAE	T
22.07.41	18.00–18.45	Pensari	30	FR-148	I-16		T
22.07.41	18.00–18.45	Pensari	30	FR-148	I-153		T
27.09.41	09.25–10.25	Rukajärvi	10	FR-150	I-153		T
05.11.42	11.55–13.40	Seesjärvi	14	MS-326	LaGG-3		T
16.03.43	14.25–14.35	Jeljärvi 5 km E	14	MS-326	I-15bis	839 IAP	T
23.03.43	08.40–08.45	Kirasjärvi 7 km SW	14	MS-326	½ I-16	197 IAP	T
13.04.44	13.45–14.15	Tungutjärvi	14	MS-622	LaGG-3		T
26.05.44	04.00–05.50	Kompakkajärvi	14	MS-622	LaGG-3	435 IAP	T
02.06.44	17.50–19.00	Kuutsjärvi 5 km S	14	MS-622	LaGG-3		T
02.06.44	17.50–19.00	Ontajärvi	14	MS-622	LaGG-3		T

Kalima flew Morane MS-326 for one year as the deputy leader of 1/LeLv 14, seen here at Tiiksjärvi in summer 1943.

Left: 2/TLeLv 14 leader Capt Kalima in the cockpit of his Morane MS-622 at Tiiksjärvi, after shooting down a LaGG-3 on 26 May 1944.

Kalima by the tail of his MS-622 at Tiiksjärvi in early June 1944. His combats were over as he was posted to Germany on 14 June 1944 to lead a group of pilots going to night fighter training.

Morane Saulnier MS.406, MS-622 of 2/TLeLv 14 flown by the flight leader Capt Martti Kalima in June 1944, based at Tiiksjärvi. (Karolina Hołda)

KARHILA, Kyösti Keijo Ensio

32¼ victories (32¼ + 0 + 4) on 304 missions
Planes: CUw-560, MT-224 and MT-461

Kössi Karhila was born at Rauma on 02.05.21. He received flying training in 1939–41 at SOK 1 and was posted on 19.03.41 as 2nd Lieutenant to LLv 32. Karhila was ordered to LLv 30 on 10.03.42 and back to LeLv 32 on 10.05.42. He was promoted on 14.03.43 to 1st Lieutenant and transferred on 20.04.43 to LeLv 34. Karhila was posted on 15.03.44 to HLeLv 30 and two months later back to HLeLv 34. On 30.06.44 Karhila was appointed to lead 3/HLeLv 24 and further on 21.07.44 to lead 2/HLeLv 30. Karhila was demobilized on 14.11.44. In civilian life he acted as an airline captain and held the rank of Captain. Karhila died on 16.09.09. He was awarded VR 4, VR 3, EK 2, VR 3 tlk and VR 2.

Curtiss Hawk 75A-6, CUw-560 flown by 1/LeLv 32 pilot 2Lt Kyösti Karhila in June 1942, based at Nurmoila. (Karolina Hołda)

CUw-560 of LeLv 32 parked under the trees at Suulajärvi in May 1942. It was the top scoring Hawk with 18¾ claims by ten different pilots, eight by Karhila.

1/LeLv 32 pilot Karhila in front of an unidentified Curtiss Hawk at Lappeenranta, after becoming an ace on 19 September 1941. He accumulated 13 victories with the type.

Date	Time	Area	Unit	Aircraft	Enemy aircraft	Enemy Unit	Victory type
10.08.41	17.40–17.45	Kirvu	32	CU-567	½ I-16		T
13.08.41	13.45–14.00	Antrea	32	CU-561	½ I-153		T
18.08.41	14.10–14.20	Sintola	32	CU-570	I-153	7 IAP	T
03.09.41	10.55–11.10	Rajajoki	32	CU-566	I-153	7 IAP	T
17.09.41	13.05–13.10	Siestarjoki	32	CU-552	I-16	13 AP, KBF	T
19.09.41	13.20–13.30	Ohalatva	32	CU-560	MiG-3	7 IAP	E
15.06.42	13.15–14.25	Mergino	32	CU-560	Yak-1		T
28.06.42	05.50–06.50	Svir	32	CU-560	½ Pe-2		T
04.07.42	18.10–19.00	Svir power plant	32	CU-560	Pe-2		V
05.07.42	16.45–17.45	Mäkriä	32	CU-560	I-16		T
21.08.42	10.45–11.50	Ljugovitsa	32	CU-560	I-16	524 IAP	T
29.09.42	10.35–11.55	Saarimäki	32	CU-571	½ Pe-2	119 RAE	T
09.11.42	13.00–15.05	Nulbenskoje - Vonozero	32	CU-571	½ LaGG-3	415 IAP	T
09.11.42	13.00–15.05	Troitsankontu - Sotkusa	32	CU-571	¼ Pe-2	119 RAE	T
09.02.43	12.40–13.40	Saarentaka	32	CU-560	Pe-2	119 RAE	T
11.02.43	10.05–11.10	Saarentaka 5 km SW	32	CU-560	½ MiG-3	415 IAP	T
11.02.43	10.05–11.10	Savijärvi	32	CU-560	LaGG-3	415 IAP	T
11.02.43	10.05–11.10	Malkjärvi	32	CU-560	U-2	4 AP GvF	T
04.05.43	11.05–12.25	Ino	34	MT-214	2 x LaGG-3	3 GIAP, KBF	T
21.05.43	18.40–19.55	Lavansaari	34	MT-224	La-5	4 GIAP, KBF	E
19.07.43	16.35–17.30	Pyötsaari	34	MT-224	Pe-2	13 ORAP	T
20.08.43	14.55–16.00	Seiskari 20 km NW	34	MT-229	La-5		T
28.05.44	04.20–05.35	Kuutsalo 10 km NW	30	MT-403	Pe-2		T

As a member of 2/LeLv 34 Karhila scored his first two Mersu kills in MT-214 on 4 May 1943, as he clearly points out in the cockpit. The location is Utti.

3/HLeLv 24 leader 1Lt Karhila in his Mersu MT-461 at Lappeenranta on 10 July 1944. He claimed eight kills flying this Kanonenboot. (SA-kuva)

SA-Kuva

Date	Time	Area	Unit	Aircraft	Enemy aircraft	Enemy Unit	Victory type
21.06.44	12.55–14.10	Tienhaara	34	MT-405	Il-2	999 ShAP	T
30.06.44	10.50–11.50	Perojoki	24	MT-436	Yak-9	404 IAP	T
01.07.44	03.55–05.05	Tuppuransaari	24	MT-461	Il-2	7 GShAP, KBF	T
02.07.44	19.50–20.00	Hovinmaa	24	MT-461	3 x Il-2		V
03.07.44	11.50–12.40	Portinhoikka	24	MT-461	Il-2		T
04.07.44	05.50–07.15	Tuppuransaari	24	MT-460	Yak-9	13 KIAP, KBF	T
05.07.44	09.25–10.25	Tuppuransaari	24	MT-461	Yak-9		T
07.07.44	19.30–20.45	Uuras	24	MT-461	Yak-9	13 KIAP, KBF	T
09.07.44	05.40–06.50	Äyräpää	24	MT-461	Il-2		T
10.07.44	12.00–13.15	Äyräpää	24	MT-461	La-5	159 IAP	T
10.07.44	12.00–13.15	Äyräpää	24	MT-461	Yak-9		T
11.07.44	04.25–05.50	Äyräpää	24	MT-461	La-5	159 IAP	T
16.07.44	06.05–07.20	Vuosalmi	24	MT-460	Yak-9	14 GIAP	T
18.07.44	05.30–06.45	Äyräpää	24	MT-460	La-5		T

31⁷/₁₂ victories (27⁷/₁₂ + 4 + 3) on appr. 350 missions
Planes: FR-112 and BW-366

Joppe Karhunen was born at Pyhäjärvi UL on 17.03.13. On 06.06.33 he was accepted to KadK and posted on 16.05.36 as 2ⁿᵈ Lieutenant to LLv 24. Two years later followed the promotion to 1ˢᵗ Lieutenant. On 02.02.40 Karhunen was appointed to lead 2/LLv 24. On 18.06.41 Karhunen was transferred to lead 3/LLv 24 and on 04.08.41 promoted to Captain. On 01.06.43 Karhunen was appointed to command LeLv 24 and on 31.08.43 promoted to major. After the war Karhunen served as commander of HLeLv 31, HLeLv 13 and LeR 1 and 2. Lsto resigning on 13.12.55 as Lieutenant Colonel. In civilian life Karhunen became a writer and held the rank of Colonel. Karhunen died on 18.01.02. He was awarded VR 4, VR 3, VR 3 tlk, VR 2 and MHR on 08.09.42.

Brewster 239, BW-366 of 3/LLv 24 flown by the flight leader Capt Jorma Karhunen in August 1941, based at Lappeenranta. (Karolina Hołda)

3/LLv 24 leader Karhunen taxies his BW-366 at Lappeenranta in August 1941. Karhunen was assigned to this machine for two years, claiming fifteen kills with it.

Brewster BW-366 of
3/LLv 24 leader Capt Karhunen re-fuelled at Lappeenranta in August 1941. Originally a naval fighter it had
a long endurance of 4 hours,
which was an advantage in
large combats.

LeLv 24's new commander
Capt Karhunen sitting on the
tail of his Brewster BW-366
at Suulajärvi in early June
1943. The fin carries all his
31 confirmed aerial victories.
(SA-kuva)

SA-Kuva

Karhunen and his dog Peggy Brown playing in the cockpit of a Brewster at Hirvas on 1 June 1942. (SA-kuva)

Date	Time	Area	Unit	Aircraft	Enemy aircraft	Enemy Unit	Victory type
01.12.39	14.17	Tainionkoski	24	FR-112	SB	41 SBAP	E
18.12.39	14.30–15.00	Humaljoki	24	FR-112	SB	24 SBAP	E
19.12.39	15.20	Oravaniemi	24	FR-112	½ DB-3	6 DBAP	T
23.12.39	10.48	Heinjoki	24	FR-112	⅓ SB	44 SBAP	T
25.12.39	13.30	Korpijärvi – Ägläjärvi	24	FR-112	2 x SB	72 SBAP	T
01.01.40	12.55	Korpijärvi	24	FR-112	SB	18 SBAP	T
01.01.40	12.58	Korpijärvi	24	FR-112	SB		V
17.01.40	13.55	Imatra – Nuijamaa	24	FR-80	2 x SB		V
29.01.40	15.20	Kaukjärvi	24	FR-80	½ R-5	16 KAO	T
29.01.40	15.25	Muolaanjärvi	24	FR-80	¼ R-5	16 KAO	T
04.07.41	11.00–12.10	Simpele	24	BW-366	SB	205 SBAP	T
09.07.41	05.10–05.20	Lahdenpohja	24	BW-363	I-153		T
12.08.41	13.00–13.30	Antrea – Kirvu	24	BW-366	I-153		E
23.09.41	13.30	Petäjäselkä	24	BW-366	½ I-16	155 IAP	T
23.09.41	14.00–14.15	Petäjäselkä	24	BW-366	I-16	155 IAP	T
26.09.41	10.15	Derevjannoje	24	BW-366	2 x I-153	65 ShAP	T
17.12.41	09.25–09.45	Byrstjagin station	24	BW-366	Hurricane	152 IAP	T
17.12.41	09.25–09.45	Byrstjagin station	24	BW-366	½ I-153	65 ShAP	T
26.02.42	10.15–10.40	Liistepohja	24	BW-366	MiG-3		T
26.02.42	10.15–10.40	Nopsa stop	24	BW-366	MiG-3		E
02.07.42	10.30–11.30	Tipinitsa – Onega	24	BW-388	Pe-2	1 GPBAP	T
14.08.42	10.20–12.00	Tolli lighthouse	24	BW-388	Hurricane	3 GIAP, KBF	T
14.08.42	10.20–12.00	Tolli lighthouse	24	BW-388	Hurricane		E
16.08.42	17.45–18.00	Seiskari – Karavaldai	24	BW-388	3 x I-16		T
18.08.42	20.00–21.20	Krasnaja Gorka	24	BW-388	I-16		T
18.08.42	20.00–21.20	Krasnaja Gorka	24	BW-388	I-16		E
18.08.42	20.00–21.20	Tolli lighthouse	24	BW-388	Pe-2	73 AP, KBF	T
20.09.42	13.00–14.40	Peninsaari	24	BW-366	Spitfire ?		T
26.10.42	11.00–12.00	Patterilahti	24	BW-366	Spitfire ?		T
23.02.43	11.15–12.30	Lavansaari 15 km S	24	BW-366	I-16		T
02.03.43	15.10–16.40	Kreivinlahti 10 km W	24	BW-366	I-153	71 IAP, KBF	T
21.04.43	08.00–09.20	Seiskari 15 km E	24	BW-366	Yak-1		T
04.05.43	10.40–12.20	Seiskari	24	BW-366	I-153	71 IAP, KBF	T

Karu, Veikko Johannes

10¾ victories (9¾ + 1 + 0) on appr. 400 missions
Planes: FR-107 and FR-129

Veikko Karu was born at Kajaani on 08.11.10. He received flying training in 1933–34 and was accepted on 06.06.34 to KadK. Karu was transferred on 16.05.36 as 2nd Lieutenant to LLv 10. On 01.02.38 he was transferred to LLv 26 and promoted to 1st Lieutenant on 16.05.38. Karu was posted to LLv 28 on 25.01.40. On 04.05.40 he became LeR 3 adjutant. On 24.05.41 Karu was posted to lead 2/LLv 30. He was promoted on 04.08.41 to Captain. On 11.11.42 Karu was assigned to the HQ of LeR 3. On 03.11.43 Karu became the acting CO of LeLv 26. Karu was transferred on 18.02.44 to lead 2/HLeLv 34, which became three weeks later 2/HLeLv 30. On 19.04.44 Karu took over the command of HLeLv 30 and on 18.07.44 was promoted to Major. After the war Karu commanded HLeLv 31n and HLeLv 23, resigning on 24.05.49. In civilian life he owned Kymen Autolento Oy. Karu died on 30.07.91. He was awarded with VR 4, VR 3, VR 3 tlk, VR 2, EK 2, EK 1 and 06.11.42 MHR.

Fokker D.XXI, FR-129 of 2/LLv 30 flown by the flight leader Capt Veikko Karu in October 1941, based at Suulajärvi. (Karolina Hołda)

2/LLv 30 leader Capt Karu received a flak hit in the cowling of his FR-129 on 1 November 1941, seen here after a successful landing at Suulajärvi.

Left: *The tail of Karu's FR-129 showing the personal emblem: devil chasing an I-16. Rudder has three victory bars dating this to early September 1941. At right is the plane's mechanic Risto Leivo.*

Right: *The tail of FR-129 from the other side showing the ship sinking marks. There are now five victory bars dating this to late March 1942.*

FR-155 of 3/LLv 30 seen at Turku in late June 1941. It was on loan to 2/LLv 30 when Karu was shot down by flak in it over enemy territory on 16 November 1941. Karu returned on foot to the Finnish side.

Date	Time	Area	Unit	Aircraft	Enemy aircraft	Enemy Unit	Victory type
17.01.40	14.05	Kuparsaari	26	FR-107	SD	31 SBAP	T
20.02.40	14.25	Estonian coast	28	MS-321	2 x DB-3	1 DABr	T
08.07.41	06.30–07.40	Naissaari 20 km N	30	FR-129	MBR	15 AP, KBF	T
06.08.41	17.46	Malmi lighthouse	30	FR-129	½ MBR	58 OAE, KBF	T
25.08.41	14.00–14.10	Tytärsaari – Lavansaari	30	FR-129	2 x MBR	58 OAE, KBF	T
12.12.41	14.45–14.55	Koivisto – Seiskari	30	FR-160	½ I-153		T
17.12.41	12.30–12.40	Narvi	30	FR-129	½ MBR	44 OAE, KBF	T
14.03.42	07.30–07.35	Peninsaari	30	FR-129	¼ MBR	15 AP, KBF	T
28.03.42	05.45–06.10	Suursaari	30	FR-136	I-16		E
15.04.42	08.20–08.30	Lumisuo	30	FR-129	I-153		T
25.05.43	23.05	Immola	3	BW-368	He 111 H	Ob.d.L	T

KATAJAINEN, Nils Edvard

34½ victories (31½ + 3 + 6) on 198 missions
Planes: BW-368, BW-353, MT-436, MT-462 and MT-476

Nils Katajainen was born at Helsinki on 31.05.19. He received flying training in 1939–40 at RAOK 6 and was demobilized on 01.03.41. In the Continuation War mobilization on 18.06.41 Katajainen was posted as Corporal to LLv 24. He was promoted to Sergeant on 23.07.41 and Staff Sergeant on 12.04.42. Katajainen was transferred on 09.09.42 to LeLv 48 for twin-engine courses. On 19.10.42 he was transferred to LeLv 6 and on 09.04.43 back to LeLv 24. Promotion to Master Sergeant took place on 24.09.43. He was demobilized on 10.11.44. In civilian life Katajainen was Helsinki city policeman. Katajainen died on 16.01.97. He was awarded VM 2, VM 1, VR 4, VR 4 tlk, VR 3 and MHR on 21.12.44.

Katajainen sitting in the cockpit of a 3/LLv 24 Brewster at Rantasalmi in July 1941. The squadron's lynx emblem is clearly seen on the forward fuselage.

BW-368 of 3/LLv 24 was assigned to Katajainen on 18 June 1941. It is seen here at Mantsi in September 1941.

Left: The engine of BW-353 quit on landing at Suulajärvi on 29 June 1943 and Sgt Kosti Koskinen crashed in the forest. BW-353 of 3/LeLv 24 was assigned to Katajainen and carried his 13 victory bars.

Right: Katajainen sitting on the tail of his BW-368 at Mantsi. The last victory marking came on 26 September 1941.

Brewster 239, BW-368 of 3/LLv 24 flown by SSgt Nils Katajainen in April 1942, based at Kontupohja. (Karolina Hołda)

BW-368 of 3/LLv 24 was assigned to Katajainen and is seen here at Kontupohja in February 1942, wearing Katajainen's six kill markings as frontal silhouettes.

Date	Time	Area	Unit	Aircraft	Enemy aircraft	Enemy Unit	Victory type
28.06.41	06.30–07.30	Loviisa	24	BW-365	SB	57 AP	T
19.07.41	13.05–13.10	Käkisalmi	24	BW-365	I-153		T
01.08.41	16.00–16.10	Rautjärvi	24	BW-368	MiG-3		T
12.08.41	13.00–13.20	Antrea – Kirvu	24	BW-368	2 x I-153		E
19.09.41	11.25–11.35	Manga	24	BW-368	LaGG-3		V
26.09.41	10.15	Derevjannoje	24	BW-368	1½ x I-153	65 ShAP	T
26.09.41	11.30	Petäjäselkä	24	BW-368	½ I-153		T
09.03.42	14.20–14.50	Uikujärvi	24	BW-368	SB		T
09.03.42	14.20–14.50	Uikujärvi	24	BW-368	MiG-3		V
14.03.42	11.00–11.15	Juka – Nopsa	24	BW-368	MiG-3		T
25.06.42	13.15–13.35	Sekehe	24	BW-388	Hurricane		T
14.08.42	11.00–11.10	Vanha Yhimäki	24	BW-380	Hurricane		T
16.08.42	17.45–18.00	Seiskari – Karavaldai	24	BW-373	2 x I-16		T
09.05.43	11.15–11.30	Seiskari	24	BW-353	Yak-7B		T
18.05.43	14.45–15.35	Harjavallanniemi	24	BW-353	Yak-1	13 KIAP, KBF	T
20.05.43	13.20–14.40	Lavansaari	24	BW-353	La-5	4 GIAP, KBF	T
23.09.43	12.00–13.45	Sepeleva lighthouse	24	BW-368	Yak-1		T
23.09.43	12.00–13.45	Sepeleva lighthouse	24	BW-368	La-5		V
23.09.43	15.30–16.30	Seiskari	24	BW-368	½ La-5		T
04.11.43	14.45–15.30	Seiskari	24	BW-368	Yak-1		V
23.06.44	07.30–08.20	Lyykylänjärvi	24	MT-441	La-5		T
23.06.44	07.30–08.20	Lyykylänjärvi	24	MT-441	Il-2	566 ShAP	T
23.06.44	12.05–13.05	Raulampi airfield	24	MT-441	La-5		T
23.06.44	12.05–13.05	Raulampi airfield	24	MT-441	La-5		V
26.06.44	13.05–14.20	Vakkila – Mannikkala	24	MT-436	2 x Il-2	566 ShAP	T
26.06.44	13.05–14.20	Lyykylänjärvi	24	MT-436	La-5		T
28.06.44	10.25–11.05	Juustila	24	MT-436	Yak-9		T
28.06.44	10.25–11.05	Mannikkala	24	MT-436	Il-2		T
28.06.44	10.25–11.05	Mannikkala	24	MT-436	Airacobra		V
28.06.44	14.00–14.55	Tammisuo	24	MT-436	Airacobra	196 IAP	T
29.06.44	18.50–19.50	Kärstilänjärvi	24	MT-462	Il-2	15 GshAP	T
30.06.44	10.50–12.00	Juustila	24	MT-462	Pe-2	140 BAP	T
01.07.44	13.10–14.10	Portinhoikka	34	MT-462	Airacobra	196 IAP	R
03.07.44	06.10–07.10	Ihantala	24	MT-462	Yak-9	14 GIAP	T
03.07.44	06.10–07.10	Tammisuo	24	MT-462	Yak-9		R
03.07.44	11.50–12.35	Mannikkala	24	MT-462	2 x Il-2		T
05.07.44	09.25–10.15	Tuppuransaari	24	MT-476	Yak-9		T

Bf 109G-8 WNr 200041 (MT-462) was assigned to Katajainen on 28 June 1944. He flew it with 3/HLeLv 24 in German markings on several missions, taking off from Lappeenranta here two days later. Katajainen crashed the plane on 3 July 1944. (SA-kuva)

Katajainen was re-trained on bombers commencing 9 September 1942. He then took a six-month spell at LeLv 6 searching for submarines in the Gulf of Finland, regularly flying this Tupolev SB coded SB-1. It is seen here at Helsinki Malmi on 14 September 1942. (Finnish Air Force)

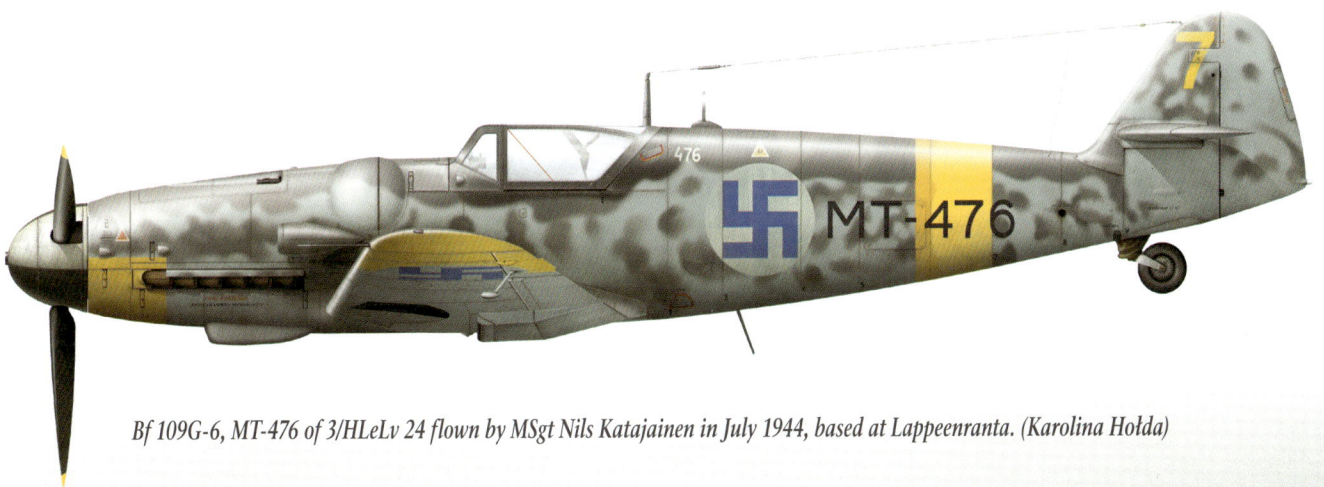

Bf 109G-6, MT-476 of 3/HLeLv 24 flown by MSgt Nils Katajainen in July 1944, based at Lappeenranta. (Karolina Hołda)

Katajainen's last assigned fighter was MT-476 (right) of 3/HLeLv 24, seen here with MT-441 at Lappeenranta on 2 July 1944. Katajainen crashed it three days later after hits in combat and was hospitalized for the rest of the war. (SA-kuva)

KAUPPINEN, Osmo Kalervo

5 victories (4 + 1 + 1) on 67 missions
Planes: BW-388

Osmo Kauppinen was born at Muhos on 02.04.16. He received flying training in 1938–39 at IRUK 8 and was accepted to KadK on 29.09.39. Kauppinen was posted on 31.12.40 as 2nd Lieutenant to LLv 24 and was promoted to 1st Lieutenant on 23.01.41. He was transferred on 29.03.42 to KoeL and was promoted on 24.05.44 to Captain. Kauppinen resigned on 16.02.48. In civilian life he was an airline captain and held the rank of Major. Kauppinen died on 17.11.00. He was awarded with VR 4, VR 3 and EK 2.

From left second lieutenants Osmo Kauppinen and Urho Sarjamo having good time with a drink between the wars, Christmas 1940

3/LLv 24 deputy leader 1Lt Kauppinen in front of his Brewster BW-388 at Kontupohja on 15 February 1942. Six weeks later he became a test pilot. (SA-kuva)

BW-388 of 3/LLv 24 was assigned to the deputy flight leader 1Lt Kauppinen on 3 October 1941. It is parked here on the ice at Kontupohja harbour on 15 February 1942. (SA-kuva)

Date	Time	Area	Unit	Aircraft	Enemy aircraft	Enemy Unit	Victory type
19.09.41	11.20–11.30	Manga	24	BW-368	½ I-16	155 IAP	T
23.09.41	14.00–14.15	Petäjäselkä	24	BW-388	½ I-153		T
17.12.41	09.25–09.45	Romantsi station	24	BW-388	MiG-3		E
26.02.42	10.15–10.20	Liistepohja	24	BW-388	MiG-3		V
26.02.42	10.15–10.20	Liistepohja	24	BW-355	MiG-3		T
09.03.42	14.20–14.50	Uikujärvi	24	BW-366	I-153		T
28.03.42	07.45–08.15	Lavansaari	24	BW-366	I-153	I-153	T

KAUPPINEN, Viljo Ilmari

7½ victories (7½ + 0 + 0) on 110 missions
Planes: BW-357

Viljo Kauppinen was born at Ruskeala on 13.11.20. He received flying training in 1940–41 at SOK 4 and was posted on 17.12.41 as Corporal to LLv 24. He was promoted to Sergeant on 29.03.42 and Staff Sergeant on 06.12.43. Kauppinen was wounded on 07.06.44 and became hospitalized. He resigned on 31.12.45. Kauppinen acted in civilian life as a businessman. He died on 20.10.03. Kauppinen was awarded with VM 2, VM 1 and VR 4.

Brewster BW-351 of 3/LeLv 24 ready for a mission from Römpötti on 1 October 1942, 2Lt Heikki Herrala at the controls this time. Flight leader Capt Karhunen and his dog look on. (SA-kuva)

Date	Time	Area	Unit	Aircraft	Enemy aircraft	Enemy Unit	Victory type
14.08.42	10.20–12.00	Tolli lighthouse	24	BW-351	Hurricane		T
23.02.43	11.15–12.30	Lavansaari 15 km S	24	BW-357	I-16		T
23.02.43	11.15–12.30	Lavansaari	24	BW-357	I-16		T
18.05.43	14.45–15.35	Harjavallanniemi	24	BW-357	Yak-1		T
20.08.43	16.30–18.00	Kronstadt	24	BW-368	½ LaGG-3		T
20.08.43	16.30–18.00	Yhinmäki	24	BW-368	LaGG-3		T
21.03.44	09.10–10.10	Valkeasaari	24	BW-357	Il-2	7 GShAP, KBF	T
02.06.44	13.45–14.35	Siestarjärvi	24	MT-241	La-5		T

BW-351 of 3/LeLv 24 returned to Römpötti on 3 October 1942 flown by the assigned pilot SSgt Leo Ahokas. Kauppinen scored his first victory in this plane. (SA-kuva)

Keskinummi, Kosti Rauni Iisakki

5½ victories (4½ + 1 + 0) on 75 missions
Planes: MT-443

Kosti Keskinummi was born at Kauhajoki on 06.01.23. He received flying training in 1940–42 at AOK 11 and was posted on 18.03.43 as Corporal to LeLv 24. Promotion to Sergeant came on 08.04.43. Keskinummi was injured in a post-combat forced landing on 28.06.44. He resigned as a Staff Sergeant on 20.08.47. In civilian life Keskinummi was a primary school teacher. He died on 18.08.18. Keskinummi was awarded with VM 2 and VM 1.

3/HLeLv 24 pilots at Lappeenranta on 18 June 1944 bound to Germany to pick up new Messerschmitts. From left SSgt Leo Ahokas, SSgt Emil Vesa, Italian transport pilot and Sgt Kosti Keskinummi. (SA-kuva)

Keskinummi scored his first victory in a 3/LeLv 24 machine, BW-363. It was assigned to 1Lt Antti Saikkonen and is seen here at Suulajärvi in early 1944.

Date	Time	Area	Unit	Aircraft	Enemy aircraft	Enemy Unit	Victory type
28.09.43	17.15–18.30	Sepeleva 10 km NW	24	BW-363	½ LaGG-3		T
20.06.44	07.15–07.50	Saarenpää	24	MT-443	LaGG-3		T
21.06.44	15.00–16.20	Tali station	24	MT-443	Airacobra		T
22.06.44	17.10–18.20	Lyykylänjärvi	24	MT-437	La-5		T
22.06.44	17.10–18.20	Kämärä	24	MT-437	La-5		R
26.06.44	16.30–17.40	Mannikkala	24	MT-443	Yak-9		T

5⅚ victories (4⅚ + 1 + 0) on appr. 200 missions
Planes: CUw-554 and CU-562

Aaro Kiljunen was born at Jyväskylä on 22.07.20. He received flying training in 1940–41 at SOK 3 and was posted on 01.03.41 as Private to LLv 32. Kiljunen was promoted to Corporal on 12.07.41, to Sergeant on 26.09.41 and to Staff Sergeant on 01.07.43. Kiljunen resigned on 31.10.44. After the war he changed his name to Aremaa. He died on 23.12.68. Kiljunen was awarded with VM 2, VM 1 and VR 4.

Curtiss Hawk 75A-2, CU-504 flown by 1/LeLv 32 pilot Sgt Aaro Kiljunen in July 1942, based at Nurmoila. (Karolina Hołda)

Curtiss CU-562 at the depot at Kuorevesi before delivery to LeLv 32, which took place on 19 February 1943. It was one of the planes frequently piloted by Kiljunen. (Finnish Air Force)

Date	Time	Area	Unit	Aircraft	Enemy aircraft	Enemy Unit	Victory type
23.12.41	13.20–15.10	Ohalatva	32	CU-555	⅓ R-5		T
08.01.42	14.10–16.00	Lumisuo	32	CU-554	½ I-15bis		T
28.03.42	05.30–07.20	Suursaari	32	CU-554	½ I-16	11 AP, KBF	T
05.07.42	14.20–15.25	Aunusjoki 30 km NW	32	CU-555	I-16		T
20.07.42	12.45–14.45	Orjenzenskojärvi	32	CU-504	MiG-1		T
25.07.42	21.20–22.00	Aunus - Nurmoila	32	CU-560	½ MBR	58 AE KBF	T
11.02.43	10.05–11.10	Saarentaka 5 km SW	32	CU-553	½ MiG-3	415 IAP	T
04.03.43	09.40–11.50	Ljugovitsa	32	CU-562	MiG-3		E
22.06.44	13.50–15.35	Kuittinen	32	CU-584	½ Il-2	257 ShAD	V

Kinnunen, Eero Aulis

22½ victories (22½ + 0 + 2) on appr. 300 missions
Planes: FR-109 and BW-352

Eero Kinnunen was born at Käkisalmi on 18.01.18. He received flying training in 1936–37 at RAOK 3 and AOK 8 and was posted as Corporal on 01.01.38 to LLv 26. Kinnunen was made Sergeant on 06.12.39 and was transferred on 29.01.40 to LLv 24. On 22.04.40 he was promoted to Staff Sergeant, on 13.07.41 to Master Sergeant and on 01.12.42 to Warrant Officer. Kinnunen was killed in action on 21.04.43, when an La-5 of 4 GIAP, KBF shot down his BW-352 at Oranienbaum. Kinnunen was awarded with VM 2, VM 1, VR 4, VR 4 tlk, VR 3 and EK 2.

Fokker D.XXI serial FR-109 was assigned to 5/LLv 24 pilot Sgt Kinnunen in the Winter War. It is seen here at Siikakangas in June 1940 belonging to LLv 32.

Brewster BW-352 of 2/LLv 24 was assigned to Kinnunen in two tours over two years. Parked here under the trees at Selänpää on 25 June 1941, the first day of the Continuation War. (SA-kuva)

Kinnunen's BW-352 under guard in the evening sun at Selänpää on 25 June 1941, after Kinnunen had destroyed 4½ Tupolev SB bombers in his first two combats. (SA-kuva)

Brewster 239, BW-352 of 2/LLv 24 flown by MSgt Eero Kinnunen in June 1942, based at Tiiksjärvi. (Karolina Hołda)

BW-352 of 2/LeLv 24 on a visit to Hirvas on 7 August 1942. The assigned pilot Kinnunen had recently been transferred to 1/LeLv 24 on the Karelian Isthmus.

Kinnunen posing in front of BW-352 at Selänpää on 25 June 1941, having claimed 4½ bombers shot down. (SA-kuva)

Kinnunen sitting on the tail of his BW-352 at Tiiksjärvi in May 1942. The victory tally is partly obscured, but shows 11½ kills, all with this plane.

Date	Time	Area	Unit	Aircraft	Enemy aircraft	Enemy Unit	Victory type
09.02.40	13.30	Summa	24	FR-109	R-5	1 KAO	T
17.02.40	13.10	Muolaanjärvi	24	FR-109	DB-3		V
18.02.40	09.05	Sirkjärvi	24	FR-109	DB-3	1 AP	T
19.02.40	15.45	Summa	24	FR-109	½ SB	24 SBAP	T
05.03.40	17.30	Vilaniemi	24	FR-109	I-16	25 IAP	T
25.06.41	07.15–07.45	Selänpää – Heinola	24	BW-352	2½ x SB	201 SBAP	T
25.06.41	10.50–11.20	Utti	24	BW-352	SB	202 SBAP	T
25.06.41	10.50–11.20	Hamina	24	BW-352	SB	2 SBAP	T
30.06.41	09.05–10.00	Kotka	24	BW-352	I-153	7 IAP	T
13.07.41	12.30–13.00	Tolvajärvi	24	BW-352	I-16	155 IAP	T
10.01.42	09.45–10.15	Konsosero	24	BW-357	MiG-3		T
13.02.42	14.30–14.45	Kangasvaara	24	BW-352	Hurricane		T
30.03.42	15.50–16.10	Pertjärvi	24	BW-352	Hurricane	152 IAP	T
30.03.42	15.50–16.10	Pertjärvi	24	BW-352	Hurricane		V
06.04.42	15.25–15.50	Rukajärvi	24	BW-352	2 x Hurricane	767 IAP	T
08.06.42	21.55–22.20	Kesä airfield	24	BW-352	Hurricane	152 IAP	T
31.08.42	18.50–20.00	Lavansaari	24	BW-382	I-153		T
26.10.42	11.00–12.00	Patterilahti	24	BW-351	DB-3	1 GMTAP, KBF	T
26.10.42	11.00–12.00	Patterilahti	24	BW-351	DB-3	1 GMTAP, KBF	T
23.02.43	11.15–12.30	Peninsaari 15 km S	24	BW-352	I-16		T
23.02.43	11.15–12.30	Peninsaari 10 km S	24	BW-352	I-16		T
21.04.43	08.00–09.00	Seiskari 15 km E	24	BW-352	½ Yak-1		T
21.04.43	08.00–09.00	Oranienbaum	24	BW-352	LaGG-3		T

9¾ victories (9¾ + 0 + 5) on 204 missions
Planes: CUc-501 and MT-402

Mauno Kirjonen was born at Viipuri on 20.04.16. He received flying training in 1939–41 at SOK 1 and was posted on 01.03.41 to LLv 32. Kirjonen was promoted to Corporal on 12.07.41 and Sergeant on 26.08.41. He started the RUK on 28.02.42 and returned on 16.07.42 as 2nd Lieutenant to LeLv 32. Kirjonen was transferred on 21.12.42 to KoeL and on 23.03.43 to LeLv 34. He was posted on 06.03.44 to HLeLv 30 and on 15.06.44 back to HLeLv 34. Kirjonen was promoted 1st Lieutenant on 06.07.44 and was transferred on 05.08.44 to HLeLv 30. He was demobilized on 14.11.44. In civilian life Kirjonen became an engineer. He died on 07.03.99. Kirjonen was awarded with VM 1, VR 4 and VR 3.

Fokker D.XXI serial FR-98 of 1/LLv 32 at Hyvinkää on 28 June 1941. It was flown by Kirjonen on 4 July 1941, when he shared with 1Lt Evinen a much faster Petlyakov Pe-2 bomber shot down.

Kirjonen scrambled in Curtiss CUc-501 of LLv 32 from Utti on 22 July 1941 encountering several I-153s. In a short combat Kirjonen downed two Chaikas, but was also hit and bailed out.

Curtiss CUw-564 of LeLv 32 on a visit to Kauhava on 2 July 1942. It was flown by eight different pilots, including Kirjonen, to claim 10 air victories.

Kirjonen at the cockpit sill of Mersu MT-227 of 2/LeLv 34 at Helsinki Malmi in August 1943. Not much aerial activity took place in the defence of the capital.

Date	Time	Area	Unit	Aircraft	Enemy aircraft	Enemy Unit	Victory type
04.07.41	09.10–09.25	Hamina	32	FR-98	½ Pe-2	58 SBAP	T
08.07.41	08.05–08.10	Ristiniemi	32	FR-100	½ DB-3		T
22.07.41	03.25–03.35	Kaukala – Haukkajärvi	32	CU-501	2 x I-153	7 IAP	T
01.08.41	19.55–20.05	Kauppila	32	CU-561	MiG-3		T
18.08.41	14.00–14.20	Sintola	32	CU-566	I-153	7 IAP	T
27.10.42	13.05–13.50	Ala–Sotkusa	32	CU-503	Pe-2		V
09.11.42	13.00–15.05	Nulbenskoje – Vonozero	32	CU-564	½ LaGG-3	415 IAP	T
09.11.42	13.00–15.05	Troitsankontu – Sotkusa	32	CU-564	¼ Pe-2	119 RAE	T
05.05.43	16.40–17.35	Haapasaari	34	MT-201	Boston	15 ORAP, KBF	T
05.05.43	16.40–17.35	Haapasaari	34	MT-201	La-5		V
21.05.43	14.35–15.40	Lavansaari 50 km SW	34	MT-230	I-153	71 KIAP, KBF	T
17.07.43	06.00–07.00	Jääski – Sakkola	34	MT-230	DB-3F		V
20.06.44	07.10–08.05	Ristiniemi	34	MT-402	Il-2	35 ShAP, KBF	T
28.06.44	19.50–21.00	Tammisuo	34	MT-423	Airacobra		V
02.07.44	20.00–20.40	Lappeenranta	34	MT-402	Il-2	448 ShAP	T
02.07.44	20.00–20.40	Lappeenranta	34	MT-402	Il-2		V

Bf 109G-2, MT-201 of 2/LeLv 34 flown by 2Lt Mauno Kirjonen in May 1943, based at Utti. (Karolina Hołda)

Mersu MT-201 of 2/LeLv 34 was assigned to the squadron CO Maj Luukkanen, here taking off from Utti in June 1943. It was also flown by many other pilots. Kirjonen scored his first Mersu kills in this machine.

Кокко, Pekka Johannes

13⅚ victories (12⅚ + 1 + 2) on appr. 150 missions
Planes: FR-86 and BW-379

Pekka Kokko was born at Jyväskylä on 03.09.18. He received flying training in 1936–37 at IRUK 6 and was accepted on 29.05.37 to KadK and was posted on 16.05.39 as 2nd Lieutenant to LLv 24. Kokko was promoted to 1st Lieutenant on 01.04.40. He was transferred on 24.11.41 to KoeL, where was promoted on 25.07.42 to Captain. Kokko was killed on a flying accident on 19.02.44, when his PY-32 hit the ground in poor weather at Lempäälä. Kokko was awarded with VR 4, VR 3 and EK 2.

From left mechanic Toivo Asikainen, 3/LLv 24 deputy leader 1Lt Pekka Kokko and an armourer, in front of Kokko's Brewster BW-379 at Rantasalmi in July 1941.

Kokko landed his BW-379 on 2 September 1941 at Immola, where the flight took a fortnight's rest, after flying top cover for the conquest of the Karelian Isthmus. (SA-kuva)

Brewster BW-379, of 3/LLv 24 flown by deputy flight leader 1Lt Pekka Kokko in September 1941, based at Immola. (Andrzej M. Olejniczak)

BW-379 of 3/LLv 24 at Immola in mid-September 1941. This plane carries one of very few personal markings, the pilot's first name "Pekka" behind the engine.

Date	Time	Area	Unit	Aircraft	Enemy aircraft	Enemy Unit	Victory type
01.12.39	14.20	Ristniemi	24	FR-87	SB	41 SBAP	E
19.12.39	11.30	Muolaanjärvi	24	FR-90	SB		V
23.12.39	10.50	Muolaanjärvi	24	FR-78	I-16		V
03.02.40	16.00	Korppoo	24	FR-86	⅓ DB-3	3/10 Abr	T
03.02.40	16.00	Korppoo	24	FR-86	2 x DB-3	3/10 Abr	T
29.06.41	10.40–12.05	Suursaari	24	BW-379	2 x MBR-2	15 AP	T
06.07.41	10.00–10.13	Parikkala	24	BW-379	½ MiG-3		T
08.07.41	14.20	Enso	24	BW-379	I-153		T
09.07.41	05.10–05.20	Lahdenpohja	24	BW-379	I-153		T
21.07.41	10.15–10.20	Käkisalmi	24	BW-363	I-153	65 ShAP	T
05.08.41	14.30–14.35	Ilmee	24	BW-379	I-153		T
12.08.41	13.00–13.45	Antrea – Kirvu	24	BW-379	I-153		E
23.09.41	13.30	Petäjäselkä	24	BW-364	½ I-16	155 IAP	T
23.09.41	14.00–14.15	Petäjäselkä	24	BW-364	I-16	155 IAP	T
26.09.41	10.15	Derevjannoje	24	BW-363	I-153	65 ShAP	T

Koskelainen, Arvo Ilmari

5 victories (3 + 2 + 3) on 140 missions
Planes: MT-237 and MT-455

Arvo Koskelainen was born at Vahviala on 24.01.17. He received flying training in 1940–41 at SOK 2 and was posted on 24.03.42 as Sergeant to LLv 30. Koskelainen was transferred on 17.06.43 to LeLv 24, where he was promoted on 26.10.44 to Staff Sergeant. He was demobilized on 10.11.44. In civilian life Koskelainen became a farmer. He died on 29.09.99. Koskelainen was awarded with VM 2, VM 1 and VR 4.

Bf 109G-6, MT-455 of 1/HLeLv 24 flown by Sgt Arvo Koskelainen in June 1944, based at Lappeenranta. (Karolina Hołda)

Mersu MT-455 of 1/HLeLv 24 at the edge of Lappeenranta airfield on 29 June 1944. It was assigned to Koskelainen, but he claimed only one of his five victories flying it. (SA-kuva)

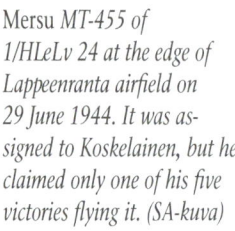

Date	Time	Area	Unit	Aircraft	Enemy aircraft	Enemy Unit	Victory type
10.06.44	09.55–11.00	Särkijärvi	24	MT-231	La-5		R
10.06.44	09.55–11.00	Särkijärvi	24	MT-231	La-5		V
17.06.44	13.30–14.30	Säiniö	24	MT-237	La-5		R
21.06.44	15.00–16.35	Pero	24	MT-244	Airacobra		T
29.06.44	07.40–09.05	Sommee	24	MT-455	Pe-2	58 BAP	R
02.07.44	19.30–20.45	Viipuri	24	MT-456	Il-2	448 ShAP	T
03.07.44	19.40–20.40	Juustila	24	MT-440	Pe-2		V
08.07.44	10.50–11.55	Tali – Juustila	24	MT-472	Airacobra		V

11⅓ victories (8⅓ + 3 + 2) on appr. 350 missions
Planes: CUw-551 and LG-3

Eino Koskinen was born at Nakkila on 23.08.07. He received flying training in 1932 at AOK 2. At Winter War mobilization Koskinen was posted on 09.10.39 as Master Sergeant to LLv 12 and was transferred on 06.02.40 to LLv 10. Koskinen was posted on 27.03.40 to KoeL, where he was promoted to Warrant Officer on 06.12.40. At the Continuation War mobilization Koskinen was posted on 20.06.41 to LLv 32. He was killed accidentally on 11.01.45. Koskinen was awarded VM 1, VM 2, VR 4, VR 3 and EK 2.

Curtiss Hawk 75A-2, CU-503 flown by 2/LeLv 32 pilot WO Eino Koskinen in July 1942, based at Nurmoila. (Karolina Hołda)

Curtiss CUw-551 of LLv 32 taxies at Suulajärvi in March 1942. It was frequently flown by Koskinen, who was one of nine pilots claiming a total of ten victories with it.

Date	Time	Area	Unit	Aircraft	Enemy aircraft	Enemy Unit	Victory type
16.07.41	14.45–14.50	Nurmi station	32	CU-553	½ I-153		T
15.09.41	15.45–17.25	Lempaalanjärvi	32	CU-560	DB-3F	1 AP, KBF	T
21.09.41	15.40–18.00	Ollila	32	CU-558	I-153	235 ShAP	T
28.03.42	17.40–18.00	Lavansaari – Suursaari	32	CU-558	I-16	71 IAP, KBF	T
07.04.42	15.20–16.00	Lumisuo	32	CU-568	2 x I-16		T
07.07.42	15.40–17.20	Lotinanpelto	32	CU-503	Pe-2		T
13.08.42	10.45–11.30	Savijärvi	32	CU-560	MiG-3		E
05.09.42	11.20–13.15	Lotinanpelto	32	CU-551	½ Pe-2		T
05.09.42	11.20–13.15	Lotinanpelto 10 km SW	32	CU-551	LaGG-3		E
05.09.42	11.20–13.15	Lotinanpelto	32	CU-551	LaGG-3		E
05.09.42	11.20–13.15	Ljugovitsha	32	CU-551	LaGG-3		V
14.11.42	08.20–10.50	Kuidosuo	32	CU-552	Pe-2		V
16.02.44	10.45–11.35	Uusi-Segesha 5 km E	32	LG-1	LaGG-3	415 IAP	T
08.05.44	09.10–10.50	Mäkriä	32	CU-587	⅓ La-5	415 IAP	T

LaGG-3, LG-1 flown by 2/HLeLv 32 pilot WO Eino Koskinen in February 1944, based at Nurmoila. (Karolina Hołda)

Koskinen taxies LaGG-3 serial LG-3 of LeLv 32 at Nurmoila in March 1943. Only the LaGG-3s had assigned pilots, Koskinen to LG-3 while WO Viljo Ikonen was assigned to LG-1. The aim was to catch fast Pe-2 bombers, but as it runed out, without success.

10¼ victories (10 ¼ + 0 + 1) on 187 missions
Planes: CUw-563, MT-227 and MT-225

Kullervo Lahtela was born at Nastola on 02.04.13. He received flying training in 1932–33 at IRUK 3 and was accepted to KadK on 06.06.33. He was posted on16.5.35 as 2nd Lieutenant to LLv 12. Lahtela was posted on 13.07.38 as 1st Lieutenant to ISK as an instructor. He was transferred on 25.02.40 to lead 2/LLv 22. On 29.03.40 he was posted to LLv 32, where was promoted to Captain on 23.01.41. Lahtela became 2/LLv 32 leader on 18.06.41 and 1/LLv 32 leader on 01.11.41. He was posted on 11.02.43 to LeLv 34 heading the 2nd Flight. On 16.02.44 he became 2/HLeLv 30 leader. On 23.06.44 Lahtela was posted to command HLeLv 32 and was promoted to Major on 18.07.44. After the was Lahtela acted as HLeLv 11 commander and resigned as a Lieutenant Colonel on 10.11.48. In civilian life Lahtela was a managing director. He died on 01.08.86. Lahtela was awarded with VR 4, VR 3, VR 3 tlk and EK 2.

Curtiss CUw-563 was pre-ferred by the flight leaders, including 2/LLv 32 leader Capt Lahtela. It is seen here at Lappeenranta in September 1941. Patched bullet holes were marked by white rings, after a well-known WW I practise.

Bf 109G-2, MT-227 of 2/LeLv 34 flown by the flight leader Capt Kullervo Lahtela in May 1943, based at Utti. (Karolina Hołda)

Lahtela bellied his MT-227 at Utti on 21 May 1943, after forgetting to lower the landing gear. It took twelve days to fix the plane, chang-ing the radiators and propeller.

MT-227 of 2/LeLv 34 again in working order at Utti in July 1943, now assigned to SSgt Niilo Erkinheimo, while Lahtela took over MT-225.

MT-227 of 2/LeLv 34 in its specified parking area at Utti in May 1943, as marked by the sign on the tree. On the wing stands MT-227's assigned mechanic, Esko Laiho.

Date	Time	Area	Unit	Aircraft	Enemy aircraft	Enemy Unit	Victory type
12.08.41	15.50–17.05	Ylikuunu	32	CU-569	I-153		T
14.08.41	05.00–06.00	Leinjärvi	32	CU-565	I-153	7 IAP	T
15.09.41	16.00–17.00	Valkjärvi	32	CU-563	DB-3	1 AP, KBF	T
16.10.41	16.00–17.00	Kivennapa	32	CU-570	I-16		T
05.11.41	13.45–15.00	Hanko	32	CU-562	¼ I-16		T
03.04.42	18.00–19.10	Seiskari	32	CU-556	I-153		T
25.07.42	21.20–22.00	Aunus – Nurmoila	32	CU-552	½ MBR	58 AE KBF	T
21.05.43	18.45–20.00	Lavansaari	34	MT-227	Il-2	7 GShAP, KBF	T
23.05.43	15.30–16.20	Tytärsaari – Lavansaari	34	MT-225	½ I-16	71 KIAP, KBF	T
02.06.43	14.40–15.45	Karavaldai	34	MT-230	LaGG-3		V
17.08.43	10.05–10.45	Someri	34	MT-225	Il-2	7 GShAP, KBF	T
23.08.43	16.35–17.20	Tallinna 15 km NW	34	MT-225	Boston	15 ORAP, KBF	T
10.10.43	14.10–14.45	Porkkala – Kallbåda	34	MT-225	Pe-2	15 ORAP, KBF	T

10 victories (9 + 1 + 0) on 75 missions
Planes: MT-441

Ahti Laitinen was born at Joroinen on 08.10.22. He received flying training in 1940–-43 at AOK 11 and UK 12 and was posted on 14.04.43 as 2nd Lieutenant to LeLv 24. Laitinen was promoted on 14.03.44 to 1st Lieutenant. He became a POW on 29.06.44, when an La-5 of 159 IAP shot down his MT-439, bailing out at Ihantala. Laitinen returned on 25.12.44 to become hospitalized and then demobilized on 13.11.45. In civilian life Laitinen acted in the Court of Appeal and held the rank of Captain. He died on 14.09.85. Laitinen was awarded with VR 4, VR 3 and VR 3 tlk.

Mersus *after the war at Utti in September 1944. At left is MT-431 of 2/HLeLv 24 and at right MT-441 of 3/HLeLv 24. The latter was assigned to Laitinen before he was shot down and captured in MT-439 on 29 June 1944.)*

Bf 109 G-6, MT-441 of 3/IILeLv 24 flown by 1Lt Ahti Laitinen in June 1944, based at Lappeenranta. (Artur Juszczak)

Date	Time	Area	Unit	Aircraft	Enemy aircraft	Enemy Unit	Victory type
31.08.43	15.45–16.30	Sepeleva 10 km W	24	BW-393	La-5		T
04.11.43	14.45–15.30	Seiskari	24	BW-366	Yak-1		T
09.06.44	06.15–07.20	Termola	24	MT-241	Airacobra		T
20.06.44	07.15–08.00	Koivisto – Seiskari	24	MT-441	Yak-9		T
20.06.44	10.40–11.25	Säiniö – Kämärä	24	MT-441	2 x Il-2	943 ShAP	T
20.06.44	19.05–20.20	Muolaanjärvi	24	MT-441	Pe-2		T
22.06.44	17.10–18.10	Tali station	24	MT-441	Pe-2		R
28.06.44	14.00–14.55	Lyykylänjärvi	24	MT-441	Airacobra	196 IAP	T
28.06.44	16.25–17.35	Karisalmi	24	MT-461	La-5	159 IAP	T

Lakio, Vilppu Mikael

5 victories (5 + 0 + 0) on appr. 150 missions
Planes: BW-382

Vilppu Lakio was born at Sortavala on 18.10.14. He received flying training in 1937–38 at IRUK 7. In the Winter War the training continued at T-LentoR 2 and he was demobilized on 26.07.40. Lakio recruited on 03.06.41 as 2nd Lieutenant to LLv 32. He was transferred on 11.08.41 to LLv 24, where he was promoted to 1st Lieutenant on 22.02.42. Lakio was posted to on 01.08.43 to VL and transfer back to HLeLv 24 came on 15.04.44. After the war Lakio served with LeV, SatLsto and IlmavV, resigning on 17.10.74 as Engineer Lieutenant Colonel. He died on 02.10.81. Lakio was awarded with VR 4.

1/LeLv 24 pilot 1Lt Lakio stands in front of his Brewster at Suulajärvi in June 1942.

Above right: *Lakio flew jointly with WO Rimminen 1/LLv 24 plane BW-382, hidden here under the trees at Nurmoila in November 1941.*

Lakio stands on the wing of 1/HLeLv 24 Mersu MT-506 at Lappeenranta after the hostilities with the Russians, which ended on 4 September 1944.

Date	Time	Area	Unit	Aircraft	Enemy aircraft	Enemy Unit	Victory type
18.08.41	16.25–16.35	Paakkola	24	BW-382	½ I-153	65 ShAP	T
06.10.41	14.30–16.30	Vosnesenja	24	BW-374	½ SB		T
18.08.42	19.20–21.40	Kronstadt	24	BW-376	Pe-2		T
18.08.42	19.20–21.40	Kronstadt	24	BW-376	I-16		T
30.10.42	11.30–11.50	Karavaldai lake	24	BW-373	I-16		T
20.05.43	09.15–10.30	Seiskari 10 km S	24	BW-356	Yak-1	21 IAP, KBF	T

14 victories (11 + 3 + 2) on 268 missions
Planes: BW-354 and MT-235

Heimo Lampi was born at Hollola on 29.02.20. He received flying training in 1940–41 RAOK 6 and was posted as Corporal on 01.03.41 to LLv 24. Lampi was promoted on 13.07.41 to Sergeant and on 16.01.42 to Staff Sergeant. On 08.01.43 he began the RUK and returned on 13.06.43 to LeLv 24, where he was promoted to 2nd Lieutenant on 18.08.43. After the war Lampi served with HLeLv 23 resigning on 20.10.46. In civilian life Lampi became a lawyer and acted as the president of Kouvola Court of Appeal. He held the rank of Major. Lampi died on 01.06.98. He was awarded with VM 1, VR 4 and VR 4 tlk.

Brewster 239, BW-354 of 2/LeLv 24 flown by SSgt Heimo Lampi in September 1942, based at Tiiksjärvi. (Andrzej M. Olejniczak)

A quartet of 2/LeLv 24 Brewsters over Rukajärvi in late September 1942. From the camera are BW-354, 352, 357 and 356. BW-354 was assigned to Lampi on 18 June 1941 and he flew it over an 18 month period.)

BW-354 of 2/LLv 24 parked at Tiiksjärvi, shortly after Lampi had scored his latest two air victories on 30 March 1942.

Lampi sitting in his BW-354 at Tiiksjärvi in late March 1942. He became an ace on 6 August 1942.

Date	Time	Area	Unit	Aircraft	Enemy aircraft	Enemy Unit	Victory type
25.06.41	07.15–07.45	Selänpää - Heinola	24	BW-354	2½ x SB	201 SBAP	T
30.03.42	15.50–16.10	Rukajärvi	24	BW-354	Hurricane	152 IAP	T
30.03.42	15.50–16.10	Ideljärvi	24	BW-354	Hurricane		E
06.08.42		Ontajärvi	24	BW-354	½ Tomahawk		T
02.04.44	16.05–16.15	Ino	24	BW-382	LaGG-3		T
17.05.44	11.10–12.30	Kulesjärvi 6 km SW	32	MT-232	La-5		T
17.06.44	06.30–07.40	Vammeljärvi	24	MT-235	La-5		T
20.06.44	08.30–09.40	Oinala	24	MT-235	Airacobra	196 IAP	R
20.06.44	10.45–12.00	Kämärä station	24	MT-235	Il-2	872 ShAP	T
20.06.44	10.45–12.00	Kämärä station	24	MT-235	La-5		R
26.06.44	12.00–13.10	Ihantala	24	MT-454	2 x Yak-9		V
30.06.44	10.45–11.55	Kärstilä	24	MT-464	Yak-9	404 IAP	T
02.07.44	19.30–20.45	Valkjärvi	24	MT-464	Il-2	448 ShAP	T
10.07.44	12.00–13.25	Vuosalmi	24	MT-477	La-5	159 IAP	T

LÄNSIVAARA, Osmo Ilmari

5 victories (3 + 2 + 2) on 268 missions
Planes: FA-31, MT-425 and MT-453

Osmo Länsivaara was born at Turku on 18.02.21. He received flying training in 1940–41 at SOK 3. Länsivaara was posted on 02.07.41 to LLv 26, where he was promoted Corporal 10 days later. On 14.12.41 came promotion to Sergeant. On 24.03.43 Länsivaara was transferred to LeLv 34 and was promoted on 25.07.43 to Staff Sergeant. He was demobilized on 11.11.44. Länsivaara recruited on 01.02.45 to LeR 1 and resigned as a Warrant Officer on 21.08.59. In civilian life Länsivaara acted as an operations manager. He died on 06.12.10. Länsivaara was awarded with VM 2, VM 1 and VR 4.

Bf 109G-6/R6, MT-453 of 1/HLeLv 34 flown by SSgt Osmo Länsivaara in July 1944, based at Taipalsaari. (Karolina Hołda)

Mersu *MT-453 of 1/HLeLv 34 seen at Taipalsaari in July 1944. This* Kanonenboot *was assigned to Länsivaara, who became an ace in it on 4 July 1944 after downing an Il-2* Shturmovik.

Date	Time	Area	Unit	Aircraft	Enemy aircraft	Enemy Unit	Victory type
24.08.42	16.40–17.20	Morje	26	FA-31	I-16		T
26.08.42	11.05–12.00	Konevitsa	26	FA-21	Il-2	57 AP, KBF	T
26.02.44	14.00–14.10	Sepeleva lighthouse	34	MT-237	Pe-2		T
06.03.44	17.10–17.35	Haapasaari	34	MT-209	La-5		V
20.06.44	08.50–10.00	Kämäränjärvi	34	MT-425	Yak-9		R
03.07.44	11.05–12.20	Ihantala – Viipuri	34	MT-469	Il-2		V
04.07.44	04.50–06.05	Pölläkkälä	34	MT-453	Il-2		R

Lautamäki, Lauri Johannes

6½ victories (5½ + 1 + 0) on appr. 250 missions
Planes: GL-253 and FA-18

Lasse Lautamäki was born at Ylistaro on 28.10.09. He received flying training in 1928–29. Lautamäki was transferred on 31.05.38 as Master Sergeant to LLv 26. Promotion to Warrant Officer came on 31.03.39. Lautamäki was transferred on 08.05.43 to KoeL as a test pilot. He resigned on 24.08.49. Lautamäki acted in civilian life as a professional pilot. He died on 10.03.70. Lautamäki was awarded with VM 1 and VR 4.

FIAT G.50, FA-26 of 1/LLv 26 flown by WO Lauri Lautamäki in September 1941, based at Lunkula. (Karolina Hołda)

Due to plane shortages, FA-26 was flown jointly by two warrant officers, Oiva Tuominen and Lauri Lautamäki. The horizontal kill bars denotes Lautamäki's claim on 13 August 1941. (SA-kuva)

During the Winter War Lautamäki was assigned to Gladiator GL-253 of 2/LLv 26. It is seen here at Helsinki Malmi in summer 1940 belonging to ErLLv.

Date	Time	Area	Unit	Aircraft	Enemy aircraft	Enemy Unit	Victory type
11.02.40	13.35	Jalovaara	26	GL-253	I-16	49 IAP	E
13.02.40	14.00–15.30	Jänisjärvi	26	GL-253	½ SB	39 SBAP	T
26.02.40		Heinjoki	26	GL-253	I-16		T
13.08.41	13.45–14.45	Aunuksenjoki	26	FA-26	I-153		T
15.08.41	17.40–18.25	Troitsankontu	26	FA-26	I-153		T
24.08.42	12.30–13.10	Saunasaari	26	FA-18	I-16		E
26.08.42	11.05–12.00	Konevitsa	26	FA-13	Hurricane	3 GIAP, KBF	T

Lehtovaara, Urho Sakari

41½ victories (26½ + 15 + 10) on appr. 400 missions
Planes: MS327, MS-304, MT-213 and MT-404

Urho Lehtovaara was born at Pyhäjärvi OL on 27.10.17. He received flying training in 1938–39 at RAOK 4 and AOK 9 and was posted as Corporal on 01.09.39 to LentoR 4. On 10.02.40 he was transferred to LLv 28. On 23.03.40 Lehtovaara was promoted to Sergeant, on 23.07.41 to Staff Sergeant. And on 11.09.42 to Master Sergeant. He was transferred on 28.03.43 to LeLv 34, where on 26.04.44 became a Warrant Officer. Lehtovaara was posted on 15.09.44 to HLeLv 28. After the war Lehtovaara served with HLeLv 21 and resigned on 20.11.46. In civilian life he managed a cinema theatre, but died on 05.01.49. Lehtovaara was awarded VM 2, VM 1, VR 4, VR 4 tlk, VR 3 and MHR on 09.07.44.

Lehtovaara in front of his MS327 of 2/LLv 28 at Joensuu on 23 July 1941. He had just shot down two Tupolev SB bombers and been promoted to staff sergeant. Behind is MS-606.

From left mechanic Jukka Paajanen and Lehtovaara in front of MS327 at Karkunranta on 9 September 1941, when Lehtovaara made his 10th confirmed kill. To the right is 2/LLv 28 pilot SSgt Urho Lehtovaara in the cockpit of his Morane MS327.

Date	Time	Area	Unit	Aircraft	Enemy aircraft	Enemy Unit	Victory type
02.03.40	14.35	Kiikala	28	MS-326	SB	35 LBAP	T
03.07.41	10.00–12.15	Ilomantsi	28	MS-327	DB-3		T
09.07.41	14.40–15.00	Räisälä	28	MS-327	MiG-3	7 IAD	T
09.07.41	14.40–15.00	Räisälä – Lumivaara	28	MS-327	2 x SB	202 SBAP	E
17.07.41	10.05–10.15	Jänisjärvi	28	MS-327	I-16		V
23.07.41	19.15–19.30	Nuosjärvi	28	MS-314	2 x SB	72 SBAP	T
17.08.41	18.15–18.25	Sortavala	28	MS-327	I-16		E
09.09.41	10.00–10.10	Svir	28	MS-304	2 x I-16		T
09.09.41	10.00–10.10	Svir	28	MS-304	I-16		E
15.08.42	04.50–04.55	Limosaari	28	MS-323	Pe-2		V
12.01.43	12.35–14.45	Ala–Voljärvi	28	MS-627	Hurricane		R
05.03.43	13.55–15.00	Liistepohja	28	MS-641	Pe-2		T
05.03.43	13.55–15.00	Liistepohja	28	MS-641	I-16		R

Morane Saulnier MS.406, MS327 of 2/LLv 28 flown by SSgt Urho Lehtovaara in September 1941, based at Karkunranta. (Karolina Hołda)

Morane MS327 of 2/LLv 28 taxiing at Viitana in November 1941. It was assigned to Lehtovaara on 18 June 1941 for six months. Behind is MS326.

Lehtovaara scored three victories in MS-304, before it was handed over to 3/LLv 28. It is seen here at Äänislinna on 22 February 1942. The leading edge of the fin carries six victory bars. (SA-kuva)

Date	Time	Area	Unit	Aircraft	Enemy aircraft	Enemy Unit	Victory type
19.04.43	05.30–06.40	Lavansaari	34	MT-216	Pe-2		T
02.05.43	10.15–10.55	Lavansaari	34	MT-216	La-5		T
16.05.43	04.50–05.45	Peninsaari 10 km N	34	MT-213	LaGG-3		T
16.05.43	04.50–05.45	Peninsaari 20 km SW	34	MT-213	LaGG-3		T
20.05.43	19.50–20.50	Peninsaari 10 km NW	34	MT-218	I-153		E
20.05.43	19.50–20.50	Peninsaari 10 km NW	34	MT-218	I-153		V
05.07.43	20.20–21.15	Suursaari 16 km SW	34	MT-209	Boston	15 ORAP, KBF	E
01.08.43	16.05–17.10	Tytärsaari 15 km NW	34	MT-223	Pe-2	73 BAP, KBF	T
01.08.43	16.05–17.10	Tytärsaari 25 km NW	34	MT-223	½ Pe-2	73 BAP, KBF	T
08.09.43	12.05–12.50	Tytärsaari 10 km NW	34	MT-216	Yak-7B	13 KIAP, KBF	T
04.11.43	12.40–13.40	Someri 15 km SW	34	MT-209	La-5		R
04.11.43	12.40–13.40	Someri 15 km SW	34	MT-209	La-5		R
26.02.44	15.40–15.45	Suursaari	34	MT-216	Pe-2		V
06.03.44	13.50–14.30	Someri	34	MT-235	Pe-2	12 GPBAP, KBF	T
06.03.44	13.50–14.30	Someri	34	MT-235	2 x La-5		V
06.03.44	17.10–17.50	Hamina	34	MT-235	La-5		T
07.04.44	07.35–08.20	Lavansaari	34	MT-235	Pe-2		V
17.05.44	10.30–11.15	Hamina	34	MT-404	Pe-2	12 GBPAB, KBF	T
17.05.44	10.30–11.15	Hamina	34	MT-404	Yak-9		T
17.06.44	05.20–06.35	Ino	34	MT-417	Airacobra	102 GIAP	T
17.06.44	05.20–06.35	Ino	34	MT-417	Yak-9		R
19.06.44	20.00–20.55	Koivisto	34	MT-406	2 x Pe-2		T
19.06.44	20.00–20.55	Kämärä	34	MT-406	2 x Airacobra	196 IAP	T
22.06.44	17.05–18.25	Kämärä – Leipäsuo	34	MT-432	2 x DB –3F		V
23.06.44	12.15–13.05	Viipuri	34	MT-432	La-5		R
01.07.44	13.10–14.05	Vahviala	34	MT-404	2 x Pe-2		R
02.07.44	20.05–20.55	Viipuri	34	MT-448	3 x Pe-2		R
25.07.44	11.20–12.25	Someri	34	MT-405	Airacobra		R
25.07.44	11.20–12.25	Someri	34	MT-405	Il-2		V

3/LeLv 34 Mersus MT-216 and 213 collided at Utti on 20 May 1943. The latter was assigned to Lehtovaara until this accident. He also scored four victories with MT-216.

Leino, Hemmo Kullervo

10 victories (4 + 6 + 1) on 251 missions
Planes: FR-146, MS-319 and MT-423

Hemmo Leino was born at Helsinki on 08.04.21. He received flying training in 1939–41 at SOK 1 and was posted on 16.09.41 as Corporal to LLv 10 and on 01.11.41 further to LLv 30. Leino was promoted to Sergeant on 06.03.42 and was transferred on 01.08.42 to LeLv 14. He was posted on 19.04.43 to LeLv 34, where he was promoted on 18.10.43 to Staff Sergeant. After the war Leino served with LeR 3, resigning on 15.05.45. In civilian life he acted as an airline captain. Leino died in 2013. He was awarded with VM 1, VR 4 and VR 4 tlk.

Fokker FR-146 of LtueKäär/LLv 14 in gun harmonization at Tiiksjärvi in early 1942. It was assigned to Leino, who shot down a Hurricane with it on 9 January 1942.

Morane MS-319 of 1/LeLv 14 taxies at Tiiksjärvi in summer 1943. It was assigned to Leino, whose two victory bars of the 16 March 1943 combat can just be seen on the rudder.

Date	Time	Area	Unit	Aircraft	Enemy aircraft	Enemy Unit	Victory type
09.01.42	12.15–12.45	Ontrosenvaara	14	FR-146	Hurricane		E
19.01.42	08.10–09.10	Ontajärvi	14	FR-146	½ R-5	669 AP	T
05.11.42	11.55–13.40	Voijärvi	14	MS-313	LaGG-3		T
16.03.43	14.25–14.35	Kotskoma 5 km SW	14	MS-319	2 x I-15bis		E
10.10.43	09.25–09.55	Luumäki	34	MT-204	½ Boston	1 GMTAP, KBF	T
16.05.44	04.30–05.10	Kiuskeri	34	MT-418	Boston	859 ORAP, KBF	T
14.06.44	13.35–14.40	Kuuterselkä	34	MT-422	Il-2	566 ShAP	T
18.06.44	07.35–08.50	Koivisto	34	MT-424	Yak-9		R
20.06.44	07.10–08.10	Vatnuori	34	MT-423	Yak-9		R
05.07.44	09.40–10.55	Vatnuori	34	MT-402	Yak-9		R
05.07.44	09.40–10.55	Vatnuori	34	MT-402	Yak-9		V

Bf 109G-6, MT-423 of 1/HLeLv 34 flown by SSgt Hemmo Leino in June 1944, based at Kymi. (Karolina Hołda)

Leino sitting on the wing of 2/LeLv 34 Mersu MT-225 at Helsinki Malmi in August 1943.

Mersu MT-423 of 1/HLeLv 34 parked at Kymi in mid-June 1944. It was assigned to Leino who scored one victory with it. The rudder bears the new squadron emblem, an eagle fledgling.

LINDBERG, Kim Konrad

5 victories (4 + 1 + 1) on 52 missions
Planes: BW-365

Kim Lindberg was born at Stuttgart on 01.09.16. He received flying training in 1939–40 at IRUK 9 and was demobilized on 20.06.40. At the Continuation War mobilization Lindberg was posted on 30.06.41 as 2nd Lieutenant to LLv 24. He was posted to Germany on 01.09.41 returning on 20.03.43 to LeLv 24 as 1st Lieutenant. On 01.09.43 followed another assignment to Germany and return on 29.06.44 to LeR 3 liaison officer. Lindberg was transferred on 17.09.44 to TLeLv 16 and demobilized on 13.11.44. In civilian life he became and aviation inspector in Germany. Lindberg was awarded with VR 4 twice.

Lindberg in front of his Brewster BW-365 of 3/LLv 24 at Rantasalmi in July 1941.

The pilots of 3/LLv 24 in front of BW-365 at Rantasalmi on 8 August 1941. From left Katajainen, Mellin, Huotari, Karhunen, Kokko, Juutilainen, Strömberg and Lindberg. All but Strömberg became aces. (SA-kuva)

Date	Time	Area	Unit	Aircraft	Enemy aircraft	Enemy Unit	Victory type
01.08.41	16.00–16.10	Rautjärvi	24	BW-362	MiG-3		E
01.08.41	16.00–16.10	Kirvu	24	BW-362	MiG-3		V
12.08.41	13.00–13.30	Antrea - Kirvu	24	BW-365	I-153	65 ShAP	T
14.04.43	06.45–08.15	Seiskari 10 km SW	24	BW-373	La-5		T
14.04.43	06.45–08.15	Seiskari 15 km SW	24	BW-373	Spitfire (?)		T
18.04.43	17.05–18.35	Seiskari – Sepelava	24	BW-393	Yak-1	21 IAP, KBF	T

LINNAMAA, Aarre Päiviö

6 victories (6 + 0 + 3) on appr. 100 missions
Planes: MS-308 and MS-607

Aarre Linnamaa was born at Pori mlk on 29.06.18. He received flying training in 1937–38 at IRUK 7. In the Winter War mobilization Linnamaa was posted on 09.10.39 as 2nd Lieutenant to T-LentoR 2 and was transferred on 15.02.40 to LLv 28. He was demobilized on 06.06.40. Linnamaa recruited on 03.06.41 back to LLv 28. Linnamaa was killed on 24.04.42, when he made a forced landing in his MS-621 to enemy side south of Lake Onega. Between the lines he walked in a mine field and shot himself to avoid capture. He was awarded with VR 4 and VR 4 tlk.

Morane MS-308 of 1/LLv 28 at Joensuu in July 1941, assigned to Linnamaa.

Linnamaa in front of his MS-308 at Joensuu in July 1941. He flew it until 5 September 1941 when it was damaged in an air raid.

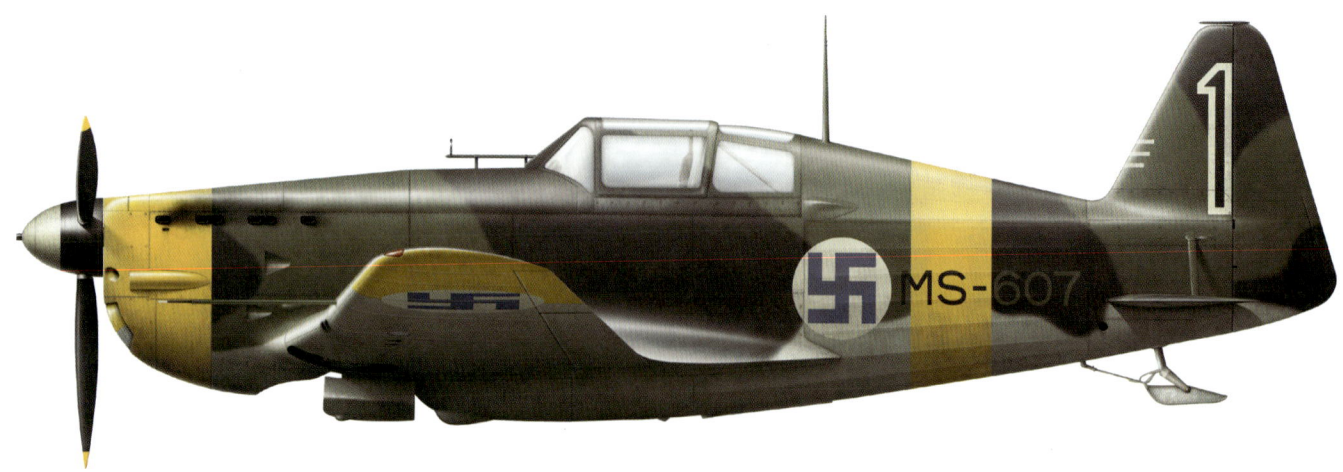

Morane Saulnier MS.406, MS-607 of 1/LLv 28 flown by 2Lt Aarre Linnamaa in March 1942, based at Viitana. (Karolina Hołda)

Morane MS-607 of 1/LLv 28 camouflaged under the trees at Viitana on 17 March 1942, assigned to Linnamaa. Two of the victory bars on the fin belong to Linnamaa and one to Hyrkki. (SA-kuva)

Date	Time	Area	Unit	Aircraft	Enemy aircraft	Enemy Unit	Victory type
11.03.40	15.15	Loviisa	28	MS-306	DB-3	7 DBAP	T
12.08.41	08.15–09.45	Vieljärvi	28	MS-611	I-15bis		T
12.08.41	08.15–09.45	Vieljärvi	28	MS-611	I-15bis		V
12.09.41	07.10–08.30	Pyhäjärvi – Onega	28	MS-607	2 x SB	72 SBAP	T
12.09.41	07.10–08.30	Pyhäjärvi – Onega	28	MS-607	SB		V
04.10.41	09.50–11.30	Suopohja	28	MS-317	I-16		V
05.02.42	09.45–09.50	Osta – Kuzra	28	MS-621	2 x R-5		T

LUMME, Kaarlo Aulis

16½ victories (15½ + 1 + 1) on 287 missions
Planes: BW-370

Aulis Lumme was born at Nurmijärvi on 08.01.17. He received flying training in 1937–38 at RAOK 3 and IRUK 7. The training continued on 23.10.39 at T-LentoR 2 and he was demobilized on 31.05.40. At the Continuation War mobilization Lumme was posted on 02.07.41 as 2nd Lieutenant to LLv 24. He was promoted to 1st Lieutenant on 22.02.42. Lumme became 2/LeLv 24 leader on 02.05.43 for four months. He was demobilized on 15.11.44. In civilian life Lumme became a headmaster of a vocational school and held a rank of Captain. He died on 04.06.92. Lumme was awarded with VR 4, VR 4 tlk, VR 3, VR 3 tlk and EK 2.

Brewster 239, BW-370 of 4/LeLv 24 flown by the flight deputy leader 1Lt Aulis Lumme in August 1942, based at Römpötti. (Karolina Hołda)

BW-370 of 4/LeLv 24 at Römpötti in August 1942. It was assigned to Lumme on 10 March 1942 for one year. The fin carries the flight's osprey emblem.

Lumme flew several Mersus for his air victories, MT-213 of 2/HLeLv 24 was one of them. It is here at Suulajärvi in May 1944. The man with the stick is the squadron CO Maj Karhunen.

Date	Time	Area	Unit	Aircraft	Enemy aircraft	Enemy Unit	Victory type
18.08.41	16.25–16.35	Paakkola	24	BW–394	½ I–153	195 IAP	T
26.10.41	15.15	Monastirskaja	24	BW–359	MiG–3		E
04.11.41	14.10	Koukkula	24	BW–359	½ MiG–3		T
09.01.42	10.15–12.05	Uikujärvi	24	BW–377	R–5		T
29.03.42	10.20	Seltokoski	24	BW–370	½ R–5		T
12.08.42	17.20–19.30	Tolli lighthouse	24	BW–380	Il–2		T
22.11.42	09.15–09.45	Kreivinlahti	24	BW–370	Spitfire ?		T
23.11.42	10.50–12.30	Seiskari	24	BW–380	Spitfire ?		T
10.03.43	15.15–16.25	Peninsaari	24	BW–370	Pe–2		T
18.04.43	17.00–18.20	Sepeleva lighthouse	24	BW–370	Il–2	7 GShAP, KBF	T
18.04.43	17.00–18.20	Yhinmäki	24	BW–370	LaGG–3		T
04.05.43	10.40–11.55	Seiskari	24	BW–387	I–153		T
20.08.43	16.40–17.50	Harjavallanniemi	24	BW–383	La–5	4 GIAP, KBF	T
10.06.44	06.25–07.40	Siestarjärvi	24	MT–213	Pe–2		T
10.06.44	09.55–11.00	Kivennapa	24	MT–213	Pe–2		V
21.06.44	08.40–09.30	Papula	24	MT–221	Il–2	703 ShAP	T
23.06.44	13.30–14.30	Kämärä station	24	MT–221	Airacobra		T
26.06.44	10.30–11.10	Lyykylänjärvi	24	MT–206	Il–2		T
08.07.44	11.30–12.15	Viipuri	24	MT–202	Yak–9	29 GIAP	T

54 victories (44 + 10 + 12) on 441 missions
Planes: FR-108, BW-375, BW-393, MT-201, MT-417 and MT-415

Eino Luukkanen was born in Jaakkima on 04.06.09. On 29.09.30 he was accepted to KadK and transferred as 2nd Lieutenant on 07.08.33 to LAs 4. On 26.01.38 he became as 1st Lieutenant the adjutant of LentoR 1. On 06.09.39 he was appointed to lead 3/LLv 24 and on 15.02.40 was promoted to Captain. In the mobilization on 18.06.41 Luukkanen was transferred to lead 1/LLv 24. On 01.11.42 he was promoted to Major and after one week posted to command LeLv 30. On 27.03.43 he was made commander of LeLv 34. After the war Luukkanen acted as commander of HLeLv 33. On 13.02.48 he was promoted to Lieutenant Colonel and given command of LeR 2. On 08.11.51 he was relieved of duty. Luukkanen died on 10.04.64. He was awarded VR 4, VR 3, VR 3 tlk, VR 2, EK 2, EK 1 and MHR on 18.06.44.

FR-108 of LLv 32 about to take off from Siikakangas in June 1940. A few months earlier in the Winter War it was flown in the same markings by 3/LLv 24 leader Capt Luukkanen.

At the beginning of the Continuation War Luukkanen was assigned to BW-375, now heading 1/LLv 24. The machine is seen here at Nurmoila in mid-September 1941.

Date	Time	Area	Unit	Aircraft	Enemy aircraft	Enemy Unit	Victory type
01.12.39	13.10	Koljolan station	24	FR-104	SB	24 SBAP	T
23.12.39	10.15	Leipäsuo	24	FR-108	½ R-5	16 KAO	T
06.01.40	13.10	Uomaa	24	FR-108	SB		T
08.07.41	04.45–05.00	Saari	24	BW-375	I-153		V
08.07.41	05.10–05.25	Jaakkima	24	BW-375	I-153		E
18.08.41	16.25–16.35	Paakkola	24	BW-375	½ I-153	65 ShAP	T
27.09.41	17.00–17.05	Derevjannoje	24	BW-390	½ I-153	152 IAP	T
06.10.41	14.30–16.30	Vosnesenja	24	BW-373	½ SB	72 SBAP	T
15.10.41	15.05–15.20	Osta – Vytegra	24	BW-375	SB	72 SBAP	T
07.11.41	08.30–10.30	Lotinanpelto	24	BW-375	LaGG-3	415 IAP	T
01.05.42	16.25–18.55	Monastirskaja	24	BW-390	I-153		E
20.07.42	04.30–06.00	Pietari	24	BW-376	Pe-2		E
06.08.42		Seiskari	24	BW-393	I-16		T
31.08.42	19.50–20.05	Lavansaari	24	BW-393	I-153		T
13.10.42	16.25–16.40	Kreivinlahti	24	BW-393	Il-4	1 GMTAP, KBF	V
22.10.42	09.30–09.45	Kreivinlahti	24	BW-393	2 x I-16		T
30.10.42	11.30–11.50	Karavaldai lake	24	BW-393	Spitfire (?)		T
30.10.42	11.30–11.50	Karavaldai lake	24	BW-393	I-16		T
19.05.43	19.35–20.35	Peninsaari	34	MT-224	I-153		E
21.05.43	19.10–19.20	Seiskari 10 km SW	34	MT-201	La-5		E
21.05.43	19.10–19.20	Seiskari 10 km SW	34	MT-201	La-5		V
22.05.43	10.50–10.55	Lavansaari	34	MT-201	La-5	4 GIAP, KBF	T
24.06.43	05.50–17.00	Krasnaja Gorka	34	MT-215	LaGG-3		T
24.06.43	05.50–17.00	Krasnaja Gorka	34	MT-215	Boston		V
16.07.43	16.29	Seiskari 20 km E	34	MT-201	Pe-2	15 ORAP, KBF	E
20.07.43	05.00–05.50	Suursaari 20 km SW	34	MT-201	2 x LaGG-3	13 KIAP, KBF	T
04.09.43	13.05–13.50	Seiskari	34	MT-201	La-5		T
08.09.43	12.05–12.50	Tytärsaari 30 km NW	34	MT-201	Yak-7B	13 KIAP, KBF	T
27.10.43	09.50–10.45	Lavansaari 10 km W	34	MT-201	Il-2		R
27.10.43	14.10–15.05	Suursaari – Someri	34	MT-201	La.5		V
04.11.43	12.50–13.40	Kotka – Someri	34	MT-201	2 x La-5		V
06.03.44	13.45–14.15	Kilpisaari	34	MT-201	Pe-2	12 GPBAP, KBF	T
06.03.44	13.45–14.15	Narvi	34	MT-201	Pe-2	12 GPBAP, KBF	R
06.03.44	13.45–14.15	Narvi	34	MT-201	La-5		T
26.04.44	11.55–12.45	Kotka	34	MT-236	La-5		V
08.05.44	07.25–07.50	Someri – Suursaari	34	MT-417	Pe-2	12 GPBAP, KBF	T
08.05.44	13.00–13.30	Haapasaari	34	MT-417	Yak-9		V
17.05.44	10.30–11.00	Hamina – Tammio	34	MT-417	Pe-2	12 GBPAB, KBF	T
17.05.44	10.30–11.00	Hamina – Tammio	34	MT-417	Yak-9		V
14.06.44	10.50–11.50	Kuuterselkä – Terijoki	34	MT-422	2 x La-5	159 IAP	T
14.06.44	17.15–18.15	Kuuterselkä	34	MT-417	Airacobra		T
15.06.44	09.45–10.55	Valkjärvi	34	MT-417	Pe-2	140 BAP	V
17.06.44	11.10–12.20	Perkjärvi	34	MT-417	Il-2	943 ShAP	T
17.06.44	11.10–12.20	Kaukjärvi	34	MT-417	La-5		R
20.06.44	17.50–19.00	Säiniö	34	MT-415	Pe-2		T
21.06.44	12.55–14.00	Tali	34	MT-415	Il-2	999 ShAP	T
23.06.44	12.15–12.50	Tali station	34	MT-415	Il-2	703 ShAP	T
23.06.44	12.15–12.50	Ristseppälä – Raulampi	34	MT-415	2 x Il-4		R
28.06.44	15.00–16.10	Jäppilä	34	MT-415	U-2		T

Date	Time	Area	Unit	Aircraft	Enemy aircraft	Enemy Unit	Victory type
28.06.44	15.00–16.10	Jäppilä	34	MT-415	La-5		V
30.06.44	20.05–21.10	Tali – Juustila	34	MT-415	La-5		T
30.06.44	20.05–21.10	Tali – Juustila	34	MT-415	Airacobra		T
01.07.44	10.15–11.25	Uuras	34	MT-415	Yak-9		V
03.07.44	11.05–12.20	Tammisuo	34	MT-415	Pe-2		T
03.07.44	11.05–12.20	Uuras	34	MT-415	Il-2		R
05.07.44	09.45–10.40	Tuppura	34	MT-415	Il-2	13 AAE, KBF	T
05.07.44	09.45–10.40	Tuppura	34	MT-415	LaGG-3	11 IAP, KBF	T
09.07.44	11.30–12.40	Pölläkkälä	34	MT-415	Yak-9	14 GIAP	T
15.07.44	09.55–10.55	Vuosalmi	34	MT-415	Yak-9		V
18.07.44	13.50–15.10	Vuosalmi	34	MT-415	Airacobra		T
25.07.44	11.20–12.35	Someri	34	MT-415	Yak-9		T
25.07.44	11.20–12.35	Ulko–Tammio	34	MT-415	Il-2	8 GShAP, KBF	R
05.08.44	12.45–13.35	Narvi	34	MT-451	Yak-7	12 KIAP, KBF	T

Brewster 239, BW-393 of 1/LeLv 24 flown by the flight leader Capt Eino Luukkanen in October 1942, based at Römpötti. (Karolina Hołda)

Luukkanen's last Brewster was BW-393, here parked at Römpötti with the rest of his flight in early October 1942. The squadron's lynx emblem appeared shortly afterwards on the front fuselage.

Luukkanen in a departure photo in front of his BW-393 at Römpötti on 8 November 1942, taking over the command of LeLv 30 at the same base. The victory markings are beer bottle labels.

Bf 109G-2, MT-201 of 2/LeLv 34 flown by the squadron commander Maj Eino Luukkanen in June 1943, based at Utti. (Karolina Hołda)

MT-201 was assigned to 2/LeLv 34, but was mostly flown by the squadron CO Luukkanen. It is seen here at Utti in summer 1943. Below 2Lt Ture Mattila taking off for an intercept mission on 2 June 1943.

Bf 109G-6, MT-415 of 3/HLeLv 34 flown by the squadron commander Maj Eino Luukkanen in June 1944, based at Lappeenranta. (Karolina Hołda)

HLeLv 34 commander Maj Luukkanen sitting in his first Bf 109G-6, serial MT-417, at Immola on 15 June 1944. (SA-kuva)

Luukkanen's second Bf 109G-6 was MT-415, seen here at Lappeenranta. HLeLv 34 was based there from 16 to 23 June 1944.

Lyly, Erik Edvard

8 victories (7 + 1 + 4) on 188 missions
Planes: BW-374 and MT-451

Erkki Lyly was born at Loimaa on 05.08.14. He received flying training in 1940–41 at SOK 4 and was posted on 19.01.42 as Sergeant to LLv 24. Lyly was transferred on 08.03.43 to LeLv 34. He was promoted to Staff Sergeant on 26.04.44 and demobilized on 11.11.44. Lyly acted in civilian life as a professional pilot and held the rank of Master Sergeant. He died on 05.02.90. Lyly was awarded with VM 2, VM 1, and VR 4.

Mersu MT-451 of 1/HLeLv 34 at Taipalsaari in July 1944, after the removal of the wing cannons. It was assigned to Lyly, though he did not score any victories in it. The rudder bears the squadron's eagle fledgling emblem.

Bf 109G-6/R6, MT-451 of 1/HLeLv 34 flown by SSgt Erik Lyly in June 1944, based at Taipalsaari. (Karolina Hołda)

Date	Time	Area	Unit	Aircraft	Enemy aircraft	Enemy Unit	Victory type
08.07.42	18.40–20.20	Someri	24	BW-374	Yak-1	21 AP, KBF	T
16.08.42	17.45–18.00	Seiskari – Karavaldai	24	BW-374	I-16		T
10.08.43	16.25–17.25	Ino 10 km S	34	MT-217	La-5		V
23.09.43	13.10–13.50	Sepeleva – Tolli	34	MT-217	2 x La-5		V
19.11.43	14.20–15.20	Seivästö 10 km SW	34	MT-217	LaGG-3		T
06.03.44	17.10–17.40	Someri	34	MT-229	La-5		T
14.06.44	10.50–11.50	Vammelsuu	34	MT-424	Airacobra		T
19.06.44	20.00–21.15	Uuras	34	MT-426	2 x Il-4	836 BAP	R
28.06.44	13.55–15.05	Ihantalanjärvi	34	MT-453	Il-2		R
30.06.44	10.50–12.15	Noskuanselkä	34	MT-435	2 x Pe-2		V

5½ victories (5½ + 0 + 0) on 158 missions
Planes: FR-99 and BW-380

Eka Magnusson was born at Yli-Tornio on 08.12.02. He was accepted on 01.12.23 to KadK. Magnusson was promoted to 2nd Lieutenant on 30.09.25 and was assigned to ErMeLL. On 14.05.27 he was promoted to 1st Lieutenant and on 06.10.29 began a three-year assignment at IlmavE. On 15.05.31 followed promotion to Captain and on 30.11.32 Magnusson was transferred to MLE. On 31.01.36 Magnusson was posted to lead 2/LLv 24. On 21.11.38 he was made commander of LLv 24. On 06.12.39 Magnusson was promoted to Major and on 10.11.41 to Lieutenant Colonel. On 27.05.43 Magnusson took the command of LeR 3 and promoted to full Colonel on 27.08.44. On 17.03.46 he resigned and became a bank manager. Magnusson was promoted to Major General on 04.06.93. Magnusson died on 27.12.93. He was awarded with VR 4, VR 3, VR 2, VR 1 and MHR on 26.06.44.

Fokker D.XXI, FR-99 of 4/LLv 24 flown by the squadron commander Maj Gustaf Magnusson in January 1940, based at Joutseno. (Karolina Hołda)

Pilots of 4/LLv 24 at Joutseno in January 1940. From left Sgt Martti Alho, Danish volunteer 1Lt Frits Rasmussen, 2Lt Tapani Harmaja and flight leader 1Lt Per Sovelius. Behind is Magnusson's Fokker FR-99.

Date	Time	Area	Unit	Aircraft	Enemy aircraft	Enemy Unit	Victory type
01.12.39	14.15–14.20	Ruokolahti	24	FR-99	SB		T
23.12.39	12.00	Lempaalanjärvi	24	FR-99	SB	44 SBAP	T
03.02.40	15.10	Joutseno	24	FR-92	DB-3	42 DBAP	T
19.02.40	14.10	Joutseno	24	FR-99	DB-3	21 DBAP	T
28.06.41	12.35–12.45	Loviisa	24	BW-380	½ SB	73 AP	T
08.07.41	10.50–11.15	Ristlahti	24	BW-380	DB-3F	1 AP	T

Brewster BW-380 of 4/LLv 24 in July 1941, above at Joensuu and below at Rantasalmi. It was assigned to the squadron CO Maj Magnusson, until he was grounded on 9 September 1941, being too valuable to be lost.

5 victories (5 + 0 + 0) on 40 missions
Planes: MS318 and MS314

Pauli Massinen was born at Viipuri on 12.03.15. He received flying training in 1935–36 at IRUK 5. At the Winter War mobilization Massinen was placed on 10.10.39 to T-LentoR 2 and transferred on 05.12.39 as 2nd Lieutenant to LLv 28. Massinen was assigned on 27.03.40 to ISK and demobilized on 16.07.40. He was promoted to 1st Lieutenant on 28.10.40. In the mobilization of the Continuation War he was assigned on 19.06.41 to LLv 28 and transferred on 07.11.41 to LeSK. Massinen was sent on 08.05.43 for air surveillance tasks to Viipuri and relieved on 20.07.44. He worked later as a department manager. Massinen died on 10.03.89. He was awarded with VR 4, VR 4 tlk and VR 3.

Morane MS318 of 2/LLv 28 at Säkylä in March 1940. Massinen shot down on 2 March 1940 one DB-3 bomber and it was marked with a silver star on the rudder.

Morane MS314 of 2/LLv 28 assigned to Massinen. It is seen here at Karkunranta in September 1941. A couple of months later he became an instructor at the Air Fighting School.

Date	Time	Area	Unit	Aircraft	Enemy aircraft	Enemy Unit	Victory type
02.03.40	13.25	Nauvo	28	MS-318	DB-3	1 AP	T
17.07.41	10.05–10.30	Loimolanjärvi	28	MS-314	SB	72 SBAP	T
31.07.41	13.30–13.40	Simpele	28	MS-314	SB	117 RAE	T
17.08.41	19.00–19.15	Lahdenpohja	28	MS-304	MBR-2	AEN Ladoga	T
21.08.41	15.20	Säämäjärvi	28	MS-314	SB	72 SBAP	T

Mattila, Ture Allan Nestor

5¾ victories (5¾ + 0 + 5) on 296 missions
Planes: FR-125 and MT-447

Ture Mattila was born in Hartola on 15.03.16. He received flying training in 1939–41 at SOK 1 and was assigned on 05.06.41 as 2nd Lieutenant to LLv 30. Mattila was transferred on 24.03.43 to LeLv 34, where was promoted on 31.07.43 to 1st Lieutenant. He was demobilized on 10.11.44. Mattila became later a merchant. He died on 03.08.00. Mattila was awarded with VR 4, VR 4 tlk and VR 3.

Fokker D.XXI, FR-125 of 2/LLv 30 flown by 2Lt Ture Mattila in July 1941, based at Hyvinkää. (Karolina Hołda)

Fokker FR-125 of 2/LLv 30 takes off from Hyvinkää on 16 July 1941 with Mattila at the controls. The specified 50 cm yellow fuselage band is exceptionally wide. (SA-kuva)

FR-125 of 2/LLv 30 parked next to a Luftwaffe *Junkers Ju 88A of KGr 806 at Helsinki Malmi in July 1941. This German unit had been laying mines around the Soviet naval base at Kronstadt.*

Mattila leaning on Mersu MT-205 of 1/LeLv 34 in April 1943. This fighter was assigned to the flight deputy leader 1Lt Lauri Pekuri.

Mersu MT-447 of 1/HLeLv 34 assigned to Mattila, seen here at Selänpää in the latter half of September 1944, after the hostilities ended.

Date	Time	Area	Unit	Aircraft	Enemy aircraft	Enemy Unit	Victory type
08.07.41	06.30–07.40	Haissaari 20 km N	30	FR–125	MBR	15 AP, KBF	T
14.07.41	14.45–15.25	Tallinna	30	FR–127	MBR	44 OAE, KBF	T
06.08.41	17.46	Malmi lighthouse	30	FR–157	½ MBR	58 OAE, KBF	T
14.03.42	07.30–07.35	Peninsaari	30	FR–131	¼ MBR	15 AP, KBF	T
02.06.43	14.40–15.45	Karavaldai – Kronstadt	34	MT–201	LaGG–3		V
19.11.43	14.20–15.20	Seivästö 10 km SW	34	MT–206	LaGG–3		T
19.11.43	14.20–15.20	Seivästö 10 km SW	34	MT–206	LaGG–3		V
09.06.44	16.10–17.10	Tienhaara	34	MT–423	Airacobra		V
02.07.44	20.00–20.40	Suomenvesi	34	MT–425	Il–2	448 ShAP	R
02.07.44	20.00–20.40	Suomenvesi	34	MT–425	Il–2		V
05.08.44	12.40–13.55	Narvi	34	MT–447	Yak–7B	12 KIAP, KBF	T
05.08.44	12.40–13.55	Narvi	34	MT–447	Il–2		V

MELLIN, Paavo Kullervo

5½ victories (4½ + 1 + 0) on 89 missions
Planes: BW-355 and BW-362

Paavo Mellin was born at Kajaani on 06.08.19. He received flying training in 1939–40 at RAOK 6 and was assigned on 01.11.40 as Corporal to LLv 24. Mellin was promoted to Sergeant on 23.09.41. He was captured on 06.03.42, when LaGG-3s of 609 IAP shot down his BW-362 in Liistepohja area. Mellin was returned on 25.12.44, promoted two days later to Staff Sergeant and demobilized on 31.01.45. He acted later as a district foreman. Mellin died on 03.05.92. He was awarded with VM 2, VM 1,VR 4 and VR 4 tlk.

Mellin at and in the cockpit of his first assigned Brewster BW-355 of 3/LLv 24 at Rantasalmi in July 1941. Nokia Oy donated the funds for this fighter and it received the inscription NOKA.

Right: Mellin flew BW-362 of 3/LLv 24 later, here revving up at Kontupohja before he was shot down and captured in this plane on 6 March 1942.

Date	Time	Area	Unit	Aircraft	Enemy aircraft	Enemy Unit	Victory type
06.07.41	10.00–10.13	Parikkala	24	BW-355	½ MiG-3		T
01.08.41	16.00–16.10	Rautjärvi	24	BW-355	MiG-3		T
19.09.41	11.25–11.30	Manga	24	BW-363	SB	72 SBAP	T
23.09.41	13.30	Petäjäselkä	24	BW-363	½ I-16	155 IAP	T
23.09.41	14.00–14.15	Petäjäselkä	24	BW-363	½ I-16	155 IAP	T
17.12.41	09.25–09.45	Romantsi station	24z	BW-363	I-153	65 ShAP	T
26.02.42	08.50–11.00	Nopsa stop	24	BW-362	MiG-3		E

METSOLA, Johannes Kai Kalevi

10 victories (10 + 0 + 2) on 296 missions
Planes: BW-390 and MT-231

Kai Metsola was born in Helsinki on 09.01.15. He received flying training in 1936–37 at IRUK 6. In the Winter War the training continued at T-LentoR 1 and T-LentoR 2 and he was demobilized on 20.06.40. Metsola recruited on 03.06.41 as 2nd Lieutenant to LLv 32 and was transferred on 02.07.41 to LLv 24. He was promoted on 22.02.42 to 1st Lieutenant and demobilized on 15.11.44. He acted later as a main porter and held the rank of Captain. Metsola died in October 1984. He was awarded with VR 4, VR 4 tlk and VR 3.

Brewster 239, BW-390 of 1/LLv 24 flown by 1Lt Kai Metsola in April 1942, based at Nurmoila. (Karolina Hołda)

Brewster BW-390 of 1/LLv 24 assigned to Metsola, seen here at Nurmoila in April 1942. The machine was destroyed in an air raid on Nurmoila on 29 May 1942.

Bf 109G-2, MT-231 of 1/HLeLv 24 flown by 1Lt Kai Metsola in June 1944, based at Lappeenranta. (Karolina Hołda)

Mersu *MT-231 of 1/HLeLv 24 in field overhaul at Lappeenranta on 29 June 1941. It was assigned to Metsola on 8 April 1944 and he flew it for four months*

Date	Time	Area	Unit	Aircraft	Enemy aircraft	Enemy Unit	Victory type
18.08.41	16.25–16.35	Paakkola	24	BW-390	½ I-153	65 ShAP	T
06.09.41	11.50–12.05	Aunus – Mäkriä	24	BW-390	SB		T
15.10.41	15.05–15.20	Osta-Vytegra	24	BW-390	SB	72 SBAP	T
31.08.42	19.50–20.05	Lavansaari	24	BW-375	½ I-153		T
30.10.42	11.30–11.50	Karavaldai lake	24	BW-375	I-16		T
18.04.43	17.05–18.35	Karavaldai – Seiskari	24	BW-356	Yak-1	21 IAP, KBF	T
09.11.43	14.00–15.30	Levashovo	24	BW-367	Yak-7B		T
02.06.44	13.45–14.35	Levashovo airfield	24	MT-247	La-5		T
07.06.44	10.55–11.50	Muolaanjärvi	24	MT-229	2 x La-5		V
17.06.44	06.25–07.25	Hämeenkylä	24	MT-231	Il-2	448 ShAP	T
20.06.44	16.00–17.15	Viipuri	24	MT-238	Airacobra	103 GIAP	T
10.07.44	18.40–19.55	Äyräpää	24	MT-477	La-5		T

MYLLYLÄ, Paavo Urho Johannes

22 victories (20 + 2 + 10) on 420 missions
Planes: MS-317, MT-229 and MT-406

Paavo Myllylä was born at Helsinki on 26.03.18. He received flying training in 1938–39 at IRUK 8. Myllylä recruited on 08.09.39 as 2ⁿᵈ Lieutenant to LLv 26, was transferred on 29.01.40 to Osasto Räty and demobilized on 08.07.40. In the Continuation War mobilization Myllylä was posted on 17.06.41 to LLv 28, where was promoted on 25.07.42 to 1ˢᵗ Lieutenant. On 09.02.43 Myllylä was transferred to LeLv 34 and was demobilized on 10.11.44. He obtained later a degree in engineering and headed the machine department at Kesko Oy. Myllylä died on 01.01.70. He was awarded with VR 4, VR 4 tlk, VR 3, VR 3 tlk.

Morane MS-317 of 1/LLv 28 at Lunkula in September 1941. It was assigned to Myllylä for eighteen months. The victory bars on the fin are for four different pilots

Bf 109G-2, MT-229 of 3/LeLv 34 flown by 1Lt Paavo Myllylä in May 1943, based at Utti. (Karolina Hołda)

Mersu MT-229 of 3/LeLv 34 parked on the platform at Utti in late May 1943. It was assigned to Myllylä for one year.

Date	Time	Area	Unit	Aircraft	Enemy aircraft	Enemy Unit	Victory type
12.09.41	07.10–08.20	Pyhäjärvi – Ääninen	28	MS-317	½ SB	72 SBAP	T
15.09.41	16.20–17.35	Ala – Prääshä	28	MS-325	DB-3		V
04.12.41	13.00–14.45	Liistepohja	28	MS-611	Hurricane		V
05.12.41	13.00–14.45	Romantsin asema	28	MS-317	Hurricane		E
13.04.43	08.30–09.10	Peninsaari 30 km S	34	MT-216	Pe-2	15 ORAP, KBF	T
20.05.43	19.50–20.50	Lavansaari – Peninsaari	34	MT-229	I-153		V
26.07.43	14.40–15.40	Kreivinlahti	34	MT-229	Il-2	7 GShAP, KBF	V
26.07.43	14.40–15.40	Kreivinlahti	34	MT-229	Il-2		V
01.08.43	16.05–17.10	Tytärsaari 25 km NW	34	MT-209	½ Pe-2	73 BAP, KBF	T
01.08.43	16.05–17.10	Pien – Tytärsaari	34	MT-209	Pe-2	73 BAP, KBF	T
08.09.43	12.05–12.50	Tytärsaari 15 km NW	34	MT-229	LaGG-3	13 KIAP, KBF	T
19.11.43	11.00–11.35	Suursaari 30 km E	34	MT-229	LaGG-3	13 KIAP, KBF	T
06.03.44	17.10–17.25	Kotka	34	MT-216	Pe-2		T
17.05.44	10.30–11.00	Koivuluoto	34	MT-406	Pe-2		T
17.05.44	10.30–11.00	Someri	34	MT-406	Yak-9		T
19.05.44	04.35–05.10	Raukinsaari	34	MT-406	Pe-2	12 GPBAP, KBF	T
19.05.44	04.35–05.10	Lavansaari	34	MT-406	Yak-9	21 KIAP, KBF	T
10.06.44	06.25–07.20	Valkeasaari	34	MT-406	La-5		V
17.06.44	13.20–14.30	Särkijärvi	34	MT-409	Il-2	943 ShAP	R
19.06.44	16.00–17.15	Kämärä	34	MT-406	Il-2	872 ShAP	T
20.06.44	10.45–11.45	Heinjoki – Leipäsuo	34	MT-406	2 x La-5	159 IAP	T
20.06.44	10.45–11.45	Heinjoki – Leipäsuo	34	MT-406	La-5		V
30.06.44	10.45–11.45	Lihaniemi	34	MT-405	La-5		V
01.07.44	18.45–19.45	Ylämaa	34	MT-432	Pe-2		T
01.07.44	18.45–19.45	Säkkijärvi	34	MT-432	Yak-9	29 GIAP	T
02.07.44	20.05–20.40	Kärstilänjärvi	34	MT-458	Il-2	448 ShAP	R
02.07.44	20.05–20.40	Kärstilänjärvi	34	MT-458	3 x Il-2		V
03.07.44	10.55–12.05	Juustila	34	MT-445	Il-2		T
15.07.44	16.50–18.00	Äyräpää	34	MT-406	Yak-9		T
25.07.44	11.20–12.25	Someri	34	MT-406	Il-2		R

3/HLeLv 34 pilot Myllylä stands on front of his Mersu MT-406 at Selänpää in September 1944, after the hostilities.

Myllymäki, Jouko Jalo Johannes

5 victories (4 + 1 + 2) on appr. 150 missions
Planes: MS-603 and MT-246

Jouko Myllymäki was born in Kauhajoki on 28.04.16. He received flying training in 1937–38 at IRUK 7 and was accepted to KadK on 28.05.38. Myllymäki was transferred on 09.10.39 as 2nd Lieutenant to T-LentoR 2 and further on 11.12.39 to LLv 28. He was promoted on 29.11.40 to 1st Lieutenant. On 12.12.42 Myllymäki became a tutor in LeSK and on 09.03.43 further in T-LeLv 35. He was promoted to Captain on 08.07.43 and was transferred on 11.09.43 to lead 2/LeLv 24. Myllymäki was missing in action on 25.06.44, the wreck of his MT-221 was later found at Ihantala. He was awarded with VR 4 and VR 3.

Moranes of 3/LLv 28 parked at Naarajärvi on 28 June 1941. Closest below the wing of MS315 is MS-603, which was assigned to the flight deputy leader Myllymäki. The next are MS318 and MS-325, the last being Tomminen's assigned fighter. (SA-kuva)

Date	Time	Area	Unit	Aircraft	Enemy aircraft	Enemy Unit	Victory type
09.03.40	12.20	Uusikirkko	28	MS-330	I-16		T
18.08.41	07.20–08.20	Vuoksenranta	28	MS-612	I-153		V
26.09.41	13.40–15.50	Matrossa	28	MS-603	SB	72 SBAP	T
26.09.41	13.40–15.50	Matrossa	28	MS-603	MiG-3		V
18.06.44	07.30–08.40	Valamo	24	MT-246	Yak-9		R
23.06.44	07.30–08.30	Säiniö – Kääntymä	24	MT-246	La-5		T
23.06.44	07.30–08.30	Säiniö – Kääntymä	24	MT-246	Airacobra		T

NIEMINEN, Urho Abraham

12 victories (12 + 0 + 2) on 112 missions
Planes: FR-96, FA-11 and BW-359

Aapo Nieminen was born at Kakskerta on 18.12.12. He received flying training in 1933–34 at IRUK 3 and was accepted on 06.06.34 to KadK. Nieminen was assigned on 01.06.36 as 2nd Lieutenant to LAs 6. On 06.09.39 he was transferred as 1st Lieutenant to LeR 4 and on 18.01.40 further to LLv 26 to lead the 3rd Flight. Nieminen was promoted on 04.08.41 to Captain and transferred on 02.11.41 to head KoeL. He was badly injured in a flying accident on 18.04.42, when his PY-31 experienced engine trouble and crashed at Nokia. On 22.10.42 Nieminen was posted to VL and demobilized on 02.12.44. In the civilian life he changed his last name to Mattelmäki, worked as a personnel manager and held the rank of Major. Mattelmäki died on 10.09.03. He was awarded with VR 4, VR 4 tlk, VR 3, VR 3 tlk and EK 2.

FIAT G.50 ,FA-11 of 3/LLv 26 flown by the flight leader 1Lt Urho Nieminen in June 1941, based at Joroinen. (Andrzej M. Olejniczak)

Nieminen (right) being congratulated by his mechanic Erkki Haimi at Joroinen on 25 June 1941, after downing three Tupolev SB bombers with FA-11 on 25 June 1941, the first day of the Continuation War.

3/LLv 26 leader Nieminen indicates the place of the third kill bar, which Haimi paints on the rudder of FA-11 at Joroinen on 25 June 1941. (SA-kuva)

LLv 26 also received one Brewster, BW-359, which was flown jointly by Urho Nieminen and Valio Porvari with 3/LLv 26. It is depicted here at Lunkula in early September 1941.

Nieminen was badly injured in a crash of Pyry serial PY-31 on 18 April 1942, burning his face and losing the sight from one eye, ending his flying career. Here is PY-31 repaired, or actually rebuilt, two years later with T-LeLv 35 at Vesivehmaa.

Date	Time	Area	Unit	Aircraft	Enemy aircraft	Enemy Unit	Victory type
19.12.39	10.50	Perkjärvi	26	FR-111	SB	13 SBAP	T
23.12.39	10.40–10.55	Muolaanjärvi	26	FR-96	I-16	25 IAP	T
23.12.39	10.40–10.55	Muolaanjärvi	26	FR-96	I-16		V
17.01.40	14.00–14.05	Vuoksi – Valkjärvi	26	FR-98	2 x SB		T
19.01.40	14.35	Rautu	26	FR-98	SB	24 SBAP	T
29.02.40	14.55	Outside Kotka	26	FA-13	SB	57 AP	T
29.02.40	14.55	Kouvola	26	FA-13	SB		V
25.06.41	11.55–12.15	Tuusmäki – Haukivesi	26	FA-11	3 x SB	72 SBAP	T
01.08.41	12.30–12.45	Tuulos	26	FA-11	I-153	197 IAP	T
15.08.41	09.50–10.10	Tuulos	26	BW-359	SB	AEN Ladoga	E
21.08.41	18.30–18.45	Suojärvi – Säämäjärvi	26	BW-359	I-153		T

NISSINEN, Lauri Vilhelm

30⅓ victories (30⅓ + 0 + 3) on appr. 300 missions
Planes: FR-98, BW-363, BW-384 and MT-225

Lauri Nissinen was born at Joensuu on 31.07.18. He received flying training in 1938 at RAOK 4 and AOK 7 and was assigned on 01.01.39 to LLv 24. On 16.05.39 Nissinen was promoted to Sergeant, on 01.01.40 to Staff Sergeant and on 12.03.40 to Master Sergeant. On 30.03.42 he was promoted to 2nd Lieutenant. Nissinen was accepted on 01.07.42 as a cadet to MaaSK. On 26.03.43 he was promoted to 1st Lieutenant and transferred three months later to lead 1/LeLv 24. Nissinen was killed on 17.06.44, when the wreck of Urho Sarjamo's MT-227 fell straight on Nissinen's MT-229 at Perkjärvi. Nissinen was awarded with VM 2, VR 4, VR 3 and MHR on 05.07.42.

Fokker FR-98 of LLv 32 taking off at Siikakangas in May 1941. During the Winter War it was assigned to Nissinen flying with 3/LLv 24.

Left: Nissinen leaning on the ski of his FR-98 in March 1940.

Nissinen in the cockpit of a 3/LLv 24 Brewster in summer 1941.

Brewster BW-363 of 3/LLv 24 lands at Lappeenranta on 21 August 1941. It was assigned to Nissinen on 18 June 1941 for two months. (SA-kuva)

Brewster 239, BW-384 of 2/LLv 24 flown by 2Lt Lauri Nissinen in April 1942, based at Tiiksjärvi. (Karolina Hołda)

Brewster BW-384 of 2/LeLv 24 taking off from Tiiksjärvi on 25 May 1942, this time with Sgt Urho Lehto at the controls. Usually it was piloted by Nissinen. (SA-kuva)

Nissinen's BW-384 of 2/LLv 24 parked at Tiiksjärvi in March 1942. It retained the 3rd Flight markings for many months.

Nissinen in front of his BW-384 at Tiiksjärvi in May 1942. The last three kills came on 6 April 1942. Then Nissinen became a cadet and, after one year, a regular officer.

1/HLeLv 24 leader 1Lt Nissinen in the cockpit of his Mersu MT-225 at Suulajärvi on 4 April 1944.

MT-225 was the first Mersu of HLeLv 24 and it was assigned to Nissinen. It is seen here on arrival at Suulajärvi on 4 April 1944, still wearing HLeLv 34 markings.

Date	Time	Area	Unit	Aircraft	Enemy aircraft	Enemy Unit	Victory type
01.12.39	13.10	Seivästö	24	FR-98	SB	24 SBAP	E
23.12.39	10.40	Muolaanjärvi	24	FR-98	I-16	25 IAP	T
23.12.39	10.40	Muolaanjärvi	24	FR-98	I-16		V
25.12.39	12.00	Enso	24	FR-98	½ DB-3	6 DBAP	T
27.12.39	14.55	Jääskenjärvi	24	FR-98	SB	2 SBAP	T
03.02.40	16.00	Korppoo	24	FR-98	⅓ DB-3	3/10 Abr	T
17.02.40	15.30	Turku	24	FR-98	SB		V
20.02.40	13.15	Tampere	24	FR-98	SB	53 DBAP	T
08.07.41	14.20	Enso	24	BW-353	2 x I-153		T
09.07.41	05.10–05.20	Huuhanmäki	24	BW-353	I-153	65 ShAP	T
09.07.41	05.10–05.20	Marjovaara	24	BW-353	I-153		T
19.07.41	13.05–13.10	Kaukola	24	BW-363	I-153		T
21.07.41	10.15–10.20	Käkisalmi	24	BW-355	I-153	65 ShAP	T
01.08.41	16.00–16.05	Rautjärvi	24	BW-363	MiG-3		T
05.08.41	14.30–14.45	Räisälä	24	BW-365	I-153		E
19.09.41	11.25–11.35	Manga	24	BW-384	SB	72 SBAP	T
19.09.41	11.25–11.35	Manga	24	BW-384	LaGG-3	238 IAP	E
23.09.41	13.30	Petäjäselkä	24	BW-384	I-16	155 IAP	T
26.09.41	11.30	Derevjannoje	24	BW-384	I-15bis		T
26.10.41	10.45–11.30	Kuuttilahti	24	BW-384	Pe-2		T
26.10.41	10.45–11.30	Kuuttilahti	24	BW-384	MiG-3		V
10.01.42	09.45–10.15	Purnojärvi	24	BW-384	MiG-3		T
13.02.42	14.40–14.45	Kangasvaara	24	BW-384	1½ x Hurricane		T
06.04.42	15.25–15.50	Tiiksjärvi – Ontajärvi	24	BW-384	3 x Hurricane	767 IAP	T
08.06.42	21.55–22.20	Kesän kenttä	24	BW-384	Hurricane		T
31.08.43	16.45–17.00	Koivistonsaari	24	BW-373	Yak-7B	13 KIAP, KBF	T
31.08.43	16.45–17.00	Oranienbaum	24	BW-373	Yak-7B	13 KIAP, KBF	T
14.04.44	16.10	Virojoki	24	MT-225	Ju 188 F	3.(F)/22	T
16.05.44	10.10–11.15	Svir	32	MT-235	La-5	415 IAP	T
28.05.44	09.45–11.05	Savijärvi	32	MT-235	2 x La-5		T

NUORALA, Aaro Eerikki

14½ victories (13½ + 1 + 3) on appr. 250 missions
Planes: FR-154, MS-611 and MT-416

Aaro Nuorala was born at Kalajoki on 28.10.17. He received flying training in 1939–40 at SOK 4 and was assigned on 01.03.41 as Corporal to LLv 30, where was promoted on 23.06.41 to Sergeant. Nuorala was transferred on 18.09.41 to LLv 10 and on 01.11.41 back to LLv 30. On 01.08.42 he was assigned to LeLv 14 and on 09.03.43 further to LeLv 34. Promotion to Staff Sergeant occurred on 26.10.43. After the war Nuorala served with LeR 3 and 3. Lsto, resigning on 10.07.56 as Warrant Officer. His civil occupation was a traffic foreman. Nuorala died on 09.01.97. He was awarded with VM 1, VR 4 and VR 4 tlk.

Fokker D.XXI, FR-154 of 3/LLv 30 flown by Sgt Aaro Nuorala in August 1941, based at Turku.
(Karolina Hołda)

Nuorala in front of his Fokker FR-154 at Turku in late August 1941.

FR-154 of 3/LLv 30 parked at Mikkeli in early September 1941. It was assigned to Nuorala for 18 months. The wheel spats were decorated with cartoons.

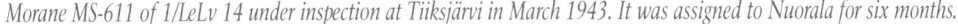

FR-154 of 3/LLv 30 on a visit to Mikkeli on 4 September 1941, piloted on this occasion by the flight deputy leader 1Lt Kalima. (SA-kuva)

Morane MS-611 of 1/LeLv 14 under inspection at Tiiksjärvi in March 1943. It was assigned to Nuorala for six months.

Date	Time	Area	Unit	Aircraft	Enemy aircraft	Enemy Unit	Victory type
14.07.41	14.45–15.25	Tallinna	30	FR-121	MBR	44 OAE, KBF	T
05.08.41	10.55–11.45	Pien-Tytärsaari 15 km SW	30	FR-125	MBR		T
28.04.42	16.15–17.30	Tunkua	14	FR-154	Hurricane		T
16.03.43	14.25–14.35	Jeljärvi	14	MS-611	I-15bis	839 IAP	E
23.03.43	08.40–08.45	Kirasjärvi 7 km SW	14	MS-640	½ I-16	197 IAP	T
05.05.43	16.35–17.35	Suursaari 20 km E	34	MT-213	La-5		E
31.07.43	20.10–21.10	Lavansaari 10 km SW	34	MT-223	2 x La-5		V
20.08.43	16.40–17.35	Harjavallanjärvi	34	MT-223	La-5		V
19.05.44	04.35–05.35	Haapasaari	34	MT-428	Yak-9		T
14.06.44	10.55–12.10	Tyrisevä	34	MT-416	La-5		T
17.06.44	05.20–06.35	Vammelsuu	34	MT-419	Airacobra	102 GIAP	T
19.06.44	20.00–20.55	Kämärä	34	MT-416	Pe-2		T
19.06.44	20.00–20.55	Kuolemanjärvi	34	MT-416	Pe-2		T
19.06.44	20.00–20.55	Härkölä	34	MT-416	La-5	401 IAP	T
22.06.44	17.05–18.25	Tali	34	MT-434	La-5		T
02.07.44	20.05–20.55	Nuijamaanjärvi	34	MT-416	Il-2	872 ShAP	T
02.07.44	20.05–20.55	Ihantala	34	MT-416	La-5	191 IAP	R

Bf 109G-6, MT-416 of 3/HLeLv 34 flown by SSgt Aaro Nuorala in July 1944, based at Taipalsaari. (Karolina Hołda)

Mersu MT-416 of 3/HLeLv 34 lands at Taipalsaari in July 1944. It was assigned to Nuorala on 30 April 1944 for four months.

5⁵⁄₆ victories (5⁵⁄₆ + 0 +1) on appr. 200 missions
Planes: CU-571

Pentti Nurminen was born at Tampere on 13.01.15. He received flying training in 1934–35 at IRUK 4 and was accepted to KadK on 03.06.36. Nurminen was assigned on 16.05.38 as 2ⁿᵈ Lieutenant to LAs 3 and transferred on 09.10.39 to LLv 10, where was promoted to 1ˢᵗ Lieutenant three weeks later. On 29.03.40 he was assigned to LLv 30 and on 01.04.41 further to LLv 32. He was appointed to lead the 3ʳᵈ Flight on 30.09.41 and was promoted to Captain on 22.05.42. Nurminen became a POW on 19.03.43, when flak shot down his CU-565 south of River Svir. Nurminen returned on 25.12.44 and served with HLeLv 11, until being relieved on 02.11.47 due to sickness. He died on 06.03.48. Nurminen was awarded with VR 4, VR 3, VR 3 tlk and EK 2.

Caudron CR.714, CA-551 of 1/LLv 30 flown by 1Lt Pentti Nurminen in September 1940. Based at Turku. (Karolina Hoda)

Curtiss CU-571 of LeLv 32 at Nurmoila in June 1942. It was regularly flown by 3ʳᵈ Flight leader Capt Nurminen. His war ended in capture on 19 March 1943, flying CU-565.

Date	Time	Area	Unit	Aircraft	Enemy aircraft	Enemy Unit	Victory type
10.08.41	17.40–17.45	Kirvu	32	CU-570	½ I-16		T
18.08.41	14.10–14.20	Sintola	32	CU-563	I-153	7 IAP	T
22.08.41	19.35–19.45	Yskjärvi	32	CU-563	I-153	7 IAP	T
10.12.41	13.50–13.55	Rajajoki	32	CU-556	⅓ SB		T
08.01.42	14.10–16.00	Lumisuo	32	CU-556	½ I-15bis		T
28.03.42	17.40–18.00	Lavansaari – Suursaari	32	CU-571	I-16	71 IAP, KBF	T
28.03.42	17.40–18.00	Lavansaari – Suursaari	32	CU-571	I-153		T
04.07.42	23.50–00.30	Lotinanpelto 10 km S	32	CU-552	Pe-2		V
07.11.42	12.15–14.00	Savijärvi	32	CU-552	½ MiG-3	415 IAP	T

Nyman, Atte Eirik Olavi

6 victories (6 + 0 + 0) on 150 missions
Planes: MT-465

Atte Nyman was born at Kouvola on 18.01.22. He received flying training in 1940–41 at AOK 11 and UK 12 and was assigned on 30.04.43 as 2nd Lieutenant to LeLv 24. Nyman was promoted to 1st Lieutenant on 14.03.44 and was demobilized 11.11.44. After the war he held managing positions in the paper industry as a diploma engineer. Nyman died on 22.07.14. He was awarded with VR 4.

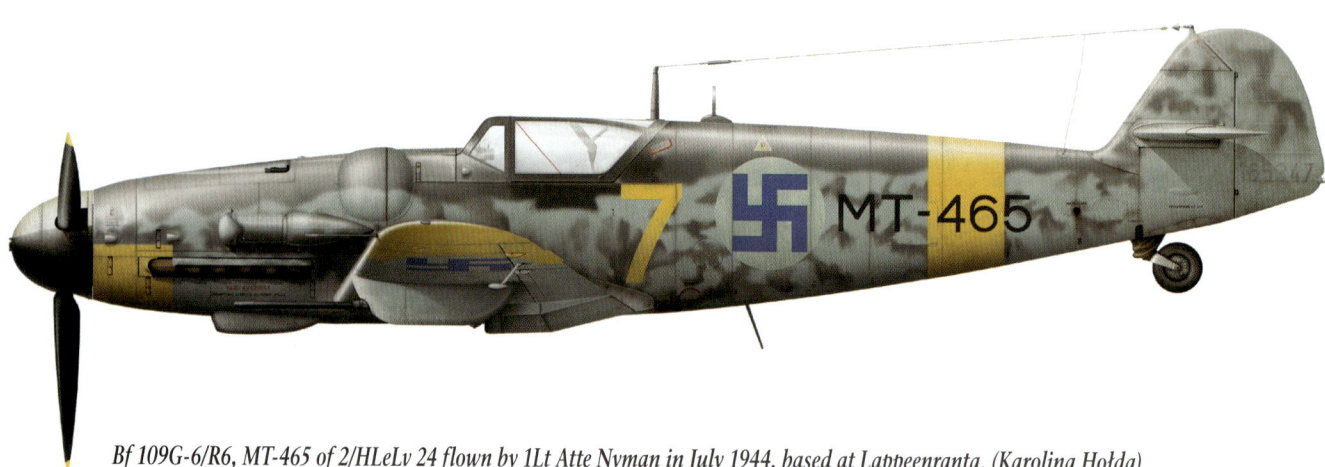

Bf 109G-6/R6, MT-465 of 2/HLeLv 24 flown by 1Lt Atte Nyman in July 1944, based at Lappeenranta. (Karolina Hołda)

Mersu MT-465 of 2/HLeLv 24 at Lappeenranta in July 1944. This Kanonenboot was assigned to Nyman on 28 June 1944 for the rest of the war. The wing cannons were soon stripped off.

Date	Time	Area	Unit	Aircraft	Enemy aircraft	Enemy Unit	Victory type
07.06.44	11.10–11.50	Rajajoki	24	MT-213	La-5		T
09.06.44	16.50–17.30	Kivennapa	24	MT-221	Il-4	55 BAP	T
10.06.44	06.15–07.35	Rajajoki	24	MT-221	Pe-2	58 BAP	R
20.06.44	19.15–20.25	Muolaanjärvi	24	MT-221	La-5		T
28.06.44	09.00–10.05	Suurpero	24	MT-201	Il-2		T
29.06.44	18.30–19.40	Tammisuo	24	MT-465	Il-2	15 GShAP	T

PALLASVUO, Yrjö Armas

12¾ victories (11¾ + 1 + 2) on appr. 250 missions
Planes: CUw-554 and MT-206

Kreivi Pallasvuo was born at Helsinki on 12.02.19. He received flying training in 1939–41 at RAOK 6 and was transferred on 01.06.41 as Corporal to LLv 32. Pallasvuo was promoted on 17.07.41 to Sergeant and on 18.09.41 to Staff Sergeant on. He started on 28.02.42 the RUK and returned on 17.07.42 as 2nd Lieutenant to LeLv 32. Pallasvuo was posted on 26.03.43 to LeLv 34 and was transferred on 06.03.44 to HLeLv 30 and on 11.06.44 back to HLeLv 34. He was killed on 03.07.44, when a Yak-9 of 29 GIAP shot down his MT-409 at Juustila. Pallasvuo was awarded with VM 2, VM 1, VR 4 and VR 3.

Curtiss Hawk 75A-1, CUw-568 flown by 2/LLv 32 pilot Sgt Yrjö Pallasvuo in August 1941, based at Lappeenranta. (Karolina Hołda)

Curtiss CUw-568 of LLv 32 parked at Lappeenranta in August 1941. Eight different pilots, including Pallasvuo, scored ten victories with this plane.

CUw-564 of LLv 32 parked next to Lappeenranta barracks in September 1941. Again, eight different pilots went on to score ten air victories with it.

Pallasvuo of 2/LeLv 34 standing by his regular Mersu MT-206 at Utti in summer 1943. He scored his squadron's 100th victory, claiming two Yak-9s on 4 October 1943.

Date	Time	Area	Unit	Aircraft	Enemy aircraft	Enemy Unit	Victory type
21.08.41	14.15–15.50	Rautu airfield	32	CU-566	I-153	7 IAP	T
23.08.41	09.45–10.30	Muolaanjärvi	32	CU-564	2 x I-153		T
29.10.41	16.20–16.40	Kivennapa	32	CU-568	I-153		T
10.12.41	13.50–13.55	Rajajoki	32	CU-552	⅓ SB		T
24.08.42	15.45–17.15	Lotinanpelto 20 km S	32	CU-554	MiG-3		E
24.08.42	15.45–17.15	Lotinanpelto 20 km S	32	CU-554	MiG-3		V
05.09.42	11.25–13.00	Ljugovitsa 7 km SW	32	CU-554	MiG-3		T
05.09.42	11.25–13.00	Ylä-Sotkusha 10 km E	32	CU-554	I-16		T
15.09.42	08.20–10.20	Mergino	32	CU-560	LaGG-3		T
29.09.42	10.35–11.55	Saarimäki	32	CU-555	½ Pe-2	119 RAE	T
04.05.43	11.05–12.30	Karavaldai lake	34	MT-206	La-5	4 GIAP, KBF	E
29.07.43	09.55–10.45	Peninsaari	34	MT-204	La-5		T
04.10.43	09.45–10.55	Ristisaari	34	MT-233	Yak-9	15 ORAP, KBF	T
04.10.43	09.45–10.55	Someri – Lavansaari	34	MT-233	Yak-9	15 ORAP, KBF	T
17.06.44	05.20–06.15	Ino	34	MT-409	Yak-9		V

10½ victories (10½ + 0 + 1) on 316 missions
Planes: FA-6, FA-33 and MT-448

Onni Paronen was born at Viipuri 02.01.18. He received flying training in 1938–39 at RAOK 4 and AOK 8 and was posted as Corporal on 01.09.39 to LLv 26. On 06.12.39 followed promotion to Sergeant and on 23.03.40 to Staff Sergeant. Paronen was transferred on 31.08.40 to LLv 32 and on 03.11.40 back to LLv 26. He was promoted on 13.07.41 to Master Sergeant. Paronen was posted on 24.03.43 to LeLv 34, where he was promoted on 26.10.44 to Warrant Officer. Paronen served after the war with HLeLv 33, until resigned on 16.07.55. In civilian life he acted as a transport manager. Paronen died on 22.08.10. He was awarded with VM 2, VM 1, VR 4, VR 3 and EK 2.

Paronen stands in front of his Fokker FR-90 belonging to 2/LLv 24 in the Winter War. He was one of many LLv 26 pilots on a commission to LLv 24.

FIAT FA-6 of 3/LLv 26 at Lunkula in early September 1941. It was assigned to Paronen, but this time the cockpit is occupied by Sgt Tage Bergman.

FIAT G.50, serial FA-6 of 3/LLv 26 flown by MSgt Onni Paronen in September 1941, based at Lunkula. (Karolina Hołda)

Paronen in the cockpit of his assigned FIAT FA-33 of 3/LeLv 26, as seen in a blast pen at Kilpasilta on 3 September 1942. (SA-kuva)

Date	Time	Area	Unit	Aircraft	Enemy aircraft	Enemy Unit	Victory type
05.01.40	11.50	Kiviniemi	26	FR-105	SB	54 SBAP	T
29.02.40	09.20	Outside Kotka	26	FA-5	DB-3		T
25.06.41	11.55–12.20	Haukivesi	26	FA-6	SB	72 SBAP	T
03.09.41	11.50–12.40	Nurmoila	26	FA-6	½ I-16		T
03.09.41	11.50–12.40	Nurmoila	26	FA-6	I-153		T
12.08.42	19.20–19.40	Kronstadt	26	FA-33	I-16		T
30.08.42	11.40–11.50	Vuoleenjärvi	26	FA-33	I-16		T
21.05.43	10.00–10.45	Lavansaari	34	MT-229	I-153	71 KIAP, KBF	T
19.05.44	04.35–04.55	Lavansaari	34	MT-414	Yak-9		T
26.06.44	10.50–11.30	Kähäri	34	MT-468	Yak-9		V
26.06.44	12.55–14.15	Juustila	34	MT-448	Il-2	566 ShAP	T
15.07.44	09.55–11.10	Vuosalmi	34	MT-448	Yak-9		T

PASILA, Mikko

9 victories (8 + 1 + 2) on appr. 200 missions
Planes: BW-388 and MT-477

Mikko Pasila was born at Iitti on 23.07.21. He received flying training in 1940–41 at SOK 3 and was transferred on 13.09.41 as 2nd Lieutenant to LLv 30. On 17.12.41 Pasila was posted to LLv 24, where he was promoted on 21.12.43 to 1st Lieutenant. He was demobilized on 10.11.44. In civilian life Pasila became a doctor of medicine and surgeon. He died on 14.02.91. Pasila was awarded with VR 4 and VR 3.

Brewster BW-382 of 1/LeLv 24 taxiing at Nurmoila in September 1941. Three months later it was assigned to Pasila, who flew it regularly for two years.

Above: Pasila in the cockpit of a Brewster of 1/LeLv 24 at Suulajärvi, in August 1943.

Left: Pasila in the cockpit of a Bf 109 G-2 of 1/HLeLv 24 at Suulajärvi in April 1944.

Date	Time	Area	Unit	Aircraft	Enemy aircraft	Enemy Unit	Victory type
08.07.42	06.15–07.25	Someri	24	BW-377	I-153		V
13.10.42	16.25–16.40	Kreivinlahti	24	BW-382	DB-3	1 GMTAP, KBF	T
22.10.42	09.30–09.40	Kreivinlahti	24	BW-382	DB-3	1 GMTAP, KBF	T
23.11.42	11.05–12.25	Oranienbaum	24	BW-388	Pe-2		E
21.04.43	08.45–09.20	Oranienbaum 10 km SW	24	BW-388	Yak-1		T
04.05.43	10.40–12.05	Vanha Yhinmäki	24	BW-374	LaGG-3		T
21.06.44	15.10–16.40	Honkaniemi	24	MT-231	Airacobra		T
28.06.44	09.00–10.15	Kämärä	24	MT-244	2 x Pe-2		T
29.06.44	07.40–08.50	Tali	24	MT-238	Il-2		V
08.07.44	10.50–11.55	Viipuri	24	MT-464	Airacobra		T

Mersu MT-216 of 1/HLeLv 24 was assigned to Pasila on 11 April 1944, seen here at that time at Suulajärvi. The assignment lasted only one month.

Pasila sitting on the cowling of his Bf 109 G-6/R6 Kanonenboot serial MT-477 of 1/HLeLv 24 in September 1944 at Utti, after the hostilities.

PEKURI, Lauri Olavi

18½ victories (17½ + 1 + 1) on 314 missions
Planes: BW-356, BW-372, MT-205 and MT-420

Lasse Ohukainen was born at Helsinki on 06.11.16. He received flying training in 1939–40 at IRUK 9 and was accepted on 06.05.40 to KadK. Ohukainen was posted on 16.05.41 as 1st Lieutenant to LeSK and was transferred on 03.09.41 to LLv 24. He changed his name in mid-1942 to Pekuri. On 09.02.43 he was posted to LeLv 34 and was transferred on 07.06.43 to lead 1/LeLv 34. Pekuri was promoted on 14.03.44 to Captain. He became POW on 16.06.44, when an Il-2 hit his MT-420, bailing out near Ino. Pekuri was returned on 23.12.44. He served thereon as HLeLv 11 and KarLsto commander, retiring on 06.11.68 as a Colonel. Pekuri was awarded with VR 4, VR 3 and VR 2.

Pekuri changed his family name from Ohukainen in July 1942. He is sitting here in the cockpit of his third Brewster, BW-384, in October 1942.

A MiG-3 punctured the port fuel tank of Ohukainen's Brewster BW-351 on 10 January 1942. He made a good forced landing outside Karhumäki in this 2/LLv 24 machine.

Date	Time	Area	Unit	Aircraft	Enemy aircraft	Enemy Unit	Victory type
04.10.41	09.55–12.00	Karhumäki	24	BW-354	I-153		E
17.12.41	09.25–09.45	Romantsi station	24	BW-351	I-153	65 ShAP	T
10.01.42	09.45–10.00	Matkajärvi	24	BW-358	MiG-3		T
24.01.42	10.15–10.30	Kiimasjärvi	24	BW-356	½ R-5		T
24.01.42	10.15–10.30	Kiimasjärvi	24	BW-356	I-15bis		T
01.02.42	12.00–12.15	Kuutsajärvi	24	BW-352	LaGG-3		T
30.03.42	15.50–16.10	Pertjärvi	24	BW-372	Hurricane	152 IAP	T
06.04.42	15.25–15.50	Tiiksjärvi – Rukajärvi	24	BW-372	3 x Hurricane	767 IAP	T
02.06.43	14.40–15.45	Yhinmäki	34	MT-224	Yak-7B	13 KIAP, KBF	T
02.06.43	14.40–15.45	Yhinmäki	34	MT-224	Yak-7B		V
31.08.43	15.35–16.25	Saarenpää 10 km S	34	MT-205	La-5		T
06.02.44	11.50–12.40	Siestarjärvi	34	MT-205	La-5	3 GIAP, KBF	T
14.06.44	10.50–11.50	Vammelsuu	34	MT-420	Airacobra		T
14.06.44	17.15–18.15	Kuuterselkä	34	MT-420	Airacobra		T
16.06.44	18.55–19.45	Ylijärvi	34	MT-420	Il-2	703 ShAP	T

Brewster 239, BW-372 of 2/LeLv 24 flown by the flight deputy leader 1Lt Lauri Ohukainen in June 1942, based at Tiiksjärvi. (Andrzej M. Olejniczak)

Ohukainen takes off from Tiiksjärvi in his BW-372 of 2/LeLv 24 on 25 May 1942. Exactly one month later he was shot down in this plane on the enemy side and walked back to the friendly side.

From left former fighter and present test pilot Capt Pekka Kokko and 1/LeLv 34 deputy leader 1Lt Lauri Pekuri exchange views in front of Pekuri's Mersu MT-205 at Helsinki Malmi on 5 April 1943. Kokko did then speed and climb trials with MT-215.

1/LeLv 34 pilots at Utti on 2 June 1943. From left SSgt Urho Lehto, SSgt Eino Peltola, 1Lt Lauri Pekuri and Sgt Lauri Mäittälä. Behind is Juutilainen's MT-222.

Pekuri's Mersu MT-205 parked at Helsinki Malmi in April 1943. He was assigned to it for one year.

147

PELTOLA, Eino Iisakki

11 victories (10 + 1 + 2) on appr. 200 missions
Planes: BW-389, BW-379 and MT-213

Eino Peltola was born at Lahti on 17.02.20. He received flying training in 1939–40 at RAOK 6 and was posted on 12.08.40 as Corporal to LLv 32. On 01.11.40 he was transferred to LLv 24 and demobilized on 01.03.41. The the Continuation War mobilization Peltola was posted on 18.06.41 to LLv 24. He was promoted on 23.07.41 to Sergeant and on 06.05.42 to Staff Sergeant. Peltola was transferred on 08.02.43 to LeLv 34. He was killed on 02.04.44, when an La-5 of 3 GIAP, KBF shot down his MT-226 at Koivisto. Peltola was awarded with VM 2, VM 1 and VR 4.

From left mechanic Paavo Vesanen and Peltola in front of his assigned Brewster BW-389 of 2/LLv 24 at Rantasalmi in July 1941. The assignment lasted until a flying accident on 29 January 1942.

Date	Time	Area	Unit	Aircraft	Enemy aircraft	Enemy Unit	Victory type
30.06.41	09.50–10.55	Lavansaari	24	BW-389	SB		V
28.02.42	09.40–09.50	Kamenskoje	24	BW-352	Hurricane		T
30.03.42	15.50–16.10	Ideljärvi	24	BW-379	2 x Hurricane		T
06.04.42	15.25–15.50	Tiiksjärvi	24	BW-379	Hurricane	609 IAP	T
06.04.42	15.25–15.50	Rukajärvi	24	BW-379	Hurricane		E
18.05.42	11.10–11.20	Tungutjärvi	24	BW-356	Pe-2		T
08.06.42	21.55–22.20	Kesä airfield	24	BW-357	Hurricane		T
10.08.43	15.55–16.45	Seiskari 15 km E	34	MT-205	LaGG-3		T
23.09.43	12.35–13.50	Sepeleva 10 km NW	34	MT-206	LaGG-3		T
23.09.43	12.35–13.50	Sepeleva 10 km NW	34	MT-206	LaGG-3		V
19.11.43	14.20–15.15	Seivästö 10 km SW	34	MT-205	Il-2		T
06.02.44	11.50–12.40	Siestarjoki	34	MT-213	La-5	3 GIAP, KBF	T
06.02.44	11.50–12.40	Siestarjoki	34	MT-213	La-5		V

Bf 109G-2, MT-215 of 1/LeLv 34 flown by SSgt Eino Peltola in May 1943, based at Helsinki Malmi. (Karolina Hołda)

Mersu MT-215 of 1/LeLv 34 in the hangar at Helsinki Malmi in late May 1943. It was assigned to Peltola on 9 March 1943 for four months.

Pokela, Väinö Nikolai

5 victories (4 + 1 + 2) on 209 missions
Planes: BW-381 and MT-424

Väiski Pokela was born at Helsinki on 27.11.18. He received flying training in 1939–40 at SOK 1 and was accepted to KadK on 03.01.41. Pokela was transferred on 05.09.41 as 2nd Lieutenant to LLv 24, where he was promoted to 1st Lieutenant on 22.06.42. He was transferred on 09.02.43 to LeLv 34 and appointed to lead 1/HLeLv 34 on 03.07.44. After the war Pokela served with HLeLv 33 and resigned on 13.02.48 as a Captain. In civilian life he was an airline captain and held the rank of Major. Pokela died on 01.01.14. He was awarded with VR 4 and VR 3.

Brewster BW-381 of 2/LeLv 24 was assigned to Pokela, seen here at Tiiksjärvi shortly before being shot down on 25 June 1942. Sgt Kalevi Anttila bailed out and then walked back through the wilderness.

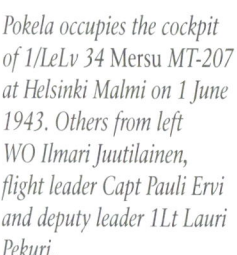

Pokela occupies the cockpit of 1/LeLv 34 Mersu MT-207 at Helsinki Malmi on 1 June 1943. Others from left WO Ilmari Juutilainen, flight leader Capt Pauli Ervi and deputy leader 1Lt Lauri Pekuri.

Date	Time	Area	Unit	Aircraft	Enemy aircraft	Enemy Unit	Victory type
13.02.42	14.40–14.45	Kangasvaara	24	BW-381	½ Hurricane		T
04.10.42	16.10–17.15	Ontrosenvaara	24	BW-357	½ I-153		T
02.06.43	14.15–15.15	Karavaldai 10 km NW	34	MT-214	Pe-2	73 BAP, KBF	T
11.09.43	14.10–14.55	Sepeleva lighthouse	34	MT-205	La-5		T
08.05.44	12.50–13.25	Haapasaari	34	MT-424	Yak-9		V
10.06.44	06.20–07.30	Terijoki	34	MT-427	Airacobra		R
18.06.44	05.15–06.20	Tiurinsaari – Kiuskeri	34	MT-422	La-5	159 IAP	R
05.07.44	09.45–10.40	Teikari	34	MT-426	Il-2		V

PORVARI, Valio Valfrid

7½ victories (7½ + 0 + 2) on 251 missions
Planes: GL-264 and FA-20

Valtsu Porvari was born at Sippola on 10.11.13. He received flying training in1936–37 at RAOK 2 and AOK 5 and was posted on 01.09.36 as Corporal to LLv 26. Porvari was promoted on 29.06.37 to Sergeant, on 08.01.40 to Staff Sergeant, on 01.03.41 to Master Sergeant and on 23.07.41 to Warrant Officer. After the war he served with LeR 3, until resigning on 16.05.47. In civilian life he was a farmer. Porvari died on 02.04.93. He was awarded with VM 1 and VR 4.

Date	Time	Area	Unit	Aircraft	Enemy aircraft	Enemy Unit	Victory type
27.12.39	12.00	Käkisalmi	26	BU-68	I-16	68 IAP	E
15.02.40	15.15	Suulajärvi	26	GL-264	DB-3		T
16.02.40	13.15	Kämärä	26	GL-264	I-16	9 LShAP	T
25.06.41	11.55–12.15	Tuusmäki – Haukivesi	26	FA-20	3½ x SB	72 SBAP	T
15.07.41	07.35–07.45	Jalovaara	26	FA-20	I-16		V
19.08.41	09.40–10.00	Säämäjärvi	26	BW-359	Pe-2		V
23.08.41	14.15–14.50	Ladoga – Vitele	26	BW-359	I-153	65 ShAP	E

FIAT FA-20 of 3/LLv 26 on its nose at Lunkula on 1 August 1941, then piloted by 2Lt Sakari Kokkonen. Porvari was the assigned pilot until this.

3/LLv 26 FIAT's in line at Joroinen on 25 June 1941. The nearest is FA-17. Porvari's FA-20 is next and then FA-5, 6, 11 and 1. The flight shot down ten Tupolev SB bombers on this day. (SA-kuva)

PUHAKKA, Risto Olli Petter

46 victories (42 + 4 + 10) on 401 missions
Planes: FR-117, FA-1, FA-25, MT-204, MT-419 and MT-433

Olli Puhakka was born at Sortavala on 11.04.16. He received flying training in 1935–36 at IRUK 5. Puhakka recruited on 01.06.39 to LLv 26 as 2nd Lieutenant. On 05.02.40 followed promotion to 1st Lieutenant and on 02.03.40 Puhakka appointed to lead 3/LLv 26. On 03.01.41 he was accepted to KadK. In the mobilization on 18.06.41 Puhakka was posted to LLv 26. On 07.11.41 Puhakka was transferred to LLv 30. On 17.11.41 commenced transfer to LeSK and on 25.02.42 further to LLv 28. On 05.06.42 Puhakka returned to lead 3/LeLv 26. On 29.12.42 he was promoted to Captain. On 09.02.43 Puhakka was assigned the leadership of 3/LeLv 34. After the war Puhakka served with HLeLv 33 and resigned on 25.07.46. He became later an airline captain and headed the flying operations of Finnair. Puhakka held the rank of Lieutenant Colonel. He died on 28.01.89. Puhakka was awarded with VR 4, VR 3, EK 2, VR 3 tlk, VR 2 and MHR on 21.12.44.

FR-117 of 2/LLv 24 was assigned to LLv 26 pilot Puhakka in the Winter War. It is seen here at Joroinen after running into a camouflaged barn on 8 April 1940.

3/LLv 26 deputy leader Puhakka and his wingman Sgt Ilmari Pöysti check their maps at Värtsilä on 10 July 1941. Behind is Puhakka's FIAT FA-1. (SA-kuva)

FIAT FA-1 of 3/LLv 26 at Joroinen on 25 June 1941, the first day of the Continuation War. The fighter was assigned to Puhakka for five months.

FIAT G.50, FA-1 of 3/LLv 26, flown by the flight deputy leader 1Lt Olli Puhakka in September 1941, based at Lunkula.
(Karolina Hołda)

FA-1 of 3/LLv 26 under service in the field at Värtsilä on 10 July 1941. Three days later Puhakka downed one SB bomber and two MiG-3 fighters, the second without firing a single round. just using superior aerobatics. (SA-kuva)

FA-1 of LLv 26 parked at Helsinki Malmi in March 1942. No longer assigned to Puhakka, but showing the plane's victory tally on the rudder, most by Puhakka.

FIAT G. 50, FA-25 of 3/LeLv 26 flown by the flight leader Capt Olli Puhakka in December 1942, based at Kilpasilta. (Karolina Hołda)

FA-25 of 3/LeLv 26 ready for a mission from Kilpasilta in December 1942. It was assigned to Puhakka and shows on the rudder his kills with this plane.

Date	Time	Area	Unit	Aircraft	Enemy aircraft	Enemy Unit	Victory type
05.01.40	11.30	Joutseno	26	FR-117	SB	41 SBAP	T
05.01.40	11.35	Nuijamaa	26	FR-117	SB		E
17.01.40	14.10	Kavantsaari	26	FR-117	SB	31 SBAP	E
19.01.40	14.30	Kiviniemi	26	FR-117	SB	24 SBAP	T
29.01.40	16.00	Urjala	26	FR-76	DB-3	53 DBAP	T
26.02.40	15.00	Pyhtää	26	FA-4	I-16	149 IAP	T
26.02.40	15.00	Kotka	26	FA-4	I-15bis		V
11.03.40	14.00	Elimäki	26	FA-21	DB-3		T
13.07.41	12.45–13.50	Jänisjärvi	26	FA-1	SB	72 SBAP	T
13.07.41	18.35–19.05	Värtsilä – Havuvaara	26	FA-1	2 x MiG-3	5 AP	T
20.07.41	09.45–10.35	Helylä airfoeld	26	FA-1	I-153		T
05.08.41	13.35–14.40	Tuulos	26	FA-1	I-15bis		T
13.08.41	13.45–14.45	Aunuksenjoki	26	FA-1	I-153	197 IAP	T
13.08.41	13.45–14.45	Aunuksenjoki	26	FA-1	I-153		E
09.07.42	08.25–09.10	Tappari – Miikkulainen	26	FA-25	I-15bis		T
12.08.42	19.20–19.40	Kronstadt	26	FA-25	I-16		T
12.08.42	19.20–19.40	Kronstadt	26	FA-25	I-16		V
15.08.42	16.30–17.05	Tappari	26	FA-25	Pe-2	26 ORAE, KBF	T
17.08.42	16.30–16.55	Konevitsa	26	FA-25	Hurricane		T
24.08.42	12.35–13.20	Miikkulainen	26	FA-10	I-16		T
19.04.43	05.30–06.40	Lavansaari	34	MT-204	La-5		T
03.05.43	05.00–06.10	Tytärsaari – Lavansaari	34	MT-204	Boston	15 ORAP, KBF	T
18.05.43	19.30–20.40	Lavansaari	34	MT-204	I-153	3 GIAP, KBF	V
18.05.43	19.30–20.40	Lavansaari	34	MT-204	I-16		V
20.05.43	13.40–14.45	Seiskari	34	MT-229	I-16		E
20.05.43	13.40–14.45	Kronstadt – Karavaldai	34	MT-229	Yak-1		T
20.08.43	16.40–17.35	Karavaldai lake	34	MT-216	La-5	4 GIAP, KBF	T
20.08.43	16.40–17.35	Karavaldai lake	34	MT-216	La-5		V
20.08.43	18.40–19.45	Lavansaari	34	MT-216	Il-2	7 GShAP, KBF	T
20.08.43	18.40–19.45	Lavansaari	34	MT-219	Yak-7B	13 KIAP, KBF	T
29.08.43	15.10–15.55	Lavansaari	34	MT-204	Yak-7B	12 KOAE, KBF	T
27.10.43	11.30–12.25	Suursaari	34	MT-216	Il-2	7 GShAP, KBF	T
19.11.43	10.00–10.30	Suursaari 20 km E	34	MT-204	Yak-7B	13 KIAP, KBF	T
07.04.44	07.35–08.20	Lavansaari	34	MT-204	Yak-9	21 KIAP, KBF	V
08.05.44	07.25–07.55	Haapasaari	34	MT-419	2 x La-5		V
17.05.44	10.30–11.10	Hamina	34	MT-419	3 x Pe-2	12 GBPAB, KBFT	
19.05.44	04.35–05.10	Haapasaari	34	MT-419	Yak-9		V
14.06.44	10.55–12.15	Tyrisevä	34	MT-419	La-5		T
14.06.44	19.45–20.45	Suulajärvi	34	MT-417	La-5		T
17.06.44	08.20–09.30	Jäppilä	34	MT-419	U-2	140 BAP	R
17.06.44	08.20–09.30	Perkjärvi	34	MT-419	Pe-2		R
17.06.44	11.10–12.00	Muolaanjärvi	34	MT-419	Il-2	943 ShAP	T
18.06.44	07.45–08.55	Kanneljärvi	34	MT-409	2 x Airacobra	102 GIAP	V
19.06.44	16.00–17.15	Kämärä	34	MT-415	Il-2	872 ShAP	T
26.06.44	12.55–14.15	Lyykylänjärvi	34	MT-427	La-5	159 IAP	T
28.06.44	13.40–15.00	Viipuri	34	MT-433	Airacobra	103 GIAP	T
29.06.44	07.25–08.25	Tali	34	MT-433	Pe-2	58 BAP	T
29.06.44	07.25–08.25	Tali	34	MT-433	3 x Pe-2		V
03.07.44	06.15–07.25	Ihantala	34	MT-468	Yak-9	26 GIAP	R

Mersu MT-204 of 3/LeLv 34 as seen here at Utti in April 1943. It was assigned to the flight leader Capt Puhakka for over a year.

Bf 109G-6, MT-433 of 3/HLeLv 34 flown by the flight leader Capt Olli Puhakka in July 1944, based at Taipalsaari. (Artur Juszczak)

Mersu MT-433 of 3/HLeLv 34 in flight after the end of hostilities in September 1944. It was assigned to Puhakka on 19 June 1944 for the rest of the war.

PURO, Kauko Olavi

33 victories (31 + 2 + 3) on 207 missions
Planes: IT-18, BW-387, MT-201 and MT-449

Olli Puro was born at Helsinki on 18.11.18. He received flying training in 1940–41 at UK 11 and was assigned on 22.06.42 as 2nd Lieutenant to LeLv 24 and on 25.09.42 further to LeLv 6. On 16.11.42 Puro was transferred to LeLv 30 and on 04.04.43 back to LeLv 24. He was promoted to 1st Lieutenant on 19.10.43 and was demobilized on 10.11.44. Later Puro became a data processing director in a savings bank. He died on 20.06.99. Puro was awarded with VR 4, VR 3 and VR 2.

Captured Polikarpov I-153 serial IT-18 of 3/LeLv 6 seen at Römpötti in October 1941. It was assigned to Puro, who shot down one similar I-153 fighter and one Pe-2 bomber flying it.

Chaikas of 3/LeLv 6 in line at Römpötti on 30 October 1942. From right IT-15, 20, 19 and 18. The last was Puro's assigned plane.

Date	Time	Area	Unit	Aircraft	Enemy aircraft	Enemy Unit	Victory type
04.10.42	08.10–09.25	Lavansaari	6	IT-18	I-153		T
12.11.42	10.30–11.40	Peninsaari	6	IT-18	Pe-2	73 BAP	V
21.04.43	08.00–09.30	Kovenskaja lake	24	BW-387	LaGG-3		T
21.04.43	08.00–09.30	Oranienbaum 12 km SW	24	BW-387	La-5		T
02.05.43	10.00–11.30	Someri 10 km SW	24	BW-387	LaGG-3		T
02.05.43	10.00–11.30	Lavansaari – Peninsaari	24	BW-387	LaGG-3		T
20.05.43	09.15–10.45	Seiskari 10 km E	24	BW-365	½ LaGG-3		T
05.06.43	14.30–15.50	Tolli – Kronstadt	24	BW-365	Yak-1		T
14.06.44	11.00–12.10	Vammeljoki – Tyrisevä	24	MT-246	2 x La-5		T
14.06.44	11.00–12.10	Vammeljoki – Tyrisevä	24	MT-246	Il-2	703 ShAP	T
17.06.44	06.20–07.10	Hämeenkylä	24	MT-246	½ Il-2		T
17.06.44	06.20–07.10	Hämeenkylä	24	MT-246	La-5		T
19.06.44	18.15–18.55	Hanhijärvi	24	MT-201	La-5	159 IAP	T
19.06.44	18.15–18.55	Hanhijärvi	24	MT-201	La-5		R
20.06.44	08.30–09.20	Viipuri – Römpötti	24	MT-201	Il-2	448 ShAP	T
20.06.44	08.30–09.20	Viipuri – Römpötti	24	MT-201	La-5	159 IAP	T
20.06.44	08.30–09.20	Viipuri – Römpötti	24	MT-201	LaGG-3		R
20.06.44	19.15–20.35	Heinjoki	24	MT-201	Airacobra	196 IAP	T
20.06.44	19.15–20.35	Valkjärvi	24	MT-201	Pe-2		T
23.06.44	13.30–14.35	Muolaanjärvi	24	MT-449	Il-2		T
23.06.44	13.30–14.35	Muolaanjärvi – Suulajärvi	24	MT-449	2 x La-5	11 GIAP	T
26.06.44	14.50–15.50	Tali	24	MT-454	La-5		T
28.06.44	14.00–14.30	Tali	24	MT-449	Airacobra		T
10.07.44	18.35–19.35	Äyräpää	24	MT-479	2 x La-5		T
10.07.44	18.35–19.35	Äyräpää	24	MT-479	La-5		V
10.07.44	18.35–19.35	Äyräpää	24	MT-479	Warhawk	191 IAP	T
15.07.44	09.25–10.50	Äyräpää	24	MT-452	2 x Yak-9		T
19.07.44	04.45–06.10	Äyräpää	24	MT-465	Yak-9		T
22.07.44	03.00–04.25	Seiskari – Narvi	24	MT-461	Il-2		T
22.07.44	03.00–04.25	Seiskari – Narvi	24	MT-461	La-5		T
22.07.44	03.00–04.25	Seiskari – Narvi	24	MT-461	La-5		V
22.07.44	03.00–04.25	Seiskari – Narvi	24	MT-461	Il-2		V
23.07.44	12.40–14.05	Heinjoki	24	MT-461	La-5		T

BW-367 fitted with a Fairchild F. 24 camera for oblique photography was Puro's plane on several mission over the Karelian Isthmus. It is seen here at Suulajärvi in April 1944.

Brewster BW-367 of 1/HLeLv 24 was fitted with a camera for reconnaissance work. Puro as a seasoned recce pilot flew this plane often over the front in spring 1944.

BW-365 of 2/LeLv 24 in flight over the Gulf of Finland in summer 1943. It was assigned to Puro for almost a year.

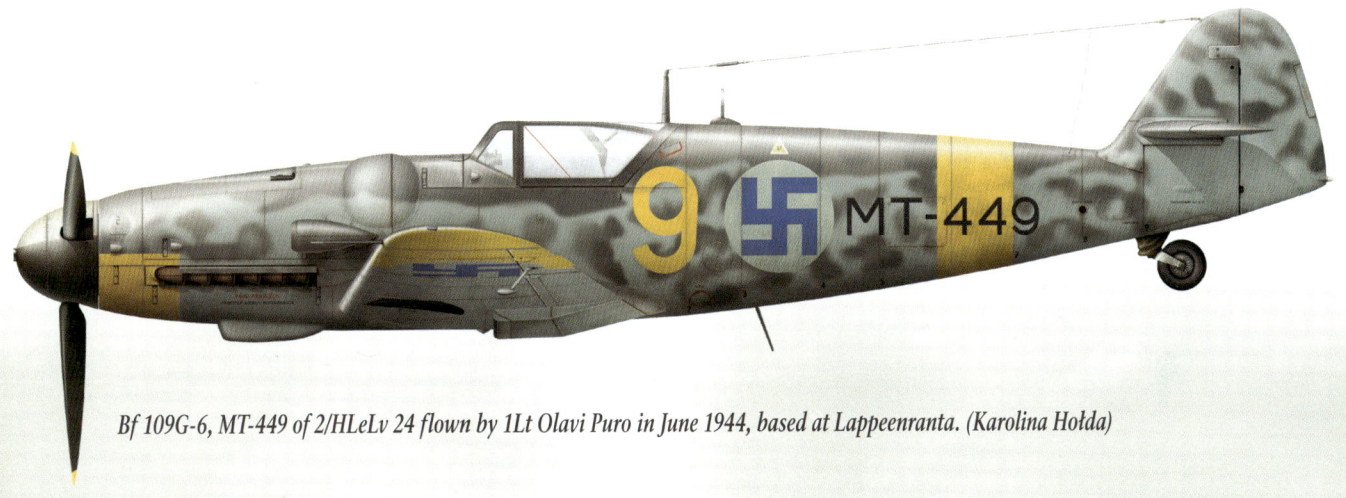

Bf 109G-6, MT-449 of 2/HLeLv 24 flown by 1Lt Olavi Puro in June 1944, based at Lappeenranta. (Karolina Hołda)

Mersu MT-449 of 2/HLeLv 24 parked at Lappeenranta in late June 1944. It was assigned to Puro for a week, before he was wounded on 28 June 1944.

Pyötsiä, Viktor

21⅓ victories (17⅓ + 4 + 6) on 437 missions
Planes: FR-110, BW-376, BW-371 and MT-244

Vikki Pyötsiä was born at Valkeala on 06.01.09. He received flying training in 1931–32 at AOK 2. Pyötsiä served as Sergeant with MLE and was transferred as Master Sergeant on 01.04.38 to ISK. On 18.03.39 he was assigned to LLv 24 and was promoted Warrant Officer on 31.03.39. Pyötsiä resigned on 15.07.45 and became an air traffic controller. He died on 13.04.96. Pyötsiä was awarded with VM 1, VR 4, VR 4 tlk, VR 3, EK 2 and EK 1.

Fokker D.XXI, FR-110 of 3/LLv 24 flown by WO Viktor Pyötsiä in March 1940, based at Lemi. (Karolina Hołda)

FR-110 of 3/LLv 24 seen at Joroinen on 8 April 1940. The port ski came loose in the air and 2Lt Olli Mustonen made a good forced landing. The plane was assigned to Pyötsiä, showing his tally on the fin.

Brewster BW-376 of 1/LeLv 24 parked at Suulajärvi in June 1942. This plane was assigned to Pyötsiä a year before, until it was shot down on 30 October 1942 killing Sgt Paavo Tolonen.

Pyötsiä was then assigned Brewster BW-371 of 1/LeLv 24, which is seen here at Suulajärvi in summer 1943 surrounded by mechanics. The plane was, unusually, fitted only with the fuselage guns.

Pyötsiä in the cockpit of a Mersu *at Suulajärvi in April 1944, possibly his assigned plane MT-244.*

Date	Time	Area	Unit	Aircraft	Enemy aircraft	Enemy Unit	Victory type
01.12.39	13.10	Korpilahti	24	FR-110	SB		V
19.12.39	11.20	Antrea	24	FR-110	SB		V
19.12.39	11.20	Muolaanjärvi	24	FR-110	SB	44 SBAP	T
23.12.39	10.48	Heinjoki	24	FR-110	⅓ SB	44 SBAP	T
27.12.39	14.30	Korpiselkä – Aittojoki	24	FR-110	2 x I-15bis		T
01.01.40	12.50	Kitelä	24	FR-110	SB	72 SBAP	T
01.01.40	12.50	Kitelä	24	FR-110	SB		V
20.01.40	14.20	Korpiselkä	24	FR-110	2 x SB	18 SBAP	T
02.02.40	11.40	Hämekoski	24	FR-110	SB	24 SBAP	T
30.06.41	16.55–17.05	Villahti	24	BW-376	Ar-2	2 SAD	T
08.07.41	04.45–05.00	Parikkala	24	BW-376	I-153	65 ShAP	T
08.07.41	04.45–05.00	Jaakkima	24	BW-376	I-153		E
18.08.41	16.25–16.35	Paakkola	24	BW-376	½ I-153	3 OIAE	T
27.09.41	17.00–17.05	Derevjannoje	24	BW-376	½ I-153	152 IAP	T
06.10.41	14.30–16.30	Vosnesenja	24	BW-376	½ SB		T
15.10.41	15.05	Osta – Vytegra	24	BW-376	SB		E
29.04.42	07.20–08.35	Alehovtshina	24	BW-376	MiG-3		V
18.04.43	17.05–18.35	Seiskari – Karavaldai	24	BW-382	Yak-1		T
18.04.43	17.05–18.35	Yhinmäki	24	BW-382	La-5	4 GIAP, KBF	T
09.03.44	18.05–19.20	Seivästö	24	BW-375	Li-2		R
17.06.44	06.20–07.10	Hämeenkylä	24	MT-244	½ Il-2		T
17.06.44	06.25–07.25	Kaukjärvi	24	MT-244	Il-2	448 ShAP	T
17.06.44	11.05–12.00	Leipäsuo	24	MT-244	Il-2	943 ShAP	V
20.06.44	08.30–09.30	Hanhijoki	24	MT-244	Airacobra	196 IAP	T
28.06.44	14.00–15.00	Tali	24	MT-238	Airacobra	196 IAP	T
30.06.44	10.45–11.55	Juustila	24	MT-238	Yak-9		V
03.07.44	10.50–11.50	Portinhoikka	24	MT-235	Il-2		R
03.07.44	10.50–11.50	Portinhoikka	24	MT-235	Il-2		V

Riihikallio, Eero Juhani

16½ victories (16½ + 0 + 1) on 110 missions
Planes: BW-374, MT-213 and MT-454

Eero Riihikallio was born at Helsinki on 20.03.19. He received flying training in 1940–41 at SOK 3 and was assigned on 14.09.41 as 2nd Lieutenant to LLv 30. On 18.12.41 Riihikallio was transferred to LLv 24, where was promoted to 1st Lieutenant on 19.10.43 and was demobilized on 10.11.44. After the war he became a master of forestry and worked as a forest manager in Canada. Riihikallio held the rank of Captain. He died on 06.09.99. Riihikallio was awarded with VR 4 and VR 3.

Aces of 2/HLeLv 24 at Suulajärvi on 12 May 1944 from left Eero Halonen (17 victories), Tapio Järvi (25½ victories), Olavi Puro (33 victories), Jorma Saarinen (23 victories) and Eero Riihikallio (16½ victories). (SA-kuva)

Brewster BW-374 of 2/HLeLv 24 at Suulajärvi on 8 May 1944, on the way to HLeLv 26. It was assigned to Riihikallio a year before, on 15 February 1943. (SA-kuva)

Bf 109G-2, MT-213 of 2/HLeLv 24 flown by 1Lt Eero Riihikallio in May 1944, based at Suulajärvi. (Karolina Hołda)

Mersu MT-213 taking off for an intercept mission from Suulajärvi on 12 May 1944, with Riihikallio at the controls of his assigned fighter. (SA-kuva)

Date	Time	Area	Unit	Aircraft	Enemy aircraft	Enemy Unit	Victory type
23.11.42	10.50–12.30	Lavansaari	24	BW-377	Tomahawk		T
10.03.43	15.15–16.25	Seiskari	24	BW-387	LaGG-3		T
18.04.43	17.00–18.20	Karavaldai	24	BW-374	Il-2		T
18.04.43	17.00–18.20	Oranienbaum	24	BW-374	LaGG-3		T
21.04.43	08.00–09.30	Lubeuskoje lake	24	BW-374	LaGG-3		T
20.05.43	09.15–10.45	Seiskari 10 km E	24	BW-374	½ LaGG-3		T
20.05.43	13.30–14.50	Karavaldai lake	24	BW-374	LaGG-3	13 KIAP, KBF	T
17.06.44	06.20–07.20	Vammeljärvi	24	MT-213	La-5		T
21.06.44	08.40–09.30	Tali	24	MT-241	La-5		V
23.06.44	13.30–14.30	Säiniö – Näykkijärvi	24	MT-213	Il-4		T
23.06.44	13.30–14.30	Säiniö – Näykkijärvi	24	MT-213	La-5		T
28.06.44	09.00–10.05	Liimatta	24	MT-454	La-5	159 IAP	T
28.06.44	11.00–12.00	Lyykylänjärvi	24	MT-454	Il-2		T
28.06.44	14.00–15.00	Perojoensuu	24	MT-454	Il-2		T
05.07.44	09.45–10.55	Tuppuransaari	24	MT-475	Il-2		T
09.07.44	14.00–15.20	Äyräpää	24	MT-463	Yak-9		T
20.07.44	16.10–17.30	Äyräpää	24	MT-481	2 x La-5		T

RIMMINEN, Toivo Veikko Johannes

6 victories (6 + 0 + 1) on 190 missions
Planes: FR-86, BW-367 and BW-382

Veka Rimminen was born at Kuhmoinen on 01.06.08. He received flying training in 1931–32 at AOK 2. Rimminen served as Sergeant at LAs 1 and was transferred on 31.12.38 as Staff Sergeant to LLv 24. He was promoted on 31.12.39 to Master Sergeant and demobilized on 13.07.40. Promotion to Warrant Officer followed on 06.08.40. At the Continuation War mobilization Rimminen was posted on 18.06.41 to LLv 24. On 15.09.42 he was transferred to LeSK and on 16.04.43 further to T-LeLv 35. On 04.10.43 Rimminen became the air traffic controller at Malmi and he was demobilized on 14.11.44. After the war Rimminen was a commercial pilot in the USA. He died on 10.07.86. Rimminen was awarded with VM 1, VR 4 and VR 4 tlk.

Date	Time	Area	Unit	Aircraft	Enemy aircraft	Enemy Unit	Victory type
05.01.40	11.35	Noskua	24	FR-86	SB	54 SBAP	T
21.02.40	12.00	Simola	24	FR-87	½ SB	54 SBAP	T
30.06.41	09.50–10.55	Lavansaari	24	BW-367	SB		T
13.07.41	12.30–13.00	Tolvajärvi	24	BW-367	I-16	155 IAP	T
29.07.41	11.20–11.30	Aunus – Alavoinen	24	BW-356	I-153		V
01.08.41	13.15–13.25	Rautjärvi	24	BW-356	½ I-153		T
06.09.41	11.50–12.05	Aunus – Mäkriä	24	BW-382	1½ x SB		T
08.09.41	14.30–14.40	Svir	24	BW-382	½ Il-4		T

BW-367 of 2/LLv 24 taking off from Rantasalmi in early July 1941, with the assigned pilot Rimminen at the controls. This plane and pilot were transferred to 1/LLv 24 on 19 September 1941.

Another Brewster of 2/LLv 24 was BW-356, in which Rimminen gained successes. It is taxiing here at Rantasalmi in August 1941, assigned to Sgt Martti Lehtovaara, the younger brother of Mannerheim Cross winner Urho Lehtovaara.

Saarinen, Jorma Kalevi

23 victories (22 + 1 + 1) on 139 missions
Planes: BW-377, MT-221, MT-452 and MT-478

Jotte Saarinen was born at Tampere on 13.04.19. He received flying training in 1940–41 at UK 11 and was assigned on 28.05.42 as 2nd Lieutenant to LeLv 24. He was promoted to 1st Lieutenant on 19.10.43. Saarinen was killed on 18.07.44, when an La-5 of 159 IAP got hits in his MT-478 and in a forced landing the plane crashed into a road bank at Antrea. He was awarded with VR 4 and VR 3.

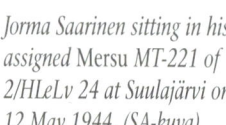

Jorma Saarinen sitting in his assigned Mersu MT-221 of *2/HLeLv 24 at Suulajärvi on 12 May 1944. (SA-kuva)*

Date	Time	Area	Unit	Aircraft	Enemy aircraft	Enemy Unit	Victory type
16.04.43	18.30–19.30	Karavaldai lake	24	BW-380	LaGG-3		T
18.04.43	17.00–18.35	Tolli lighthouse	24	BW-380	LaGG-3		T
20.05.43	13.30–14.50	Karavaldai lake	24	BW-377	LaGG-3	13 KIAP, KBF	T
05.06.43	14.30–15.50	Tolli lighthouse	24	BW-374	Il-2		T
20.08.43	16.40–17.40	Tolli lighthouse	24	BW-386	La-5	4 GIAP, KBF	T
07.06.44	11.05–11.45	Tipuna	24	MT-221	La-5		T
08.06.44	07.25–08.05	Lempaalanjärvi	24	MT-221	Airacobra		V
09.06.44	12.05–12.45	Valkeasaari	24	MT-221	Airacobra		T
14.06.44	11.00–12.05	Tyrisevä	24	MT-227	Il-2	703 ShAP	T
14.06.44	11.00–12.05	Raivola	24	MT-227	Airacobra		T
17.06.44	13.25–14.25	Suulajärvi	24	MT-221	La-5		T
19.06.44	20.20–21.10	Johannes	24	MT-246	Airacobra	103 GIAP	R
26.06.44	12.45–13.45	Juustila	24	MT-450	Il-2	872 ShAP	T
26.06.44	12.45–13.45	Tali	24	MT-450	Il-2		R
28.06.44	14.00–15.00	Karisalmi – Tali	24	MT-452	3 x Il-2	566 ShAP	T
28.06.44	16.10–17.00	Juustila	24	MT-452	La-5	159 IAP	T
05.07.44	12.20–13.25	Tiurinsaari	24	MT-463	LaGG-3		T
06.07.44	07.30–08.35	Viipuri	24	MT-470	Airacobra	196 IAP	T
10.07.44	12.00–13.25	Äyräpää	24	MT-478	La-5	159 IAP	T
11.07.44	11.05–12.25	Äyräpää	24	MT-478	Yak-9	29 GIAP	T
15.07.44	08.50–10.15	Äyräpää	24	MT-478	Yak-9		T
18.07.44	05.30–06.45	Paakkola	24	MT-478	La-5		T

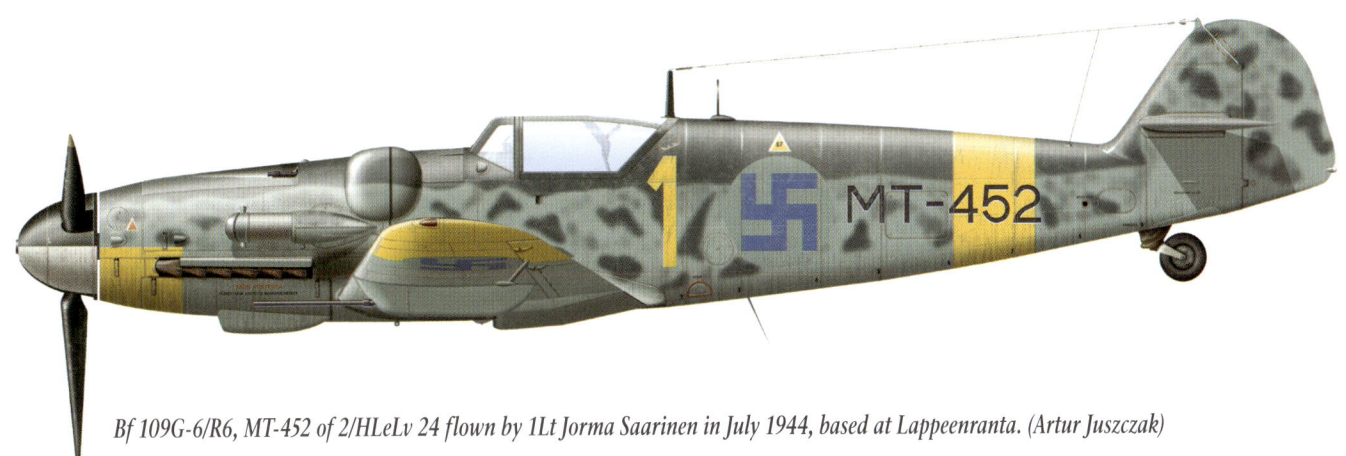

Bf 109G-6/R6, MT-452 of 2/HLeLv 24 flown by 1Lt Jorma Saarinen in July 1944, based at Lappeenranta. (Artur Juszczak)

Mechanic about to wind the inertial starter crank of MT-221 belonging to 2/HLeLv 24 at Suulajärvi on 12 May 1944. Saarinen waits in the cockpit of his assigned aircraft. (SA-kuva)

Saarinen ready to take off in his MT-221 of 2/HLeLv 24 at Suulajärvi on 12 May 1944. (SA-kuva)

SALMINEN, Paul Erik

6 victories (5 + 1 + 0) on appr. 250 missions
Planes: CUw-568

Pauli Salminen was born at Valkeala on 06.11.17. He received flying training in 1938 at RAOK 4 and AOK 7 and was posted on 01.01.39 as Corporal to Er.LLv. Salminen was promoted on 23.05.39 to Sergeant and on 07.05.40 to Staff Sergeant. On 31.08.40 he was assigned to LLv 32, where was promoted to Master Sergeant on 31.07.42 and Warrant Officer on 15.07.44. Salminen resigned on 30.09.45. He died on 24.08.84. He was awarded with VM 2, VM 1, VR 4 and VR 4 tlk.

Curtiss CUw-558 was one of several in which 2/LeLv 32 pilot Salminen claimed victories. It is seen here between missions at Nurmoila in early August 1942.

CU-503 arrived with LeLv 32 on 7 May 1942, after spending eleven months at the factory in engine trials and engine change from Cyclone to Twin Wasp. It was flown by eight pilots to gain 9¾ victories.

Date	Time	Area	Unit	Aircraft	Enemy aircraft	Enemy Unit	Victory type
03.07.41	07.35–08.20	Vehkalahti	32	FR-108	½ Pe-2	58 SBAP	T
27.03.42	15.55–17.55	Suursaari	32	CU-568	½ I-16	71 AP, KBF	T
19.07.42	12.00–13.30	Kuuttilahti	32	CU-554	I-16	524 IAP	E
02.08.42	18.45–20.20	Svir	32	CU-503	LaGG-3		E
05.09.42	11.20–13.15	Mergino 10 km N	32	CU-558	LaGG-3		T
05.09.42	11.20–13.15	Lotinanpelto 10 km W	32	CU-558	I-16	524 IAP	T
05.09.42	11.20–13.15	Lotinanpelto 5 km E	32	CU-558	I-16		T

SALOVAARA, Martti Johannes

5½ victories (4½ + 1 + 1) on appr. 100 missions
Planes: BW-364

Martti Salovaara was born at Savonlinna on 19.05.17. He received flying training in 1939–40 at SOK 1 and was accepted to KadK on 03.01.41. Salovaara was assigned on 05.09.41 as a cadet to LLv 24, where was promoted to 1st Lieutenant on 22.06.42. He was killed in a take-off accident on 12.05.44, when MT-242 rammed from behind his MT-236 at Suulajärvi. Salovaara was awarded with VR 4 and VR 3.

When Juutilainen was posted in February 1943 to LeLv 34, Salovaara was assigned his Brewster BW-364 with 3/LeLv 24, seen here at Immola after an overhaul on 26 April 1943, the fin still carrying Juutilainen's full score of 36 kills.

Date	Time	Area	Unit	Aircraft	Enemy aircraft	Enemy Unit	Victory type
26.09.41	10.15	Derevjannoje	24	BW-353	½ I-153	65 ShAP	T
06.02.42	15.00–15.10	Tiiksjärvi	24	BW-379	Hurricane		E
06.02.42	15.00–15.10	Tiiksjärvi	24	BW-379	Hurricane		V
28.02.42	09.40–09.50	Kamenskoje	24	BW-379	Hurricane		T
20.05.43	09.00–10.30	Kreivinlahti	24	BW-364	Yak-1	21 IAP, KBF	T
23.09.43	12.00–13.45	Sepelava lighthouse	24	BW-364	Yak-1		T
23.09.43	15.30–16.30	Seiskari	24	BW-364	La-5		T

3/HLeLv 24 pilot Salovaara revving up the engine of his MT-236 at Suulajärvi on 12 May 1944, creating a dust cloud. Ahokas behind in MT-242 thought that he had taken off and opened the throttle, hitting and jumping over MT-236, killing Salovaara. (SA-kuva)

Sarjamo, Urho Kaarlo

Urkki Sarjamo was born at Ii on 27.02.17. He received flying training in 1936–37 at IRUK 8. Sarjamo was assigned at the Winter War mobilization as 2nd Lieutenant on 09.10.39 to T-LentoR 2 and was accepted on 06.05.40 to KadK. In the Continuation War mobilization he was assigned on 17.06.41 to LLv 24, where was promoted to 1st Lieutenant on 20.08.41. Sarjamo was killed on 17.06.44, when an La-5 of 159 IAP shot down at Perkjärvi his MT-227, which fell on Lauri Nissinen's MT-229. Sarjamo was awarded with VR 4 and VR 3.

4/LLv 24 pilot 1Lt Sarjamo (centre) and his mechanics at the tail of BW-380. The plane belonged to the squadron CO Maj Magnusson, but he was grounded in early September 1941.

BW-386 of 2/LeLv 24 after overhaul at Immola on 18 March 1943. When the assigned pilot Ikonen was posted six weeks earlier to LeSK as an instructor, this plane was assigned to Sarjamo.

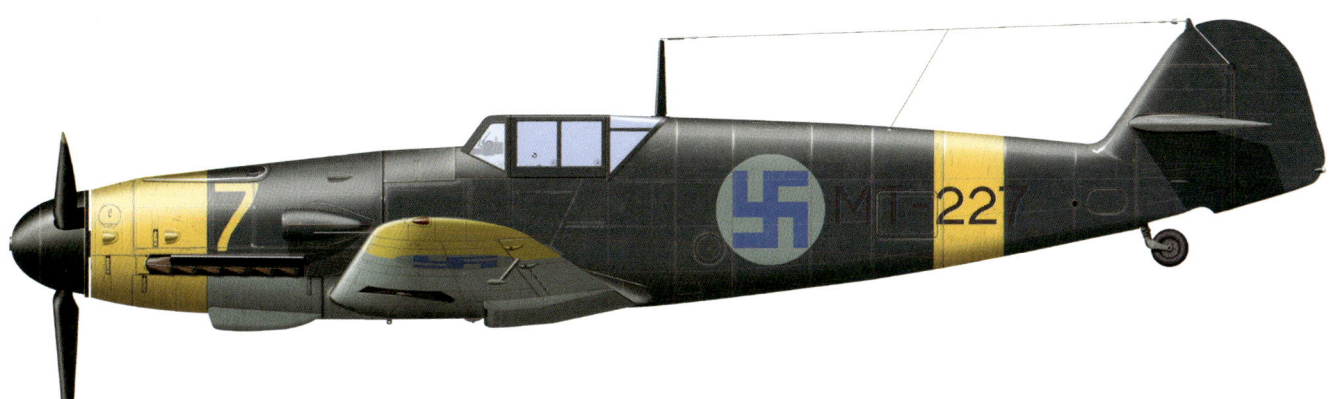

Bf 109G-2, MT-227 of 2/HLeLv 24 flown by the flight deputy leader 1Lt Urho Sarjamo in May 1944, based at Suulajärvi. (Artur Juszczak)

Mersu MT-227 of 2/HLeLv 24 seen at Suulajärvi on 8 May 1944. It was assigned to the flight deputy leader Sarjamo. He was shot down and killed flying this fighter on 17 June 1944. (SA-kuva)

Date	Time	Area	Unit	Aircraft	Enemy aircraft	Enemy Unit	Victory type
04.11.41	14.10	Koukkula	24	BW-378	½ MiG-3		T
16.04.43	18.30–19.30	Oranienbaum	24	BW-386	LaGG-3		T
18.04.43	17.00–18.20	Seiskari 10 km E	24	BW-386	LaGG-3		T
21.04.43	08.00–09.30	Yhinmäki/Viro	24	BW-386	LaGG-3		T
02.05.43	10.00–11.30	Peninsaari	24	BW-386	LaGG-3		T
20.05.43	09.15–10.45	Tolli lighthouse	24	BW-386	LaGG-3		T
12.12.43	12.15–12–45	Mustamäki 25 km E	24	BW-386	Li-2	102 AP	T
07.06.44	11.05–11.25	Muolaanjärvi	24	MT-227	Airacobra		T
10.06.44	06.25–07.40	Retunkylä	24	MT-227	Pe-2		T
10.06.44	09.55–11.00	Vaskelovo station	24	MT-227	Tu-2	12 BAP	T
10.06.44	09.55–11.00	Riihiö	24	MT-227	Airacobra		R

SARVANTO, Jorma Kalevi

16⅚ victories (16⅚ + 0 + 5) on 251 missions
Planes: FR-97, FR-80, BW-357 and BW-373

Jorma Sarvanto was born at Turku on 22.08.12. He received flying training in 1934–35 at IRUK 4 and was accepted to KadK on 04.06.35. Sarvanto was assigned on 16.05.37 as 2ⁿᵈ Lieutenant to LAs 1. On 29.04.39 he was transferred to LLv 24, where was promoted to 1ˢᵗ Lieutenant on 16.05.39. Sarvanto was promoted to Captain on 04.08.41. A post at IlmavE commenced on 19.10.41 and transfer to lead KoeL came on 08.05.42. Sarvanto was posted to Germany on 17.07.42 and he returned to LeLv 24 on 16.01.43 to lead the 1ˢᵗ Flight. On 09.07.43 Sarvanto was transferred to LeSK and on 22.06.44 further to T-LeLv 35. After the war Sarvanto commanded HLeLv 21, was a military attaché in London and commanded KarLsto resigning on 08.06.60 as Lieutenant Colonel. He died on 16.10.63. Sarvanto was awarded with VR 3 and VR 2.

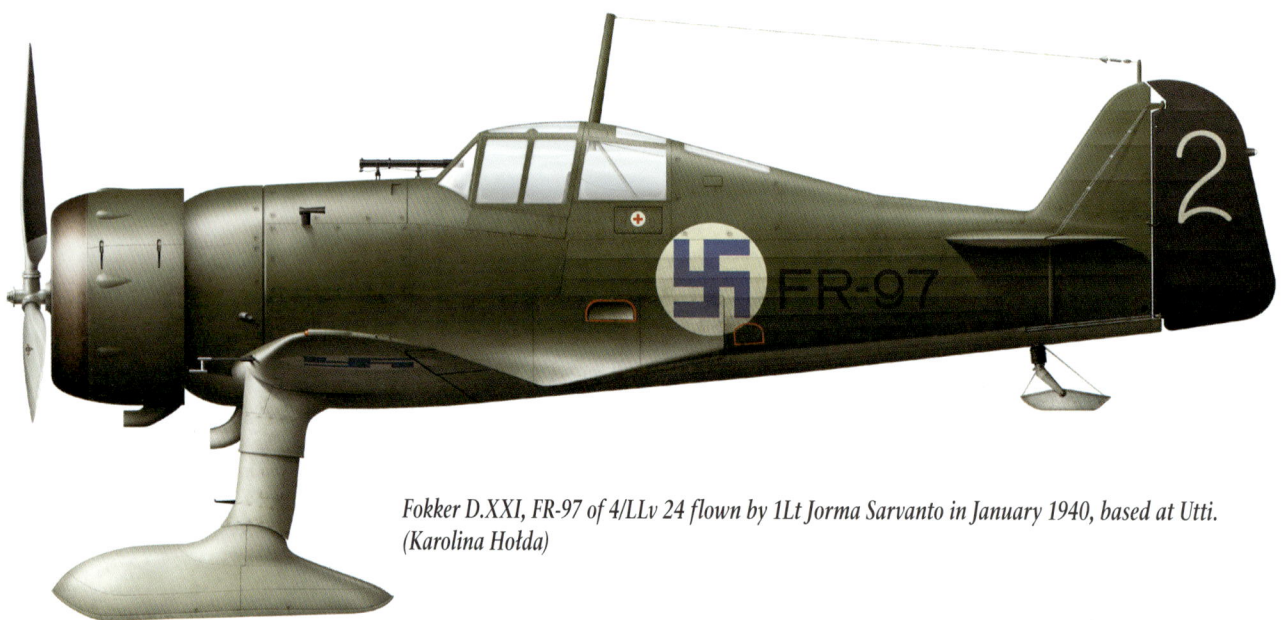

Fokker D.XXI, FR-97 of 4/LLv 24 flown by 1Lt Jorma Sarvanto in January 1940, based at Utti. (Karolina Hołda)

Sarvanto in the cockpit of his FR-97 of 4/LLv 24 at Utti on 6 January 1940, after the victorious combat in which he downed six Ilyushin DB-3 bombers in just four minutes. His plane had 23 bullet holes.

172

Brewster 239, BW-357 of 2/LLv 24 flown by the flight deputy leader 1Lt Jorma Sarvanto in June 1941, based at Selänpää. (Karolina Hołda)

Sarvanto's Brewster BW-357 of 2/LLv 24 at Rantasalmi in early July 1941. He was the top scorer in the Winter War with 13 victories. These are marked in white on the rudder while the two Continuation War victories are in yellow, the last gained on 29 June 1941.

Date	Time	Area	Unit	Aircraft	Enemy aircraft	Enemy Unit	Victory type
23.12.39	11.08	Ristseppälä	24	FR-97	SB	44 SBAP	T
23.12.39	11.15	Noisniemi	24	FR-97	SB	44 SBAP	E
06.01.40	12.03–12.07	Utti – Tavastila	24	FR-97	6 x DB-3	6 DBAP	T
17.01.40	14.00	Heinjoki	24	FR-99	SB	54 SBAP	T
17.01.40	14.00	Heinjoki	24	FR-99	SB		V
03.02.40	15.20	Nuijamaanjärvi	24	FR-80	DB-3	42 DBAP	T
15.02.40	11.15	Viipuri	24	FR-80	DB-3	42 DBAP	T
18.02.40	09.05	Nuijamaa	24	FR-100	DB-3		V
18.02.40	09.05	Simola	24	FR-100	DB-3	1 AP	T
19.02.40	15.45	Summa	24	FR-100	SB		V
19.02.40	15.45	Summa	24	FR-100	½ SB	24 SBAP	T
21.02.40	12.00	Simola	24	FR-100	⅓ DB-3	85 DAP	T
25.06.41	07.45–08.00	Selänpää	24	BW-357	SB	201 SBAP	T
29.06.41	12.30–13.40	Sippola	24	BW-357	Pe-2	58 SBAP	T
02.07.41	16.10	Rauha hospital	24	BW-357	I-153		V
29.07.41	11.20–11.30	Aunus – Alavoinen	24	BW-357	I-153		V
21.04.43	08.45–09.20	Sepeleva 10 km NW	24	BW-373	Yak-1		T
09.05.43	11.45–12.45	Peninsaari	24	BW-373	Yak-7B		T

SAVONEN, Joel Adiel

9 victories (9 + 0 + 4) on 313 missions
Planes: BW-361, BW-375 and MT-464

Joel Savonen was born at Lahti on 22.08.14. He received flying training in 1934–35 at IRUK 4. At Winter War mobilization Savonen was posted on 09.10.39 to T-LentoR 2 and was transferred on 20.12.39 to LLv 26. Savonen was demobilized on 15.06.40 and was promoted to 1st Lieutenant on 28.10.40. In the Continuation War mobilization on 18.06.41 he was posted to LLv 24 and on 02.08.41 further to LLv 25. Return to LLv 24 took place on 21.09.41 and transfer to LeSK on 15.12.42. Savonen was posted on 13.02.43 to LeLv 24 and led the 1st Flight for two weeks from 17.06.44. Savonen was demobilized on 10.11.44. He became later a researcher. Savonen died on 18.09.98. He was awarded with VR 4 and VR 3.

Savonen was leading a detachment to fly top cover to the General Headquarters, his BW-361 of 1/LLv 24 seen here at Mikkeli in mid-July 1941. There is a victory bar on the fin, next to tail number 8.

Mersu MT-464 was assigned to 1/HLeLv 24 deputy leader Savonen on 28 June 1944, seen here at Lappeenranta in the following month. By this time his scoring was over.

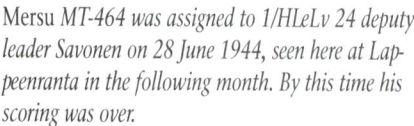

Date	Time	Area	Unit	Aircraft	Enemy aircraft	Enemy Unit	Victory type
16.07.41	03.00–03.10	Hankoniemi	24	BW-361	I-16		T
18.07.41	18.15–18.40	Sakkola	24	BW-361	MiG-3		V
06.10.41	14.30–16.30	Vosnesenja	24	BW-362	½ SB	72 SBAP	T
01.05.42	17.00–19.50	Uusi-Ladatokka	24	BW-392	I-153		V
08.07.42	18.40–20.30	Someri	24	BW-392	Il-2	57 AP, KBF	V
18.08.42	19.35–21.15	Kreivinlahti	24	BW-375	I-16		T
25.10.42	14.00–14.10	Tolli lighthouse	24	BW-375	Hurricane	3 GIAP, KBF	T
18.04.43	17.05–18.35	Seiskari 15 km E	24	BW-375	2 x Yak-1		T
19.04.43	06.10–07.40	Kronstadt	24	BW-375	½ Pe-2	73 AP, KBF	T
20.08.43	16.30–18.15	Ino 10 km S	24	BW-375	LaGG-3		T
02.04.44	15.45–16.30	Seivästö	24	BW-375	LaGG-3		V
17.06.44	13.30–14.30	Perkjärvi	24	MT-235	Yak-9D	29 GIAP	R
02.07.44	19.30–20.45	Nuijamaa	24	MT-231	Il-2		V

14¾ victories (12¾ + 2 + 4) on 257 missions
Planes: FR-92 and BW-378

Pelle Sovelius was born at Oulu on 03.06.14. He received flying training in 1933–34 at IRUK 3 and was accepted on 06.06.34 to KadK. Sovelius was assigned as 2nd Lieutenant on 16.05.36 to LLv 10 and was transferred on 28.05.37 to LLv 24. He was promoted to 1st Lieutenant on 16.05.38 and appointed to lead the 4th Flight on 18.06.41. Sovelius was promoted to Captain on 04.08.41. He was assigned on 16.02.42 to IlmavE and on 30.07.42 to command KoeLeLv. Sovelius was promoted to Major on 14.03.44 and assigned on 30.05.44 the command of HLeLv 28. He resigned on 16.05.45. Sovelius became a farmer and held the rank of Lieutenant Colonel. He died on 27.08.95. Sovelius was awarded with VR 4, VR 3, VR 3 tlk and VR 2.

Fokker FR-92 of LLv 32 parked at Siikakangas in June 1940. A couple of months earlier Sovelius gained all his Winter War victories flying this plane as 4/LLv 24 deputy leader.

Left: 4/LLv 24 deputy leader 1Lt Per Sovelius by the tail of his Fokker FR-92 during the Winter War. No victory markings applied in this conflict.

Right: 1Lt Per Sovelius (left) and his mechanic in front of Fokker FR-105 belonging to 5/LLv 24 in early December 1939.

Brewster 239, BW-378 of 4/LLv 24 flown by the flight leader Capt Per Sovelius in October 1941, based at Lunkula. (Karolina Hołda)

BW-378 of 4/LLv 24 leader Sovelius seen at Lunkula in early December 1941. He claimed all his Brewster victories in this particular plane.

Date	Time	Area	Unit	Aircraft	Enemy aircraft	Enemy Unit	Victory type
01.12.39	13.10	Rautu	24	FR-92	SB	24 SBAP	E
19.12.39	10.30	Seivästö	24	FR-92	SB	13 SBAP	T
19.12.39	10.30	Seivästö	24	FR-92	SB		V
25.12.39	12.05	Immola	24	FR-92	½ DB-3	6 DBAP	T
06.01.40	10.10	Utti	24	FR-92	DB-3	6 DBAP	T
06.01.40	12.30	Suursaari – Lavansaari	24	FR-92	DB-3		E
01.02.40	15.00–15.30	Viipuri	24	FR-92	I-16	7 IAP	E
02.02.40	11.15	Virolahti	24	FR-92	½ DB-3	6 DBAP	T
09.02.40	13.30	Summa	24	FR-92	¼ R-5		T
14.02.40	15.15	Nuijamaa	24	FR-92	½ SB	48 SBAP	T
17.02.40	13.15	Muolaanjärvi	24	FR-92	DB-3	31 DBAP	T
30.06.41	09.15–10.45	Porvoon edusta	24	BW-378	MBR-2	58 OAE	T
09.07.41	05.20	Lahdenpohja	24	BW-378	I-153		T
09.07.41	05.20	Lahdenpohja	24	BW-378	I-153		V
22.07.41	10.20–12.00	Käkisalmi	24	BW-378	MiG-3		T
10.08.41	18.35	Sairala station	24	BW-378	MiG-3		T
07.10.41	09.30–11.15	Suopohja	24	BW-378	I-16	155 IAP	T
07.10.41	09.30–11.15	Suopohja	24	BW-378	I-16		V
09.01.42	10.15–12.05	Urosjärvi	24	BW-378	R-5		T
09.01.42	10.15–12.05	Urosjärvi	24	BW-378	2 x R-5		V
06.02.42	08.05–10.20	Telekina	24	BW-378	MiG-3		E

19½ victories (19½ + 0 + 2) on 261 missions
Planes: MT-229 and MT-238

Väinö Suhonen was born at Rääkkylä on 25.11.15. He received flying training in 1937–38 at IRUK 7. At Winter War mobilization Suhonen was posted on 09.10.39 as 2nd Lieutenant to T-LentoR 2 and was demobilized on 15.06.40. In the Continuation War mobilization on 18.06.41 Suhonen was posted to LLv 25 and was transferred on 05.07.41 to LLv 24. He was promoted to 1st Lieutenant on 25.07.42. Suhonen was assigned the leadership of 3/HLeLv 24 on 21.07.44 and was demobilized on 10.11.44. He became later a lawyer and held the rank of Captain. Suhonen died on 17.04.90. He was awarded with VR 4, VR 3 and VR 3 tlk.

When landing BW-375 of 1/LeLv 24 to Mensu-vaara on 4 July 1942, 2Lt Suhonen slid off the runway and the plane flipped over. This was the assigned plane of the flight leader Capt Luukkanen.

Mersu MT-229 of 1/HLeLv 24 parked at Suulajärvi in May 1944. It was assigned to Suhonen, but was destroyed on 17 June 1944, killing the flight leader 1Lt Nissinen.

Bf 109 G-2, MT-238 of 1/HLeLv 24 flown by 1Lt Väinö Suhonen in June 1944, based at Lappeenranta. (Artur Juszczak)

MT-238 of 1/HLeLv 24 parked near the trees at Lappeenranta on 29 June 1944. It was flown jointly by Suhonen and SSgt Aimo Vahvelainen. (SA-kuva)

Date	Time	Area	Unit	Aircraft	Enemy aircraft	Enemy Unit	Victory type
08.09.41	14.30–14.40	Svir	24	BW-371	½ DB-3F		T
07.11.41	08.30–10.30	Lotinanpelto	24	BW-392	LaGG-3	415 IAP	T
25.10.42	14.00–14.10	Tolli lighthouse	24	BW-376	2 x Hurricane		T
10.08.43	16.25–17.35	Sepeleva 10 km W	24	BW-379	LaGG-3		T
30.05.44	08.55–10.15	Totleben	24	MT-229	La-5		T
06.06.44	08.35–09.25	Ohalatva	24	MT-229	La-5		T
13.06.44	09.15–10.15	Kuuterselkä	24	MT-244	Pe-2		T
13.06.44	09.15–10.15	Mainila	24	MT-244	Airacobra		T
20.06.44	20.10–20.50	Muolaanjärvi	24	MT-238	2 x Airacobra	196 IAP	T
26.06.44	10.55–12.20	Tammisuo	24	MT-244	La-5	159 IAP	T
28.06.44	09.00–10.15	Karhusuo	24	MT-238	Il-2		T
28.06.44	09.00–10.15	Karhusuo	24	MT-238	Il-2		V
28.06.44	11.05–11.50	Mannikkala	24	MT-238	Il-2		T
28.06.44	11.05–11.50	Tali	24	MT-238	Yak-9	29 GIAP	T
02.07.44	19.30–20.45	Viipuri	24	MT-235	Yak-9		V
04.07.44	13.00–14.00	Teikarsaari	24	MT-472	Il-2	47 ShAP	T
09.07.44	18.55–20.20	Vuosalmi	24	MT-461	Yak-9	14 GIAP	R
25.07.44	11.55–13.05	Lavansaari-Ulko-Tammio	24	MT-461	3 x Il-2		T

20½ victories (20½ + 0 + 5) on 272 missions
Planes: MS311, MS-619, MT-209 and MT-428

Antti Tani was born at Hämeenlinna on 03.08.18. He received flying training in 1938–39 at RAOK 5 and AOK 9 and was posted on 13.02.40 as Corporal to LLv 28. Tani was promoted to Sergeant on 23.03.40, to Staff Sergeant on 23.07.41 and to Master Sergeant on 20.11.41. He was assigned on 15.04.43 to LeLv 34. After the war Tani served with LeR 3 and 3. Lsto resigning on 10.07.56 as Warrant Officer. Thereafter he became a watchmaker. He died on 30.03.17. Tani was awarded with VM 2, VM 1, VR 4 and VR 3.

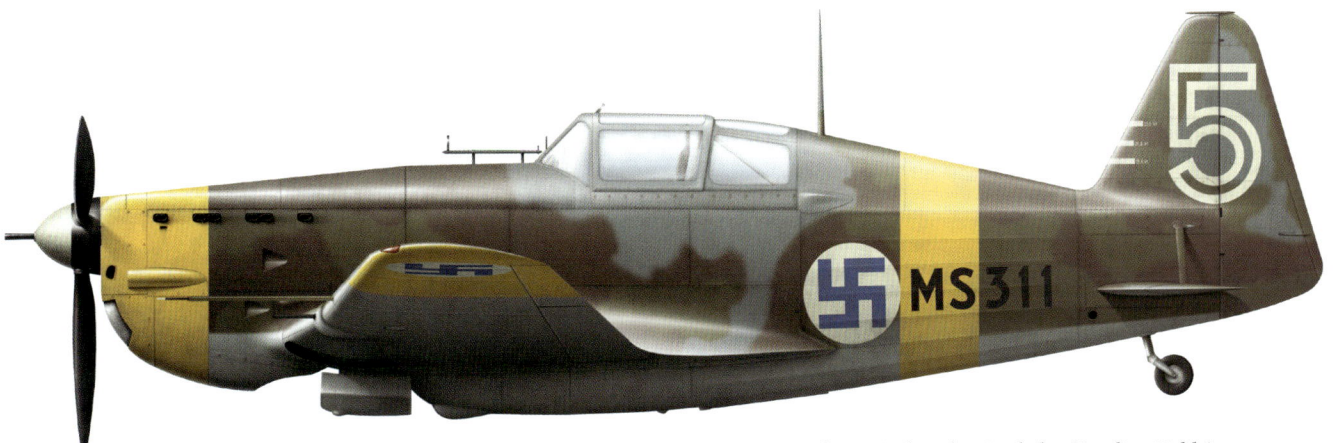

Morane Saulnier MS.406, MS311 of 1/LLv 28 flown by SSgt Antti Tani in September 1941, based at Lunkula. (Karolina Hołda)

Morane MS311 at fixed ski trials at the State Aircraft Factory at Tampere in spring 1940. It was assigned to Tani on 21 June 1941 for six months.

Tani's second Morane was MS-619, seen here in take-off for a reconnaissance mission from Viitana on 17 March 1942, with Tani at the controls. (SA-kuva)

Date	Time	Area	Unit	Aircraft	Enemy aircraft	Enemy Unit	Victory type
25.06.41	12.50–13.00	Sulkava	28	MS-311	SB	10 SBAP	T
12.08.41	08.15–09.45	Vieljärvi	28	MS-311	I-152		V
21.08.41	18.00–18.40	Suojärvi – Säämäjärvi	28	MS-311	I-153	65 ShAP	T
21.08.41	18.00–18.40	Suojärvi – Säämäjärvi	28	MS-311	I-153		V
23.08.41	14.40–14.50	Lotinanpelto	28	MS-311	I-153		V
02.09.41	13.40–15.20	Säämäjärvi	28	MS-308	½ I-16		T
12.09.41	07.10–08.20	Pyhäjärvi – Onega	28	MS-328	½ SB	72 SBAP	T
29.09.41	10.40–12.40	Vorobjeva	28	MS-317	SB	117 RAE	T
25.03.42	08.25–09.00	Äänislinna	28	MS-619	Pe-2	4 GPBAP	T
24.02.43	12.55–13.55	Kotsosero	28	MS-619	Pe-2	119 RAE	R
21.05.43	18.45–20.00	Lavansaari	34	MT-229	Il-2	7 GShAP, KBF	T
29.08.43	15.10–15.55	Lavansaari – Seiskari	34	MT-223	Jak-7B	12 KOAE, KBF	V
08.09.43	12.05–12.50	Tytärsaari 30 km NW	34	MT-223	Il-2	35 ShAP, KBF	T
10.10.43	09.25–09.55	Luumäki	34	MT-201	½ Boston		T
14.10.43	08.15–09.05	Ilmee	34	MT-204	Boston	1 GMTAP, KBF	T
06.03.44	13.45–14.20	Hamina	34	MT-209	Pe-2	12 GPBAP, KBFT	
06.03.44	13.45–14.20	Hamina	34	MT-209	Pe-2		V
06.03.44	17.10–17.45	Haapasaari	34	MT-237	La-5		T
08.05.44	07.20–07.50	Luppi – Suursaari	34	MT-414	Jak-9	21 KIAP, KBF	T
07.06.44	04.10–05.05	Vahviala	34	MT-424	Pe-2		V
16.06.44	18.55–19.45	Halilanjärvi	34	MT-428	Il-2	448 ShAP	T
20.06.44	08.50–09.50	Mieltyjärvi	34	MT-428	Airacobra	196 IAP	R
21.06.44	13.00–14.00	Ristiniemi	34	MT-435	Jak-9		T
01.07.44	14.10–15.30	Juustila	34	MT-453	3 x Il-2	566 ShAP	T
11.07.44	05.30–06.30	Paakkola	34	MT-469	Il-2		T

Morane Saulnier MS.406, MS-619 of 1/LeLv 28 flown by MSgt Antti Tani in August 1942, based at Äänislinna. (Karolina Hołda)

Morane MS-619 of 1/LeLv 28 seen at Äänislinna in August 1942. It was assigned to Tani for 18 months. He shot down two fast Pe-2 bombers in this plane, which was quite an achievement.

Tani's assigned Mersu with 1/HLeLv 34 was MT-428 (behind), but he flew MT-435 on several occasions. The latter is seen here at Lappeenranta during the third week of June 1944.

TEROMAA, Erik Uolevi

19 victories (19 + 0 + 1) on 225 missions
Planes: BW-367, BW-364 and MT-470

Erik Teromaa was born at Tampere on 30.09.17. He received flying training in 1939–40 at SOK 1 and was accepted to KadK on 04.01.41. Teromaa was posted as 2nd Lieutenant on 17.08.41 to LLv 24, where was promoted to 1st Lieutenant on 22.06.42. He was transferred on 30.05.44 to HLeLv 26 and was appointed to lead the 1st Flight on 09.06.44. Teromaa was assigned the leadership of 2/HLeLv 24 on 03.07.44. He was transferred to lead 2/HLeLv 26 on 20.08.44. After the war Teromaa served with HLeLv 23, until being killed in a flying accident on 07.02.46, when his FA-2 crashed at take-off from Rissala due to engine failure. Teromaa was awarded with VR 4, VR 3 and VR 3 tlk.

Brewster 239, BW-367 of 4/LeLv 24 flown by 1Lt Erik Teromaa in November 1942, based at Suulajärvi. (Andrzej M. Olejniczak)

Brewster BW-367 of 4/LeLv 24 taking off from Suulajärvi in November 1942. Teromaa had just become an ace in this plane and scored three more victories with it.

Teromaa was appointed to lead 1/HLeLv 26 on 9 June 1944 and assigned Brewster BW-364. It is here the closest with BW-382 behind, as seen at Mensuvaara in early July 1944.

Date	Time	Area	Unit	Aircraft	Enemy aircraft	Enemy Unit	Victory type
17.09.41	16.20–16.40	Pyhäjärvi – Prääsä	24	BW-377	MiG-3	179 IAP	E
07.10.41	09.30–11.15	Suopohja	24	BW-387	I-153		T
18.08.42	20.10–21.20	Karavaldai – Tolli	24	BW-370	2 x I-16		T
26.10.42	12.15–14.00	Kreivinlahti	24	BW-367	I-16		T
22.11.42	09.15–09.40	Tolli lighthouse	24	BW-367	Il-2		T
22.11.42	09.15–09.40	Tolli lighthouse	24	BW-367	Tomahawk		T
23.11.42	10.50–12.30	Seiskari	24	BW-367	Pe-2	73 AP, KBF	T
09.06.44	16.50–18.00	Kivennapa	26	BW-364	Il-4	836 BAP	T
09.06.44	16.50–18.00	Kivennapa	26	BW-364	Yak-7B		V
10.06.44	06.30–07.40	Suulajärvi	26	BW-364	Il-4	55 BAP	T
14.06.44	23.25–00.45	Valkjärvi	26	BW-364	Il-4	455 AP	T
18.06.44	07.45–09.05	Sakkola	26	BW-364	La-5	159 IAP	T
28.06.44	16.10–17.00	Kärstilänjärvi	24	MT-206	La-5	159 IAP	T
02.07.44	19.30–20.15	Lappeenranta	24	MT-241	Pe-2		T
09.07.44	14.00–15.20	Äyräpää	24	MT-479	Yak-9		T
19.07.44	04.45–06.10	Äyräpää	24	MT-470	2 x Yak-9		T
20.07.44	16.10–17.30	Äyräpää	24	MT-470	La-5		T
03.10.44	13.15–15.45	Ristijärvi – Lonejärvi	26	BW-361	Ju 87 D	German	T

TERVO, Altto Kalevi

**21¼ victories (15¼ + 6 + 3) on appr. 150 missions
Planes: CU-552 and MT-207**

Kale Tervo was born at Ii on 08.07.19. He received flying training in 1939–40 at IRUK 9 and was accepted to KadK on 04.01.41. Tervo was posted on 30.08.41 as 2nd Lieutenant to LLv 30 and further to LLv 24 on 04.09.41. He was assigned on 25.12.41 to LejKoulK and was transferred on 08.04.42 to LLv 32. Tervo was promoted on 08.10.42 to 1st Lieutenant. On 11.02.43 he was transferred to LeLv 34. Tervo was killed on 20.08.43, when a Yak-7 of 13 IAP, KBF shot down his MT-219 near Lavansaari. He was awarded with VR 4 and VR 3.

Curtiss Hawk 75A-6, CUw-555 flown by 2/LeLv 32 pilot 2Lt Kalevi Tervo in August 1942, based at Nurmoila. (Karolina Hołda)

Curtiss CUw-555 of LeLv 32 during an overhaul at Immola in early May 1942. Tervo flew it on several missions and gained two air victories with it.

Curtiss Hawk 75A-3, CU-552 flown by 2/LeLv 32 pilot 1Lt Kalevi Tervo in October 1942, based at Nurmoila. (Karolina Hołda)

CU-552 of LeLv 32 on the edge of Nurmoila airfield in June 1942. Here WO Eino Koskinen occupies the cockpit. Twelve different pilots accumulated 14½ victories flying this plane.

2/LeLv 32 pilot Tervo was the Curtiss top scorer with 14½ victories. Here he sits in the cockpit of CU-503 at Nurmoila in June 1942.

Mersu *MT-207 of 1/LeLv 34 in the blast pen of Suulajärvi in late July 1943. It was assigned to Tervo, who met his untimely death on 20 August 1943 flying MT-219.*

Date	Time	Area	Unit	Aircraft	Enemy aircraft	Enemy Unit	Victory type
27.09.41	17.00–17.05	Derevjannoje	24	BW-373	½ I-153	152 IAP	T
16.06.42	16.35–17.15	Nurmoila 25 km E	32	CU-556	LaGG-3		T
25.06.42	13.10–14.10	Mergino 10 km SW	32	CU-556	Pe-2		T
05.07.42	10.30–11.20	Savijärvi	32	CU-560	Pe-2	4 GPAP	E
05.07.42	10.30–11.20	Savijärvi	32	CU-560	LaGG-3		E
13.08.42	10.45–11.30	Lotinanpelto	32	CU-564	MiG-3	524 IAP	E
24.08.42	15.45–17.15	Mergino	32	CU-555	MiG-3		E
24.08.42	15.45–17.15	Mergino	32	CU-555	MiG-3		V
05.09.42	11.20–13.15	Lotinanpelto	32	CU-552	½ Pe-2		T
05.09.42	11.20–13.15	Nikonovitsina	32	CU-552	MiG-3		E
05.09.42	11.20–13.15	Ylä–Sotkusha	32	CU-552	I-16		E
11.09.42	13.00–15.40	Sjarkjärvi	32	CU-556	LaGG-3	524 IAP	V
15.09.42	08.20–09.30	Sotkusha	32	CU-554	MiG-3		T
30.09.42	11.05–12.00	Laskojärvi	32	CU-552	Pe-2		T
09.11.42	13.00–15.05	Vonozero	32	CU-552	MiG-3		E
09.11.42	13.00–15.05	Troitsankontu – Sotkusa	32	CU-552	¼ Pe-2	119 RAE	T
11.11.42	09.05–11–35	Savijärvi	32	CU-564	MiG-3		E
11.11.42	09.05–11–35	Savijärvi	32	CU-564	MiG-3		V
02.06.43	14.15–15.30	Lavansaari	34	MT-207	Pe-2	73 BAP, KBF	T
02.06.43	14.15–15.30	Seiskari	34	MT-207	La-5		T
02.06.43	14.15–15.30	Seiskari	34	MT-207	La-5		V
10.07.43	15.35–16.50	Seiskari	34	MT-207	2 x I-153		T
07.08.43	15.10–15.55	Lavansaari	34	MT-205	Pe-2	OZv, PVO	T
07.08.43	15.10–15.55	Peninsaari	34	MT-205	La-5	3 GIAP, KBF	T
20.08.43	18.55	Lavansaari	34	MT-219	LaGG-3		T

5 victories (5 + 0 + 0) on appr. 60 missions
Planes: FR-103

Pentti Tilli was born at Turku on 05.02.17. He received flying training in 1936–37 at RAOK 2 and AOK 5 and was posted on 01.01.38 as Sergeant to LLv 26. Tilli was promoted to Staff Sergeant on 31.12.39. He was killed on 20.01.40, when an I-153 of 49 IAP shot down his FR-107 at Sääksjärvi. Tilli was awarded with VR 4.

Fokker FR-103 of LLv 32 faced an undercarriage failure on landing at Siikakangas on 5 June 1941, piloted on this occasion by 2Lt Kai Metsola. During the Winter War it was assigned to 3/LLv 24 pilot Tilli, until he was shot down and killed in FR-107 on 20 January 1940.

Date	Time	Area	Unit	Aircraft	Enemy aircraft	Enemy Unit	Victory type
19.12.39	11.30	Vammelsuu	26	FR-103	I-16	7 IAP	T
23.12.39	10.55	Kämärä	26	FR-103	I-16	7 IAP	T
23.12.39	11.10	Noskuanselkä	26	FR-103	SB		T
16.01.40		Ladoga	26	FR-107	DB-3	21 DBAP	T
20.01.40	15.15	Uomaa	26	FR-107	DB-3	21 DBAP	T

TOMMINEN, TOIVO

6½ victories (5½ + 1 + 0) on appr. 50 missions
Planes: MS-325 and MS329

Toivo Tomminen was born at Kirvu on 06.10.19. He received flying training in 1939–40 at SOK 1 and was posted on 31.03.41 to LLv 28. Tomminen was promoted on 30.04.41 to Corporal and on 23.07.41 to Sergeant. He was killed on 04.12.41, when a Hurricane (Lt N.F. Repnikov) of 152 IAP collided with his MS-329 near Karhumäki. Tomminen was awarded with VM 2 and VM 1.

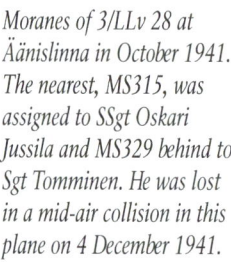

Morane MS305 of 3/LLv 28 at Joensuu in July 1941, assigned to Sgt Yrjö Pulliainen. It was one of half a dozen planes in which Tomminen made victory claims.

Moranes of 3/LLv 28 at Äänislinna in October 1941. The nearest, MS315, was assigned to SSgt Oskari Jussila and MS329 behind to Sgt Tomminen. He was lost in a mid-air collision in this plane on 4 December 1941.

Date	Time	Area	Unit	Aircraft	Enemy aircraft	Enemy Unit	Victory type
14.07.41	20.00–20.20	Suojärvi	28	MS-325	SB		E
20.08.41	06.20–06.30	Rantalahti	28	MS-318	I-153		T
02.09.41	14.40–14.45	Säämäjärvi	28	MS-324	½ I-16		T
07.10.41	10.15–10.20	Suopohja	28	MS-305	I-16		T
19.10.41	13.10–13.20	Poventsa	28	MS-315	I-153		T
19.10.41	13.10–13.20	Poventsa	28	MS-315	R-Z		T
04.12.41	13.00–13.05	Maaselkä	28	MS-329	Hurricane	152 IAP	T

TÖRRÖNEN, Iikka Veikko Santeri

10¾ victories (9¾ + 1 + 4) on appr, 260 missions
Planes: BW-385 and BW-380

Iikka Törrönen was born at Joensuu on 27.06.14. He received flying training in 1934–35 at IRUK 4. At Winter War mobilization Törrönen was posted on 09.10.39 as 2nd Lieutenant to LLv 24. He was accepted on 06.05.40 to KadK and was assigned as 1st Lieutenant on 17.06.41 to LLv 24. On 15.02.42 he was appointed to lead the 4th Flight and on 11.02.43 this became the 2nd Flight. Törrönen was promoted to Captain on 13.04.43. He was killed on 02.05.43, when a LaGG-3 (SrSgt Kutsherenko) of 10 GIAP, KBF shot down his BW-380 over Oranienbaum. Törrönen was awarded with VR 4, VR 3 and VR 3 tlk.

Brewster 239, BW-385 of 4/LLv 24 flown by the flight deputy leader 1Lt Iikka Törrönen in September 1941, based at Immola. (Andrzej M. Olejniczak)

Brewster BW-385 of 4/LLv 24 taxiing at Immola on 2 September 1941. It was assigned to the flight deputy leader 1Lt Törrönen. (SA-kuva)

Törrönen (right) and his mechanic in front of the tail of 4/LLv 24 machine BW-385 in September 1941. It was assigned to Törrönen until lost to flak on 3 December 1941.

Date	Time	Area	Unit	Aircraft	Enemy aircraft	Enemy Unit	Victory type
09.02.40	13.30	Summa	24	FR-104	¼ R-5		T
14.02.40	15.15	Lappeenranta	24	FR-104	½ SB	48 SBAP	T
28.06.41	12.35–12.45	Loviisa	24	BW-384	½ SB	73 AP	T
08.07.41	10.25–11.45	Ristlahti	24	BW-384	I-153		T
11.07.41	18.50–20.05	Jänisjärvi	24	BW-384	MiG-3		T
17.09.41	16.00–17.10	Pyhäjärvi – Prääsä	24	BW-385	MiG-3	179 IAP	T
17.09.41	16.00–17.10	Pyhäjärvi – Prääsä	24	BW-385	MiG-3		E
17.09.41	16.00–17.10	Pyhäjärvi – Prääsä	24	BW-385	MiG-3		V
26.10.41	14.00–15.45	Monastirskaja	24	BW-385	MiG-3		V
09.01.42	10.15–12.05	Uikujärvi	24	BW-383	R-5		T
09.01.42	10.15–12.05	Uikujärvi	24	BW-383	I-153		T
10.01.42	12.20–13.35	Haapaselkä	24	BW-377	2 x MiG-3		V
29.03.42	10.20	Seltokoski	24	BW-380	½ R-5		T
10.03.43	15.15–16.25	Narvi – Peninsaari	24	BW-380	LaGG-3		T
21.04.43	08.00–09.30	Oranienbaum	24	BW-380	LaGG-3		T
21.04.43	08.00–09.30	Kronstadt	24	BW-380	La-5	4 GIAP, KBF	T

4/LeLv 24 leader Törrönen is strapped into the cockpit of his Brewster BW-380 at Römpötti on 4 October 1942. In the background 1Lt Wind taxies by in Sgt Järvi's BW-377. (SA-kuva)

Törrönen bellied his BW-380 of 4/LLv 24 north of Karhumäki on 7 April 1942 due to an engine failure. After a propeller change and minor repairs it flew again after nine days.

6 victories (5 + 1 + 1) on appr. 150 missions
Planes: FA-13 and BW-367

Nils Trontti was born at Vaasa on 8.3.21. He received flying training in 1940–41 at SOK 2 and was assigned on 19.03.42 as 2nd Lieutenant to LLv 26. Trontti was promoted to 1st Lieutenant on 19.10.43. He was transferred on 25.06.44 to HLeLv 34. Trontti was captured on 26.06.44, when his MT-434 was hit over the Karelian Isthmus and he bailed out. On 10.07.44 Trontti reported back to duty at his unit and was transferred to home front. He was demobilized on 18.11.44. Thereafter Trontti became a clerk. He died on 18.10.66. Trontti was awarded with VR 4 and VR 4 tlk.

3/LeLv 26 FIATs at Kilpasilta on 20 August 1942. On the left is FA-13 assigned to 2Lt Nils Trontti while FA-29 taxies at right. It was also occasionally flown by Trontti. (SA-kuva)

Brewster BW-367 of 2/HLeLv 26 ready for take-off from Onttola on 2 October 1944. Trontti made his last air victory in this plane a couple of months earlier.

Date	Time	Area	Unit	Aircraft	Enemy aircraft	Enemy Unit	Victory type
08.07.42	18.40–19.40	Petäjämäki airfield	26	FA-33	LaGG-3	11 GIAP	T
12.08.42	19.20–19.40	Kronstadt	26	FA-13	I-16		T
17.08.42	16.30–16.55	Konevitsa	26	FA-13	Hurricane		V
26.08.42	11.20–11.35	Saunasaari	26	FA-29	Hurricane		T
22.10.42	10.50–11.50	Morje	26	FA-13	DB-3		E
02.05.43	09.50–10.25	Miikkulainen – Morje	26	FA-29	I-153		T
17.06.44	17.20–17.25	Palkeala	26	BW-367	Pe-2	34 GBAP	V

Tuomikoski, Kauko Olavi

5 victories (5 + 0 + 0) on appr. 150 missions
Planes: FA-22, FA-3 and MT-402

Kauko Tuomikoski was born at Lappeenranta on 27.04.18. He received flying training in 1937–38 at RAOK 3 and AOK 6. During the Winter War the training continued at T-LentoR 2 and he was demobilized on 31.05.40. In the Continuation War mobilization Tuomikoski was posted on 18.06.41 as Corporal to LLv 26. He was promoted on 15.12.41 to Sergeant and on 10.08.42 to Staff Sergeant. Tuomikoski was assigned on 25.06.44 to HLeLv 34. After the war he served with LeR 3 and YhtLtue, resigning on 27.04.58 as Warrant Officer. Thereafter Tuomikoski worked as a flight clearer. He died on 13.05.92. Tuomikoski was awarded with VM 2, VM 1 and VR 4.

FIAT G.50, FA-22 of 2/LLv 26 flown by Cpl Kauko Tuomikoski in June 1941, based at Joroinen. (Karolina Hołda)

Tuomikoski stands in front of FIAT FA-22 of 2/LLv 26 at Joroinen at the end of June 1941. He flew this plane jointly with the flight deputy leader 1Lt Kauko Linnamaa.

Date	Time	Area	Unit	Aircraft	Enemy aircraft	Enemy Unit	Victory type
30.07.41	08.40–09.25	Aunus	26	FA-33	MBR		T
17.06.44	05.20–06.35	Vammeljärvi	34	MT-422	Il-2	277 ShAD	T
20.06.44	10.45–11.45	Leipäsuo	34	MT-410	U-2		T
22.06.44	08.50–10.00	Heinjoki	34	MT-410	La-5		T
16.07.44	06.05–07.20	Vuosalmi	34	MT-402	Yak-9	14 GIAP	R

Tuomikoski was then assigned to 1/LLv 26 and flew FA-3, seen here at Lunkula in early September 1941. From left Tuomikoski and mechanic Kaarlo Korpi.

Tuomikoski remained with 1/LeLv 26 and again flew FA-22, which is seen here at Kilpasilta in summer 1942, with all white tactical numbers.

From left 1Lt Ilmari Joensuu and SSgt Kauko Tuomikoski in front of 3/HLeLv 34 Mersu MT-416 at Taipalsaari in July 1944. Both pilots were transferred from HLeLv 26 in late June 1944.

193

Tuominen, Oiva Emil Kalervo

47 victories (35 + 12 + 8) on appr. 400 missions
Planes: GL-255, FA-26, MT-212, MT-220 and MT-405

Oiva Tuominen was born at Kouvola on 05.03.08. He received flying training in 1933 at AOK 3 and on 01.11.33 was transferred as Sergeant to LAs 5. On 01.01.38 Tuominen was transferred as Staff Sergeant to LLv 26. On 25.04.40 came promotion to Master Sergeant and on 23.07.41 to Warrant Officer. Tuominen was transferred on 08.02.43 to LeLv 34. On 06.03.44 he was posted HLeLv 30. On 09.07.44 Tuominen was assigned to HLeLv 34. He resigned on 06.01.45. Thereafter Tuominen acted as a light plane pilot and taxi driver. He died on 28.01.76. Tuominen was awarded with VR 4 and MHR on 18.08.41.

Gladiator II, GL-255 of 2/LLv 26 flown by SSgt Oiva Tuominen in February 1940, based at Mensunkangas. (Karolina Hołda)

Gladiator GL-255 of 2/LLv 26 at Mensunkangas in February 1940. It was assigned to Tuominen, who on one mission downed three SB bombers and one I-15bis, plus sharing a fourth bomber with Lautamäki.

FIAT FA-26 of 1/LLv 26 seen at Lunkula in early September 1941. It was assigned to Tuominen, who became the first air force Mannerheim Cross winner on 18 August 1941.

FIAT G.50, FA-26 of 1/LLv 26 flown by WO Oiva Tuominen in September 1941, based at Lunkula. (Karolina Hoda)

Left: 1/LeLv 26 pilot Tuominen sits on the tail of his FA-26 at Kilpasilta, pointing to the spot for his 23rd Continuation War victory, which came on 22 October 1942.

Right: During the next winter Tuominen's FA-26 was repainted and new tactical number applied. A defect in the photo has covered three last victory markings.

FIAT G.50, FA-26 of 1/LeLv 26 flown by WO Oiva Tuominen in October 1942, based at Kilpasilta. (Karolina Hołda)

Date	Time	Area	Unit	Aircraft	Enemy aircraft	Enemy Unit	Victory type
25.12.39	12.20	Joutseno	26	FR-117	½ DB-3	6 DBAP	T
19.01.40	14.30	Igolkanniemi	26	FR-86	SB	24 SBAP	T
02.02.40	15.40	Elimäki	26	GL-258	I-16		V
02.02.40	15.40	Kotka	26	GL-258	I-16	149 IAP	T
02.02.40	16.00	Suursaari	26	GL-258	I-16		E
11.02.40	13.30	Soanlahti	26	GL-255	I-16	49 IAP	E
13.02.40	14.00–15.30	Jänisjärvi	26	GL-255	2½ x SB	39 SBAP	T
13.02.40	14.00–15.30	Roikonkoski	26	GL-255	SB		E
13.02.40	14.00–15.30	Loimola	26	GL-255	I-15bis		E
04.07.41	11.00–11.20	Pyhäselkä – Tohmajärvi	26	FA-3	4 x SB	72 SBAP	T
14.07.41	09.30–10.30	Suojärvi	26	FA-26	2 x SB	72 SBAP	T
30.07.41	08.35–09.20	Aunus	26	FA-26	2 x SB		E
01.08.41	09.30–10.30	Säämäjärvi	26	FA-26	I-16	7 IAP	T
03.08.41	17.30–17.55	Ladoga	26	FA-26	1½ x MBR	Gr Hrolenko	T
05.08.41	13.35–14.40	Tuulos	26	FA-26	I-15bis		T
03.09.41	11.45–12.35	Nurmoila	26	FA-26	1½ x I-16		T
23.05.42	15.40–16.20	Kallbådagrund	26	FA-6	Hurricane		E
05.07.42	12.00–13.15	Lavansaari	26	FA-6	MBR		E
24.08.42	12.40–13.20	Saunasaari	26	FA-32	I-16		T
24.08.42	16.40–17.20	Morje	26	FA-32	I-16		T
26.08.42	11.05–12.00	Miikkulainen – Lumisuo	26	FA-18	2 x I-153		E
21.09.42	18.50–19.40	Miikkulainen	26	FA-26	I-153		T
21.09.42	18.50–19.40	Miikkulainen	26	FA-26	I-15bis		E
22.10.42	11.40–12.50	Ladoga	26	FA-26	I-153		T
22.10.42	13.50–15.00	Ladoga	26	FA-26	I-15bis	11 AP, KBF	T
23.04.43	05.55–06.50	Karavaldai lake	34	MT-212	Pe-2	15 ORAP, KBF	T
02.05.43	09.00–09.35	Leksa, Estonia	34	MT-205	Boston	15 ORAP, KBF	T
02.06.43	14.15–14.45	Someri	34	MT-212	La-5		V
02.06.43	14.15–14.45	Someri 10 km S	34	MT-212	Pe-2	73 BAP, KBF	T
19.07.43	20.35–21.35	Someri 25 km SW	34	MT-220	LaGG-3		T
20.07.43	05.30–06.15	Seiskari	34	MT-220	Yak-7B	15 ORAP, KBF	R
24.07.43	07.55–08.55	Seiskari – Peninsaari	34	MT-220	2 x LaGG-3		T
26.07.43	07.00–07.55	Lavansaari	34	MT-204	2 x La-5		V
17.08.43	10.05–10.55	Someri 10 km E	34	MT-201	Il-2	7 GShAP, KBF	T
20.08.43	14.55–16.00	Seiskari	34	MT-209	Boston	15 RAP, KBF	V
22.08.43	14.55–15.45	Seiskari	34	MT-216	Il-2		T
07.09.43	10.45–11.40	Tytärsaari 10 km NW	34	MT-229	LaGG-3		V

Bf 109G-2, MT-212 of 1/LeLv 34 flown by WO Oiva Tuominen in June 1943, based at Utti. (Karolina Hołda)

When Tuominen was posted to 2/LeLv 34, he was assigned this MT-220. Tuominen poses on it at Utti, before it was lost by another pilot on 27 July 1943.

Date	Time	Area	Unit	Aircraft	Enemy aircraft	Enemy Unit	Victory type
07.09.43	10.45–11.40	Tytärsaari 15 km NW	34	MT-229	LaGG-3		V
03.02.44	09.30–10.50	Kallbådagrund	34	MT-232	Pe-2		V
24.05.44	16.45–17.35	Stenskär 20 km SW	30	MT-407	Pe-2		V
08.07.44	13.50–14.35	Humaljoki	34	MT-445	Pe-2	140 BAP	R
09.07.44	11.30–12.30	Pölläkkälä	34	MT-468	La-5		T
15.07.44	16.50–18.00	Äyräpää	34	MT-405	La-5		T
15.07.44	16.50–18.00	Äyräpää	34	MT-405	Yak-9	21 KIAP, KBF	T
18.07.44	13.50–15.10	Heinjoki	34	MT-405	Yak-9		R

TURKKA, Yrjö Olavi

17³⁄₄ victories (16³⁄₄ + 1 + 3) on appr. 350 missions
Planes: FR-83, BW-351, BW-357 and MT-219

"Pappa" Turkka was born at Savitaipale on 29.01.09. He received flying training in 1930–31 at AOK 1. Turkka was transferred on 28.03.34 as Sergeant to LAs 1 and as Master Sergeant on 01.01.38 to LLv 24. He was promoted on 31.03.39 to Warrant Officer. Turkka was assigned on 16.04.43 to LeLv 34 and on 10.04.44 back to HLeLv 24. After the war he served with HLeLv 31 until resigned on 16.05.47. He became later a social manager. Turkka died on 19.02.91. He was awarded with VM 1, VR 4 twice, VR 4 tlk, VR 3 and EK 2.

Fokker D.XXI, FR-83 of 1/LLv 24 flown by WO Yrjö Turkka in February 1940, based at Ruokolahti. (Andrzej M. Olejniczak)

Fokker FR-83 of LLv 32 at Siikakangas in June 1940. In the Winter War it was assigned to 1/LLv 24 pilot Turkka who scored all his victories, becoming an ace, with this plane.

Turkka by the tail of his Brewster BW-357 of 2/LeLv 24, as seen at Tiiks-järvi on 31 July 1942. (SA-kuva)

Mersu MT-219 of 1/LeLv 34 parked at Utti on 2 June 1943. It was assigned to Turkka until 20 August 1943 when 1Lt Tervo was shot down and killed in this machine.

Date	Time	Area	Unit	Aircraft	Enemy aircraft	Enemy Unit	Victory type
23.12.39	10.48	Heinjoki	24	FR-83	⅓ SB	44 SBAP	T
23.12.39	10.50	Pölläkkälä	24	FR-83	SB	44 SBAP	T
25.12.39	11.25	Lauritsala	24	FR-83	SB	41 SBAP	T
29.01.40	15.20	Kaukjärvi	24	FR-83	½ R-5	16 KAO	T
29.01.40	15.25	Muolaanjärvi	24	FR-83	¼ R-5	16 KAO	T
30.01.40	13.25	Virolahti	24	FR-83	⅓ SB	50 SBAP	T
18.02.40	09.05	Vainikkala	24	FR-83	DB-3	1 AP	T
18.02.40	09.05	Vainikkala	24	FR-83	DB-3		V
21.02.40	12.00	Simola	24	FR-83	⅓ DB-3		T
25.06.41	08.00–08.30	Virolahti	24	BW-351	2 x SB	2 SBAP	T
25.06.41	08.00–08.30	Virolahti	24	BW-351	SB		V
03.07.41	09.30–10.40	Enso	24	BW-351	I-153	7 IAP	T
13.07.41	12.30–13.00	Tolvajärvi	24	BW-351	I-16	155 IAP	T
24.01.42	10.15–10.30	Ontajärvi	24	BW-352	I-15bis	65 ShAP	T
24.01.42	10.15–10.30	Kuusjärvi	24	BW-352	½ R-5		T
28.01.42	09.45–10.00	Karasjärvi	24	BW-357	R-5	669 AP	T
01.02.42	12.00–12.15	Kuutsajärvi	24	BW-357	LaGG-3		T
30.03.42	15.50–16.10	Rukajärvi	24	BW-357	Hurricane	152 IAP	T
06.08.42		Ontajärvi	24	BW-357	½ Tomahawk		T
02.03.43	15.10–16.45	Seiskari	24	BW-355	I-153	71 IAP, KBF	T
04.05.43	10.20–11.10	Lavansaari	34	MT-203	LaGG-3		E
21.05.43	10.00–10.45	Lavansaari	34	MT-218	I-153	71 KIAP, KBF	T
26.07.43	17.50–18.30	Piisaari	34	MT-219	Il-2		T
09.06.44	12.15–12.50	Valkeasaari	24	MT-213	La-5		V

VESA, Emil Onerva

30½ victories (27½ + 3 + 1) on 198 missions
Planes: BW-357, MT-438 and MT-460

Emppu Vesa was born at Viipuri on 19.03.12. He received flying training in 1940–41 at SOK 2 and was posted on 03.12.41 as Sergeant to LLv 24. Vesa was promoted to Staff Sergeant on 22.10.42 and to Master Sergeant on 16.07.44. He was demobilized on 10.11.44. He became later a purchasing manager at Lännen Sokeri Oy. He died on 19.06.64. Vesa was awarded with VM 2, VM 1, VR 4 and VR 3.

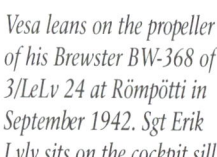

Vesa leans on the propeller of his Brewster BW-368 of 3/LeLv 24 at Römpötti in September 1942. Sgt Erik Lyly sits on the cockpit sill.

BW-357 of 3/HLeLv 24 at Suulajärvi in April 1944. It was flown jointly by Vesa and SSgt Viljo Kauppinen for one year. The squadron's lynx emblem was gone by this time.

Date	Time	Area	Unit	Aircraft	Enemy aircraft	Enemy Unit	Victory type
25.06.42	13.15–13.35	Sekehe	24	BW-353	Hurricane		E
18.08.42	20.00–21.20	Tolli lighthouse	24	BW-351	I-16		T
02.03.43	15.05–16.20	Karavaldai – Seiskari	24	BW-351	I-153	71 IAP, KBF	T
21.04.43	08.00–09.30	Karavaldai lake	24	BW-351	Yak-1		T
20.05.43	13.20–14.40	Seiskari	24	BW-357	La-5	4 GIAP, KBF	T
20.05.43	13.20–14.40	Seiskari	24	BW-357	La-5		V
30.05.43	04.30–05.45	Ystinski	24	BW-357	½ La-5		T
30.05.43	04.30–05.45	Kreivinlahti	24	BW-357	LaGG-3		T
10.08.43	16.25–17.30	Seiskari – Sepeleva	24	BW-357	Il-2	7 GShAP, KBF	V
20.08.43	16.30–18.00	Kronstadt	24	BW-357	½ LaGG-3		T
20.08.43	16.30–18.00	Yhinmäki	24	BW-357	LaGG-3		T
23.09.43	15.30–16.30	Seiskari	24	BW-357	½ La-5		T
10.11.43	13.15–14.40	Oranienbaum	24	BW-393	Yak-1		T
02.06.44	13.45–14.50	Levashovo airfield	24	MT-209	La-5		T
06.06.44	07.50–09.05	Lempaalanjärvi	24	MT-241	LaGG-3		T
13.06.44	09.00–10.05	Kaukjärvi – Raivola	24	MT-241	Pe-2		T
20.06.44	10.40–11.25	Karhusuo	24	MT-438	2 x Il-2	943 ShAP	T
20.06.44	19.05–20.15	Punnusjärvi	24	MT-438	Airacobra	196 IAP	T
21.06.44	08.40–09.30	Suomenvesi	24	MT-438	La-5		T
23.06.44	07.30–08.20	Tienhaara – Viipuri	24	MT-438	2 x Il-2	872 ShAP	T
23.06.44	07.30–08.20	Tienhaara – Viipuri	24	MT-438	Il-2		R
28.06.44	09.00–10.00	Noskuanselkä	24	MT-438	2 x Il-2	448 ShAP	T
30.06.44	10.50–11.55	Karhusuo	24	MT-460	Yak-9	404 IAP	T
30.06.44	10.50–11.55	Lyykylänjärvi	24	MT-460	Ar-2 ?		T
01.07.44	03.55–05.05	Tuppura	24	MT-460	LaGG-3	9 IAP, KBF	T
03.07.44	06.10–07.00	Viipuri	24	MT-460	Il-4		T
05.07.44	09.25–10.30	Vatnuori	24	MT-460	Il-2		T
10.07.44	12.00–13.25	Äyräpää	24	MT-460	2 x La-5		T
19.07.44	04.40–06.10	Äyräpää	24	MT-460	La-5		R

Bf 109G-6, MT-460 of 3/HLeLv 24 flown by SSgt Emil Vesa in July 1944, based at Lappeenranta. (Artur Juszczak)

Mersus of 3/HLeLv 24 at Lappeenranta on 2 July 1944. From left MT-440 (yellow 4) of Huotari, MT-460 (yellow 8) of Vesa, MT-462 (German) of Katajainen and MT-443 (Yellow 5) of Keskinummi. (SA-kuva)

Virta, Toimi Kelpo Jalmari

6 victories (5 + 1 + 1) on appr. 110 missions
Planes: FR-84

Kelpo Virta was born at Pori on 08.09.16. He received flying training in 1936–37 at RAOK 3 and AOK 6 and was posted on 01.01.38 as Corporal to LLv 24. Virta was promoted on 06.12.39 to Staff Sergeant, on 25.01.40 to Master Sergeant and on 22.04.40 to Warrant Officer. He was killed in a flying accident on 28.01.41, when his BW-391 hit the woods in low level manoeuvring at Vesivehmaa. Virta was awarded with VR 4.

Fokker FR-116 of 5/LLv 24 at Joroinen in early April 1940. Though not Virta's assigned plane, he flew it on several occasions during the Winter War.

BW-365 of LLv 22 seen at Hollola on 21 March 1940. After delivery to LLv 24 a month later it was often flown by Virta, including the last but one flight before his accidental death on 28 January 1941 flying BW-391.

Date	Time	Area	Unit	Aircraft	Enemy aircraft	Enemy Unit	Victory type
01.12.39	14.05	Johannes	24	FR-84	SB	41 SBAP	T
19.12.39	10.15–10.25	Muolaanjärvi	24	FR-84	I-16	68 IAP	T
19.12.39	10.15–10.25	Muolaanjärvi	24	FR-84	I-16		E
23.12.39	10.45	Kämärä	24	FR-84	I-16		V
25.12.39	12.05	Immola	24	FR-101	½ DB-3	6 DBAP	T
25.12.39	12.20	Joutseno	24	FR-101	½ DB-3	6 DBAP	T
17.01.40	14.30	Sakkola	24	FR-84	SB		T
04.02.40	14.20	Utö	24	FR-84	SB	57 AP	T

VIRTANEN, Väinö Johannes

8 victories (6 + 2 + 0) on appr. 280 missions
Planes: CUw-569

Väinö Virtanen was born at Pori on 26.02.21. He received flying training in 1940–41 at SOK 3 and was posted on 01.03.41 to LLv 32. Virtanen was promoted on 17.07.41 to Corporal, on 26.08.41 to Sergeant and on 28.06.43 to Staff Sergeant. After the war he served with LeR 1, 2. Lsto and SatLsto resigning as Warrant Officer on 17.10.58. He changed later his family name to Vuontila and worked as a process planner. Vuontila died on 12.02.03. He was awarded with VM 2, VM 1 and VR 4.

Curtiss Hawk 75A-2, CU-551 flown by 2/LeLv 32 pilot SSgt Väinö Virtanen in July 1943, based at Nurmoila.
(Karolina Hołda)

CU-551 of LeLv 32 ready for a mission from Nurmoila in early July 1943. It was often piloted by Virtanen, a long time member of the 2nd Flight.

Date	Time	Area	Unit	Aircraft	Enemy aircraft	Enemy Unit	Victory type
21.09.41	15.40–18.00	Siestarjoki	32	CU-552	I-153	235 ShAP	E
23.12.41	13.20–15.10	Ohalatva	32	CU-556	U-2	174 ShAP	T
28.03.42	17.40–18.00	Lavansaari – Suursaari	32	CU-554	I-153		T
07.04.42	15.20–16.00	Lumisuo	32	CU-554	I-16		E
19.07.42	13.00–14.30	Kuuttilahti	32	CU-552	I-16	524 IAP	E
19.07.42	13.00–14.30	Kuuttilahti	32	CU-552	LaGG-3	524 IAP	V
13.08.42	10.45–11.30	Savijärvi	32	CU-558	MiG-3		E
05.05.43	12.35–14.15	Mergino 7 km N	32	CU-569	MiG-3		T

VUORIMAA, Toivo Olavi

5 victories (4 + 1 + 1) on appr. 80 missions
Planes: FR-93

Toivo Vuorimaa was born at Simo on 26.01.17. He received flying training in 1936–37 at RAOK 2, AOK 5 and took the reserve officer school. Vuorimaa was transferred on 22.12.37 as a Sergeant to LLv 24. He was promoted on 31.12.39 to Staff Sergeant and on 30.01.40 to 2nd Lieutenant. Vuorimaa was assigned as instructor on 27.03.40 to T-LLv 35, where he was promoted on 22.06.42 to 1st Lieutenant. On 04.07.43 He was transferred to LeSK and on 31.10.43 back to T-LeLv 35. After the war he served with LeSK and EsikLtue resigning on 08.05.47. Vuorimaa acted later as a station service manager. He died on 11.08.04. Vuorimaa was awarded with VR 4.

Vuorimaa sitting in the cockpit of a 1/LLv 24 Fokker during the Winter War, becoming then an ace.

Fokker FR-93 seen at the factory at Tampere shortly before the hand-over to LLv 24 on 3 April 1939. On 10 October 1939 it was assigned to 1/LLv 24 and Vuorimaa.

Date	Time	Area	Unit	Aircraft	Enemy aircraft	Enemy Unit	Victory type
19.12.39	13.00	Sairala	24	FR-83	SB	44 SBAP	T
25.12.39	13.30	Korpiselkä – Äglājärvi	24	FR-93	2 x SB	72 SBAP	T
01.01.40	12.55	Uuksujärvi	24	FR-93	SB		E
20.01.40	15.15	Uomaa	24	FR-93	DB-3		T
19.02.40	14.20	Imatra	24	FR-91	I-15bis		V

Wind, Hans Henrik

74½ victories (64½ + 10 + 2) on 302 missions
Planes: BW-378, BW-393, MT-201 and MT-439

Hans Wind was born at Tammisaari on 30.07.19. He received flying training in 1939–40 at IRUK 9 and on 07.12.40 he was accepted to KadK. Wind was promoted to 1st Lieutenant on 17.06.41 and posted to LLv 25 and on 01.08.41 further to LLv 24. Wind became on 07.11.42 the deputy leader of 1/LeLv 24. On 27.05.43 he was assigned the leadership of 3/LeLv 24 and on 17.10.43 was promoted to Captain. Wind was wounded on 28.06.44, when a Yak-9 of 29 GIAP got hits in the cockpit of his MT-439. After the war Wind served with HLeLv 31 and resigned on 10.10.45. He obtained a degree in economics and held leading positions in shoe and bakery industry. Wind died on 24.07.95. He was awarded with VR 4, VR 3 tlk and MHR twice, on 31.07.43 and on 28.06.44.

Brewster BW-378 of 4/LeLv 24 parked at Vesivehmaa on 4 June 1942. It was assigned to Wind, who was a member of a detachment flying top cover to Adolf Hitler's Fw 200 during his visit to greet Marshall Mannerheim on his 75th birthday.

1/LeLv 24 leader Wind after landing his BW-393 at Suulajärvi. The 16th victory bar marked on the fin came on 24 January 1943.

Brewster 239, BW-393 of 3/LeLv 24 flown by the flight leader 1Lt Hans Wind in September 1943, based at Suulajärvi. (Karolina Hołda)

3/LeLv 24 leader Wind taxies his Brewster BW-393 for the photographer at Suulajärvi on 12 September 1943. (SA-kuva)

Wind (centre) and his mechanics in front of BW-393 at Suulajärvi on 12 September 1943. The fin has 33 victory bars and there would be six more.

Wind on the wing of his BW-393 at Suulajärvi on 26 August 1943. He had received the Mannerheim Cross four weeks earlier. (SA-kuva)

3/LeLv 24 leader Wind posing for the official photographer at the tail of his BW-393 at Suulajärvi in August 1943. A standard practise was to apply the victory markings on both sides of the fin. (SA-kuva)

Date	Time	Area	Unit	Aircraft	Enemy aircraft	Enemy Unit	Victory type
27.09.41	17.00–17.05	Derevjannoje	24	BW-367	½ I-153	152 IAP	T
07.11.41	08.30–10.30	Lotinanpelto	24	BW-373	LaGG-3	415 IAP	T
09.01.42	10.15–12.05	Kamenitsinjärvi	24	BW-393	R-5		T
10.01.42	12.20–13.35	Haapaselkä	24	BW-377	MiG-3		T
06.02.42	08.05–10.20	Telekina	24	BW-380	MiG-3		E
29.03.42	09.45–11.35	Sekehe	24	BW-378	½ R-5		T
14.08.42	12.30–13.45	Tolli lighthouse	24	BW-393	Hurricane	3 GIAP, KBF	T
14.08.42	12.30–13.45	Oranienbaum	24	BW-393	Hurricane		E
18.08.42	20.00–21.00	Kreivinlahti	24	BW-393	Hurricane		T
18.08.42	20.00–21.00	Kreivinlahti	24	BW-393	I-16		T
18.08.42	20.00–21.00	Kreivinlahti	24	BW-393	I-16		E
26.10.42	12.15–14–00	Karavaldai – Sepeleva	24	BW-377	2 x I-16		E
23.11.42	11.05–12–25	Patterinlahti	24	BW-393	I-16		T
23.11.42	11.05–12.25	Tolli lighthouse	24	BW-393	Pe-2		T
23.11.42	11.05–12.25	Tolli lighthouse	24	BW-393	Pe-2		V
24.01.43	14.45–16.00	Kronstadt	24	BW-393	I-16		T
02.03.43	15.45–17.00	Tipuna	24	BW-393	I-16		T
14.04.43	06.45–08.15	Seiskari 12 km SW	24	BW-367	Spitfire ?		T
14.04.43	06.45–08.15	Seiskari 15 km SW	24	BW-367	Spitfire ?		T
21.04.43	08.00–09.00	Seiskari 15 km E	24	BW-393	½ Yak-1		T
21.04.43	08.15–09.30	Sepeleva	24	BW-393	Yak-1		T
21.04.43	08.15–09.30	Oranienbaum	24	BW-393	Yak-1		T
04.05.43	10.40–12.20	Seiskari	24	BW-393	I-153		T
04.05.43	10.40–12.20	Seiskari – Oranienbaum	24	BW-393	2 x Il-2	7 GShAP, KBF	T
04.05.43	10.40–12.20	Tolli lighthouse	24	BW-393	Il-2		T
09.05.43	11.15–11.30	Seiskari – Oranienbaum	24	BW-366	La-5		T
20.05.43	09.00–10.30	Ystinskin niemi	24	BW-393	Yak-7B	13 KIAP,KBF	T
20.05.43	13.20–14.20	Seiskari 15 km E	24	BW-393	La-5	4 GIAP, KBF	T
20.05.43	13.20–14.20	Seiskari – Sepeleva	24	BW-393	Yak-1		T
30.05.43	04.30–05.45	Ystinski	24	BW-393	LaGG-3		T
30.05.43	04.30–05.45	Ystinski	24	BW-393	½ La-5	4 GIAP, KBF	T
05.06.43	14.40–15.50	Yhinmäki	24	BW-366	La-5		T
14.07.43	05.30–06.30	Oranienbaum	24	BW-366	Pe-2		E
17.07.43	12.30–13.30	Koivisto – Teikari	24	BW-393	LaGG-3		E
19.09.43	14.10–15.35	Tolli – Retusaari	24	BW-393	La-5		T
23.09.43	15.30–16.30	Seiskari	24	BW-393	Il-2		T
23.09.43	15.30–16.30	Seiskari	24	BW-393	La-5	4 GIAP, KBF	T
28.09.43	17.15–18.30	Seivästö	24	BW-393	½ Yak-1		T
28.09.43	17.15–18.30	Sepeleva lighthouse	24	BW-393	Yak-1		T
21.03.44	09.10–10.10	Valkeasaari	24	BW-355	Il-2	7 GShAP, KBF	T
27.05.44	06.40–07.20	Kasimovo – Perkjärvi	24	MT-201	2 x La-5		T
03.06.44	11.20–12.15	Kivennapa – Valkeasaari	24	MT-201	LaGG-3		T
04.06.44	09.15–10.00	Kronstadt	24	MT-201	La-5		T
06.06.44	07.50–09.05	Lempaalanjärvi	24	MT-201	Airacobra		T
13.06.44	09.00–10.10	Kaukjärvi – Raivola	24	MT-201	4 x Pe-2		T
13.06.44	09.00–10.10	Kaukjärvi – Raivola	24	MT-201	Airacobra		V
15.06.44	08.35–09.35	Kuuterselkä	24	MT-201	Airacobra	196 IAP	T
15.06.44	08.35–09.35	Kuuterselkä	24	MT-201	Il-2	999 ShAP	T
19.06.44	20.10–21.10	Summa	24	MT-439	2 x Pe-2		T
19.06.44	20.10–21.10	Johannes	24	MT-439	La-5		T
20.06.44	07.15–08.00	Koivisto	24	MT-439	2 x La-5	159 IAP	T
20.06.44	08.55–09.55	Koivisto – Seiskari	24	MT-439	Pe-2		T
20.06.44	08.55–09.55	Koivisto – Seiskari	24	MT-439	2 x Yak-9	29 GIAP	T
22.06.44	01.00–01.50	Viipuri – Säiniö	24	MT-439	2 x La-5		T
22.06.44	12.00–13.20	Suulajärvi	24	MT-439	Airacobra		T

Date	Time	Area	Unit	Aircraft	Enemy aircraft	Enemy Unit	Victory type
23.06.44	12.05–13.05	Säiniö – Viipuri	24	MT-439	2 x Il-4		T
23.06.44	12.05–13.05	Säiniö – Viipuri	24	MT-439	2 x La-5		T
26.06.44	14.45–15.45	Viipuri – Tali	24	MT-439	2 x Yak-9	14 GIAP	T
26.06.44	14.45–15.45	Viipuri – Tali	24	MT-439	Yak-9		R
26.06.44	16.25–17.45	Tali	24	MT-439	2 x Yak-9		T
28.06.44	09.00–10.00	Viipuri – Säiniö	24	MT-439	2 x Il-2	448 ShAP	T
28.06.44	10.30–11.30	Juustila – Viipuri	24	MT-439	Yak-9		T
28.06.44	10.30–11.30	Juustila – Viipuri	24	MT-439	2 x Yak-9		R

3/HLeLv 24 leader Capt Hans Wind boarding a Savoia-Marchetti S.81 of TGr 10 in the morning of 19 June 1944, bound to Insterburg, Germany to pick up new Bf 109 G-6s. TGr 10 was part of Gefechtsverband Kuhlmey, which came to assist repelling the major Soviet summer offensive. (SA-Kuva)

Bf 109G-6, MT-439 of 3/HLeLv 24 flown by the flight leader Capt Hans Wind in June 1944, based at Lappeenranta. (Karolina Hołda)

Capt Wind about to start for the evening mission from Lappeenranta on 19 June 1944. He is piloting his assigned plane MT-439, still in German transfer markings. Wind had earlier this day brought this fighter from Germany. In ten days he claimed 25 air victories flying MT-439. (SA-kuva)

Fighter Units

Lentolaivue 24

On 8 September 1932 the Finnish Air Force received a new commander, Colonel Jarl Lundqvist, who was previously a field artillery officer. Like his predecessors he created a new organization. While the bases remained the same the names were changed to numbered air stations (*Lentoasema*):

- *Lentoasema* 1 – previous *Maalentoeskaaderi* of Utti
- *Lentoasema* 2 – previous *Merilentoasema* of Santahamina
- *Lentoasema* 3 – previous 2. *Erillinen Merilentolaivue* of Sortavala
- *Lentoasema* 4 – previous *Merilentoeskaaderi* of Turkinsaari
- *Lentoasema* 5 – previous *Erillinen Maalentolaivue* of Suur-Merijoki
- *Lentoasema* 6 – previous 1. *Erillinen Merilentolaivue* of Viipuri

This organization was still heavily maritime oriented, as four of the six units were naval air stations.

Another updating was followed by the order of the day on 10 October 1934, which gave numbers to the squadrons (*Lentolaivue*):

- *Lentolaivue* 10 – Reconnaissance squadron of LAs 1
- *Lentolaivue* 12 – Reconnaissance squadron of LAs 5
- *Lentolaivue* 24 – Fighter squadron of LAs 1
- *Lentolaivue* 26 – Fighter squadron of LAs 5
- *Lentolaivue* 34 – Squadron of LAs 4
- *Lentolaivue* 36 – Squadron of LAs 2
- *Lentolaivue* 38 – Squadron of LAs 3
- *Lentolaivue* 44 – Squadron of LAs 6

The shift to land-based units was clear, five out of eight squadrons operated now from land bases.

1 January 1938 saw the modernization of the Finnish Air Force. Static air stations gave way to mobile aviation regiments (*Lentorykmentti*). *Lentorykmentti* 2 was formed at Utti. *Lentolaivue* 24 based already there and *Lentolaivue* 26 from *Lentoasema* 5 at Suur-Merijoki were subordinated to it. Thus the whole fighter force was concentrated in *Lentorykmentti* 2, under one commander.

Lentolaivue 24 was still flying Gloster Gamecocks from 1930 and *Lentolaivue* 26 had Bristol Bulldogs obtained in 1935. At the foundation of the aviation regiments it was clear that biplanes had become obsolete and the regiment was promised new monoplane aircraft. In summer 1938 these appeared, in the form of Dutch Fokker D.XXI fighters.

Fighter Tactics

In 1934 Major Richard Lorentz, commanding *Lentolaivue* 24, discovered that contrary to the traditional form of a lead aircraft and two wingmen, a pair was much more flexible and suited better to most tactical conditions and could without difficulty be increased to a four aircraft swarm (= finger-four) etc. Before assuming the command of *Lentolaivue* 24 on 21 November 1938 Captain Gustaf Magnusson paid a number of visits to other air forces, including a three month tour to JG 132 "Richthofen".

They had experience from the Spanish Civil War and he received valuable information how to shoot down a Soviet Tupolev SB bomber and how to deal with Polikarpov I-15*bis* and I-16 fighters. The Germans had also given up the three-plane basic form in favour of the finger-four. This convinced the top brass that the basic formations and thus tactics were sound.

Due to the lack of funds the elementary fighter pilot's flying training was complete, but the advanced training did not go through all possible modes of offensive, since it was discovered that usually two, or three at the most, types of attacks against the bombers were enough. These three ways and the associated gun-

Lentolaivue *24 and its predecessors flew the Gloster Gamecock for ten years, from December 1929 to October 1939. Here is one of the first, coded GA-45, at Utti on 5 December 1929 when the unit was* Hävittäjälaivue *of Maalentoeskaaderi (HLv/MLE). (Finnish Air Force)*

One of seven Dutch-built Fokker D.XXIs, coded FR-80, of LLv 24 running up at Helsinki Malmi on 11 July 1939. The flight is bound for Utti, the squadron's base, where this aircraft flipped over on landing. (Finnish Air Force)

nery were rehearsed throughout, and it served both the doctrine and economical limits. This was based on the assumption that the enemy bombers would arrive without fighter escort, which proved to be true when the hostilities began.

The aircraft machine guns were usually set to converge at 150 metres, but the pilots were trained to hold fire until at 50 metres. Getting there was somewhat risky, but again at that distance behind the bomber you had two major advantages: 1) you were out of the sectors of the defensive fire, as the Soviet bombers did not have tail gun positions and 2) you could not miss. It was also thought that the rifle calibre MG fire did not cause the bomber to explode, when pieces of bomber could damage the fighter.

When the war broke out Magnusson gave strict orders to avoid fighter duels, the Fokker D.XXI did not have enough manoeuvrability for this but was fit as an interceptor. Even if lacking speed it had a good rate of climb, and could always pull away in a dive. The D.XXI possessed simple construction and systems, which was an advantage in the harsh winter conditions, all aircraft being quickly made operational and remaining so.

Towards the War

After the German and Soviet occupation of Poland in September 1939 the political situation grew intense between Finland and the Soviet Union. In this crisis the Finns began on 10 October 1939 additional military rehearsals, which in fact was full-scale mobilization. The troops were concentrated on the south-east and east borders, the emphasis on the Karelian Isthmus.

The Finnish fighter defence was concentrated to *Lentorykmentti* 2 commanded by Lieutenant Colonel Richard Lorentz, under the command of the General Headquarters. *LentoR* 2 possessed two fighter squadrons: *Lentolaivue* 24 and 26, the former fully and the latter mainly equipped with the Fokker

D.XXI. At the beginning of the mobilization *Lentolaivue* 24 was dispersed to forward airfields and *Lentolaivue* 26 delivered its D.XXIs and their pilots to LLv 24 on 23 October 1939.

During the autumn the Russian had positioned troops along the whole eastern border from the Gulf of Finland to the Arctic Sea, The Russians were estimated to possess 2,000 combat panes, which recent research has proven to be 2,318 aircraft, This amount would grow by another 1,500 during the course of the Winter War.

The main front was on the Karelian Isthmus, where the ten division strong Soviet 7th Army was facing five Finnish divisions. The air forces of the 7th Army had 993 combat aircraft, supported by the 158 planes of long-distance bomber command and 219 fighters from the Leningrad air defence. In addition the 469 plane strong Baltic Fleet air force operated on the coastline from Viipuri up to Vaasa.

Winter War

The Soviet offensive began on the morning of 30 November 1939. Against this massive air armada Finland could deploy only one modern fighter squadron, *Lentolaivue* 24. Its order of battle was then:

- Squadron commander Captain Gustaf Magnusson, Immola
- 1st Flight, Captain Eino Carlsson, Immola (6 FR)
- 2nd Flight, 1st Lieutenant Jaakko Vuorela, Suur-Merijoki (6 FR)
- 3rd Flight, 1st Lieutenant Eino Luukkanen, Immola (6 FR)
- 4th Flight, Captain Gustaf Magnusson, Immola (7 FR)
- 5th Flight, 1st Lieutenant Leo Ahola, Immola (10 FR)

The air defence commander, Major General Lundqvist, had given orders for *Lentorykmentti* 2 to prevent the flying of enemy aircraft in the airspace of the Karelian Isthmus. The operational area of LLv 24 was defined as the north-western side of line between Viipuri and Antrea. The mobilization was done and needed no air protection, thus the squadron could fully concentrate on defence.

On the first day of the Winter War, 30 November 1939, Soviet air forces bombed with 200 aircraft many towns and air bases in southern Finland, while fighters cruised midway above the Karelian Isthmus. But the interceptors of LeR 2 failed to meet the invaders due to poor weather.

On first day of December 250 unescorted bombers were in the air again, attacking many of the same targets as the day before. The Fokkers of LLv 24 took off in pairs, led by the CO Capt Magnusson, for 59 sorties claiming eleven bombers destroyed in the Viipuri-Lappeenranta area, eight from 41 SBAP (41 Fast Bomber Aviation Regiment) and three from 24 SBAP.

The first fell at 1205 hours under the guns of the 2nd Flight leader 1Lt Jaakko Vuorela and the last at 1440 hours by the 5th Flight boss 1Lt Leo Ahola. Vuorela became a double scorer and the other victors were Capt Gustaf Magnusson, 1Lt Eino

Luukkanen, 1Lt Jussi Räty, 2Lt Pekka Kokko, Sgts Lasse Heikinaro, Lauri Nissinen, Lauri Rautakorpi and Kelpo Virta with one each. No combat reports exist from these first encounters, since the forms were not available until after three weeks later. However, Capt Magnusson insisted that every pilot involved in a combat should write down his experiences on a piece of paper. His report said the following:

"1.12.39 at 1410-1445 hours. Based on an announcement that a Soviet bomber formation was approaching Imatra we took off. We met the formation above Imatra. I attacked the one flying on the extreme right wing shooting first along the fuselage. When the firing did not seem to have any effect, I aimed the fire to the starboard engine, which started to smoke after a few bursts.

I had to interrupt my attack since the one on the left to my target had reduced speed being about 70 metres on my port side

Fokker FR-86 of 2/LLv 24 camouflaged under sheets at Utti on 1 December 1939. On this day the flight leader 1Lt Jaakko Vuorela scored the unit's first air victories, two Tupolev SB bombers of 24 SBAP. (SA-kuva)

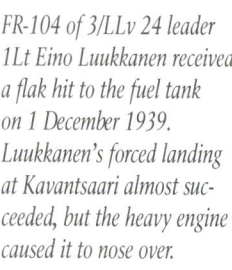

FR-104 of 3/LLv 24 leader 1Lt Eino Luukkanen received a flak hit to the fuel tank on 1 December 1939. Luukkanen's forced landing at Kavantsaari almost succeeded, but the heavy engine caused it to nose over.

with the dorsal gunner firing all the time. I slowed down behind this plane and shot it into flames. The plane crashed burning into the ground.

Since the squadron did not have other that normal bullets and tracers, it was not possible to gain results with a small amount of ammunition. 1200 rounds spent. Own machine FR-99."

This example shown by the commanding officer was much more important that anything the pilots had learnt so far. It proved that the methods used were working and nobody was willing to do less.

From 2 December 1939 poor weather kept the interceptors of LLv 24 in the ground for almost three weeks.

Main Target Bombers

On 18 December 1939 LLv 24 took off for 30 sorties. Most could not engage the enemy bombers due to poor weather, but 1Lt Jorma Karhunen claimed one SB of 24 SBAP.

LLv 24 flew 58 sorties on 19 December 1939 to the front on the Karelian Isthmus and was engaged in combat on 22 occasions between 10.50 and 15.20 hours. The Soviets lost seven SBs, six from 44 SBAP and five Ilyushin DB-3s from other regiments. However, SSgt Virta was the first in action, with 25 IAP fighters and claimed two shot down.

The ground troops observed both crashes and Virta was credited with two aircraft destroyed. The squadron adjutant interviewed every pilot engaged in combat and made a summary, where the matters were told in a combat report manner, but a bit more free hand. 1Lt Per Sovelius flew FR-92 and told this:

"19.12.39 at 0955-1105 hours, on air combat patrol I was leading the 3rd pair with Sgt Ikonen on my wing. We took off by the announcement. I flew over Antrea area when the radio informed 45 and I headed towards south-west. Somewhere near Kämärä I observed a 7-plane SB formation and started the chase. The SBs were flying toward south-west, but turned towards south. However, we did not gain on them, but we observed a bit aside three SB planes flying roughly in the same direction.

Sgt Ikonen got well behind the plane on the starboard wing and shot it into flames from a very close range at the altitude of 2,000 metres over Kipinola. I tried to get behind the port wingman, but did not have enough speed. I observed another three SB planes heading south-west, tried after them but began to loose them. The planes were throwing leaflets.

During the chase I observed again three SB planes a little below going southwards. I picked the port wingman as my target and shot first to the rear fuselage, when the dorsal gunner quit firing. After this I aimed the fire to the port engine, which began to smoke and finally caught fire. The plane fell to the starboard wing and dived towards the sea close to Seivästö about 10 kilometres from the coast.

I fired then at the starboard wingman, when its right engine started to pour smoke, but it stayed with the lead plane and kept going on. I did the return flight at 3,000 metres and while about 5 kilometres above the land I observed a-a artillery explosions. On the south coast of lake Muolaanjärvi two I-16s managed to take me by surprise coming from the sun. I woke up when bullets sounded in my plane.

I pulled instantly towards them, but noticed soon that the I-16 was more manoeuvrable than the Fokker. I tried to tighten my turns, but got only once the enemy into my sight and fired a short burst. I noticed at the same time that I had ammunition left only in one gun. Once after trying to turn as tight as possible I lost control and slipped into a spin. I continued to make all sorts of evasive actions and down at the surface I managed to shake off my pursuers. I was then nearby Heinjoki.

Both I-16s attacked simultaneously and obviously in order to avoid a collision did not get straight behind me and were forced to shoot with a small deflection. Judged by the tracers the I-16s seemed to fire all the time even when my plane was clearly out of their sights.

After the mission my plane had two hits. One in the tail plane and the other had gone in through the mg compression bottle hatch and out from the fuselage bottom."

The weather permitted flying in the morning on 21 December 1939 and 300 bombers were counted over the south-eastern Finland. LLv 24 too off for 62 intercept sorties and claimed three DB-3 bombers shot down on the Karelian Isthmus.

Bad luck hit 44 SBAP again on 23 December 1939. The Fokkers of LLv 24 downed at around eleven o'clock another six SBs on the Karelian Isthmus, 1Lt Sarvanto claiming two. Magnusson describes his shoot-down of one bomber in FR-99 this way:

Pilots of LLv 24 on New Year's Day 1940. Behind from the left: WO Yrjö Turkka, Sgt Lauri Heikinaro, 1Lt Jorma Sarvanto and 1Lt Erhard Frijs, in front from the left: Sgt Risto Heiramo, Sgt Eero Kinnunen and Sgt Tauno Kaarma. Behind is a Dutch-built Fokker with serial FR-81.

"Sgt Kinnunen flying on the left flank observed nine SB planes above Vuoksenranta. I dived after Sgt Kinnunen, who pulled away thinking I was an I-16.

I continued after the formation and caught it at Kiviniemi. I chose as my target the rearmost bomber on the left flank. First I fired in the starboard engine, which started to smoke, thereafter I shot the port engine into fire, when the plane began to descend.

Tactically the enemy unit worked well, e.g. lowering the landing gears simultaneously for speed reduction and the next to the one under attack lowered speed in order to obtain a better position for its rear gunner. At 12 o'clock the aircraft hit the ground at Lempaalanjärvi."

The engagements of the day totalled 21 and during fighter duels a further two I-16s of 7 IAP and two of 64 IAP were shot down. Sgt Pentti Tilli accounted for both of the former and 1Lt Urho Nieminen and 2Lt Heikki Ilveskorpi the latter. On the other hand one Fokker D.XXI was hit and Sgt Tauno Kaarma was injured when he crashed the plane.

By 25 December 1939 the Finns had stopped all advances of all Soviet armies on the 1,000-mile long border and a stalemate situation for five weeks began. The extremely cold winter, temperatures often below 30 and on several days below 40 degrees centigrade, favoured less the attacking Russians.

Fokkers of LLv 24 destroyed two Tupolev SB bombers from 6 DBAP (long-range bomber aviation regiment) over the Karelian Isthmus. The 3rd Flight was strengthened to Detachment Luukkanen and transferred to Värtsilä in support of the troops on the northern coast of Lake Ladoga. There it immediately sent down four SB bombers of 18 BAP, both 1Lt Jorma Karhunen and Sgt Toivo Vuorimaa claiming two.

On 27 December 1939 the Soviets lost on the Karelian Isthmus three SB bombers of 2 SBAP and WO Pyötsiä shot down in FR-110 two Polikarpov I-15bis fighters north of Lake Ladoga.

On the last day of 1939 Sgt Ilmari Juutilainen, member of LLv 24´s Detachment Luukkanen and the future top scorer with 94 kills, blasted behind 1Lt Karhunen's tail a lone I-16 on the northern shore of Lake Ladoga.

Lentolaivue 24 scored steadily and the first month of operations 54 enemy aircraft were sent down for the cost of only one Fokker lost and another damaged. The small number of available fighters permitted intercepts over the Karelian Isthmus, but did not allow the intercept of almost daily air raids of southern Finland, where numerous bombers attacked many locations.

By the beginning of 1940 it was discovered that the Soviet bombers orientated by the aid of the Finnish railway network on their way to attack targets in the inner parts of the country. North of Lake Ladoga the Fokkers of LLv 24 shot down three bombers on New Year's Day.

LLv 24 fighters carried out 35 intercept sorties on 5 January 1940 claiming four bombers shot down over the Karelian Isthmus.

6 January 1940

In the morning of 6 January 1940 seventeen Ilyushin DB-3Ms of 6 DBAP took off in two detachments from Estonia to bomb Kuopio in eastern Finland. The first nine planes bombed as planned, but the second formation of eight bombers drifted too far west and crossed the Gulf of Finland south of Utti, where 4/LLv 24 was then based. 1Lt Sovelius, who was in the air, caught the bombers at 10.10 hours at 3,000 metres and shot one down.

The remaining seven continued to Kuopio, released the bombs (not causing any serious damage) and returned the same route along the railway. 1Lt Sarvanto had meanwhile taken off and met the bombers. In his memoirs he describes those famous four minutes just after noon as follows:

"The clouds over Utti had disappeared and the sun gleamed from the light bellies of the marvellous looking row of bombers. I counted them to be seven. On the left flew an echelon of three and to the right four almost in a row. The distance between the planes was hardly one aircraft.

Pilots of LLv 24 in front of Fokker FR-95 in February 1940. From the left: 2Lt Olli Mustonen, 2Lt Iikka Törrönen, WO Yrjö Turkka, 1Lt Per Sovelius, Sgt Eero Kinnunen, MSgt Sakari Ikonen, 2Lt Toivo Vuorimaa, Sgt Martti Alho and Sgt Lauri Heikinaro.

I banked to the right to the south and continued to climb. For a moment I was in the sights of the nose gunners, but facing the sun they obviously did not see me. When I reached the altitude of the bombers I was already 500 metres behind them.

At full power I started the chase and selected the one on the left wing, although the third from the left was more behind and the fire from its rear gunner felt dangerous. At the distance of 300 metres it banged unpleasantly in my plane – I obviously flew into the stream of enemy bullets.

I opened fire at 20 meters with a short burst to the fuselage of the left machine. The tracers seemed to hit the target, and in the same way I silenced the previously dangerous rear gunner. One more good aim at the starboard engines of both bombers, as light touches as possible on the trigger, and both enemies were on fire going where they should.

I must have cheered and threw my Fokker to the right side of the formation. In the same manner I made the engines of one bomber smoke. I attacked the next at a very close range, and the planes were in flames as soon as I had hit them with two or three very short bursts. On the right I saw the one already smoking and being separated from the others diving as a fireball towards the ground.

Now I had set my goal to destroy every one. Others turned away from the formation like burning pages of a book, others pulled up steeply when the pilot met his end. The reddish January sun shone through the haze towards me all the time, except when the dark smoke of the burning planes cast a shadow.

e last but one was much tougher than the others. My wing guns were probably empty by then. Well, it finally caught fire, so I could take the last one. Its rear gunner had been silent for quite some time and I went in very close. I aimed at the engine and pulled the trigger.

The guns were quiet!

I made a couple of charging attempts but without any result. I had run out of ammunition, and the only thing to do was to return home."

All DB-3s crashed between Utti and Tavastila, a distance of 30 kilometres. Sarvanto was credited with six victories in four minutes and became the first ace of the Finnish Air Force. His FR-97 had taken 23 hits, none very serious, but was still flown to the repair shop. The foreign press were immediately interested, nothing like this had yet happened in Europe!

Frequent Combats

17 January 1940 was a very cold day, temperature below minus 40 degrees centigrade. Ten Fokkers of LLv 24 scrambled and at 13.55 hours caught 25 SBs of 54 SBAP returning in three formations via the Karelian Isthmus. 25 minutes later nine bombers had met their end and several more were damaged.

LLv 24 gained new aces on 19 January 1940. Both 1Lt Urho Nieminen (LLv 26 member) and SSgt Kelpo Virta achieved the status, shooting down an SB bomber each over the Karelian Isthmus. For two weeks the airspace over south-eastern Finland became almost free of bombers.

On 20 January 1940 Detachment Luukkanen of LLv 24 engaged, north of Lake Ladoga, SB bombers of 21 DBAP, both arriving and departing, and succeeded in destroying five of them. WO Viktor Pyötsiä scored two and SSgt Pentti Tilli (LLv 26 member) one, both becoming aces. Then the luck turned and two I-16s appeared on the scene to chase the Finnish Fokkers and Tilli was shot down, dying in the crash.

1Lt Tatu Huhanantti took off from Tampere in one Fokker D.XXI, which had been under repair at the State Aircraft factory. On the way back to the base he met three SB bombers belonging to 35 SBAP and quickly shot down two before the escort of five I-153s could interfere. The wrecks were easily found as the planes came down 60 kilometres north of Helsinki, next to the railway line.

Though up till now the Soviet fighters had shot down only two Fokkers, there had been a number of close calls. On 28 January 1940 LeR 2 commander LtCol Richard Lorentz is-

Pilots of LLv 24 in front of Fokker FR-110 in January 1940. From the left: Sgt Martti Alho, 2Lt Tapani Harmaja, 1Lt Jussi Räty, 1Lt Veikko Karu, Maj Gustaf Magnusson, WO Viktor Pyötsiä, Sgt Sakari Ikonen, 1Lt Per Sovelius, 2Lt Iikka Törrönen and a war correspondent.

sued a firm ban on searching for and engaging enemy fighters for combat, only the bombers were to be attacked. The Fokker D.XXI was no match for the much more manoeuvrable Polikarpov I-153s and I-16s, which additionally possessed a huge numerical superiority.

On 29 January 1940 Russian artillery was aiming gunnery on the Karelian Isthmus, with the aid of Polikarpov R-5 fire control aircraft. The Karelian army commander contacted LeR 2 commander, requesting them to put a stop to this and 1Lt Jorma Karhunen of LLv 24 told how the job was done:

"At 14.55 I got an order to take off with four Fokkers to Summa, either to drive away or destroy two artillery fire control R-5 planes. Five minutes later the Fokker swarm speeded on the ice and took off. WO Yrjö Turkka was my wingman in the lead pair while 2Lt Olli Mustonen and Sgt Tauno Kaarma as wingman headed the other pair.

I plotted the battle plan, which had to be foolproof. The enemy planes were called off as soon as a smallest hint of arriving Finns was received. When the Finns were gone they popped up again. I decided to fool the airspace control of the opposition properly.

We flew along the west coast of Viipurinlahti at the cloud line at 2,000 metres. We continued to the south and at Koivisto we turned towards Summa. I took the swarm inside the clouds. When approaching Summa I pulled my FR-80 out of the cloud just to make sure that the "patients" were still there. There they circled, distributing instructions to the artillery. A way to go and I returned inside the clouds. Three minutes took a long time, but I controlled myself. Then we bounced out of the clouds.

One R-5 was conveniently below the lead pair. We attacked together with "Daddy" Turkka and fired simultaneously from up ahead and behind. In a moment the wings collapsed together after

the fuselage burst into fire. The flaming ball of fire came down between the lines. Then we all fired at the other R-5 and it crashed beyond the lines. The job was done.

We also managed to get unhurt out of the fierce anti-aircraft fire back to our own side, which as such was a miracle."

At the end of January there were now 28 serviceable Fokkers in LLv 24, when ten "hired" pilots returned to their original unit *Lentolaivue 26*, which had just began receiving Gloster Gladiators. January score for the Fokkers still showed 34 enemy aircraft shot down.

On 3 February 1940 1Lt Jorma Sarvanto of LLv 24 claimed his 10th kill, shooting down an Ilyushin DB-3M bomber of 51 DBAP at Nuijamaa in south-eastern Finland. In the west four Fokkers took completely by surprise three DB-3 bombers of 10 ABr (aviation brigade) and sent all down in the Turku archipelago, 2Lt Pekka Kokko claimed a double.

One DB-3 was shot down on 4 February 1940 over the sea south-west of Turku and the Russians became cautious. The sudden appearance of a Finnish fighter unit was not expected this far west.

On 9 February 1940 the Fokkers of LLv 24 took off for 31 intercept sorties over the front at the Karelian Isthmus and claimed two R-5s shot down.

The 4th Flight of LLv 24 attacked on 10 February 1940 a large bomber formation over Lappeenranta, but the equally numerous fighter escort handled the situation and shot down one Fokker D.XXI, MSgt Sakari Ikonen being wounded. Big claims were now history for the Fokker pilots, as the daily scores were three bombers at the most. Losses also started to accumulate.

LLv 24 flew 69 sorties on 14 February 1940 protecting troop transfers on the way to the Karelian Isthmus in order to repel the Soviet 7th Army. Two bombers of a 19-aircraft strong group from 48 SBAP were shot down near Lappeenranta.

The main target of Soviet bombers on 17 February 1940 was Kouvola on several occasions. It was an important railway and road junction. LLv 24 managed to scatter two 30-aircraft strong bomber formations over the immediate rear of the Karelian Isthmus front. In 26 combats three bombers were shot down

On 18 February 1940 over 300 aircraft bombed Viipuri. LLv 24 flew 65 intercept sorties claiming three bombers shot down.

Large formations of bombers escorted by numerous fighters flew on 19 February 1940 over the Karelian Isthmus. LLv 24 took of for 60 intercept missions claiming two bombers destroyed. 25 IAP fighters shot down Danish volunteer 1Lt Erhard Frijs near Käkisalmi, killing him.

On 20 February 1940 LLv 24 received 480 observations of Soviet bombers and climbed for 59 intercept sorties. They failed to prevent the bombardment of Elisenvaara railway junction, when the transfer of a division to the Karelian Isthmus was delayed.

From 21 February 1940 onwards many fighter bases and forward landing grounds became under almost continuous attacks by patrolling Soviet fighters. LLv 24 claimed two bombers and two fighters shot down in the encounters.

On 26 February 1940 Soviet bombers attacked Immola, the main base of LLv 24, with large forces. The Fokkers took off for 47 intercept missions trying to engage formations as large as 60 aircraft. The Soviet escort fighters made it impossible to attack the bombers. Sgt Tauno Kaarma of LLv 24 escaped injury when 68 IAP fighters shot his Fokker D.XXI into flames when taxiing at Immola.

The 29th of February was the saddest day for the fighter arm, when Soviet fighters carried out a series of air raids on *Lentolaivue* 24 and 26 bases. Detachment Luukkanen of LLv 24 had three weeks earlier moved to Ruokolahti and two Gladiator flights of *Lentolaivue* 26 were there put under his command.

At noon a bomber formation was announced to be approaching Ruokolahti, but it turned out to be the 68 IAP with six *Chaikas* and eighteen *Ratas*, which took by complete surprise the Gladiators at the take-off. Three were instantly destroyed and in a low-level combat a further two Gladiators, plus a Fokker D.XXI piloted by 1Lt Tatu Huhanantti, were downed. Only one I-16 was shot down and another hit the trees while manoeuvring at low altitude.

During February LLv 24 had claimed 27 aircraft destroyed and the serviceability at the end of the month was down to 22 Fokkers.

In March the flights of LLv 24 continued to use several forward landing grounds, mostly frozen lakes, in order to stay operational.

Viipurinlahti

On 4 March 1940 The Soviet troops had managed to cross Viipurinlahti (Gulf of Viipuri) over the ice and formed a bridgehead at Vilaniemi and Häränpääniemi. Troops and columns

Four 3/LLv 24 pilots with a Fokker at Ruokolahti in February 1940. From the left: the flight leader Capt Eino Luukkanen, Sgt Jalo Dahl, MSgt Ilmari Juutilainen and Sgt Martti Alho.

Left: Three 4/LLv 24 pilots trying to occupy the cockpit of a Fokker D.XXI coded FR-92 at Joutseno in January 1940. From the left: 1Lt Per Sovelius, 2Lt Olli Mustonen and 2Lt Iikka Törrönen.

flowed across the ice from Pulliniemi and Tuppura. All regiments were thrown in to repel this extremely serious threat.

Viipurinlahti drew also the attention of the Soviet fighters. On 5 March 1940 Capt Eino Luukkanen of LLv 24 led several strafing missions and described one in the early evening as follows:

"I am breaking radio silence and order the formation (15 Fokkers) out of the clouds. Now we are playing the enemy by approaching from their direction. I bank to the left and lead the formation into a dive. There seems to be no lack of the targets, because the 4 kilometres distance between Tuppura and Vilaniemi is full of columns, cars, trucks and tanks. Above Uuras circles an I-16 squadron and another fighter unit is on the other side of the gulf over Ristiniemi. I continue the shallow dive; the sooner we hit the better for us.

The range to the nearest targets is only one kilometre, when the air around us fills with explosions and tracers of anti-aircraft fire. Both white and black explosion clouds puff close to us, which attracts the enemy fighters after us.

My first burst hits an infantry column, next in sight is a line of trucks and finally I manage to fire at two tanks. The bullets of our rifle-calibre machine guns do not seem to have any effect on the latter, at least the tracers bounced away from the armour plates. After my strafing run I look back and ascertain that the rest of the formation followed my recent example.

Immediately after the attack the return begins – first at low level to the west and then banking to the north towards our base, so that the enemy would not discover our location in case someone was tailing us."

LLv 24 flew 18 strafing sorties on 6 March 1940 against the troops, columns and vehicles on the ice of Gulf of Viipuri. Next day LLv 24 flew 43 strafing sorties to Viipurinlahti. The tasks had become very difficult and dangerous. The Soviets had placed anti-aircraft artillery units on both sides of the gulf in addition to continuous combat air patrols by large fighter formations, often the whole regiment of 50 aircraft.

In seven days the ice over the Gulf of Viipuri was swept clean, as a result of the coastal artillery and air force, causing heavy casualties to the Russians. The advance on the continent had been stopped also and Viipuri remained unconquered.

The Fokkers of *Lentolaivue* 24 flew 154 assault sorties, claimed one I-16 shot down and lost in return one Fokker.

Their last combats were fought on 11 March 1940 over southern Finland, where fighter formations as large as 200 were observed. LLv 24 was involved in the fighter duels with no results.

The 105 days long Winter War ended on 13 March 1940 at 11.00 hours in the peace treaty negotiated in Moscow.

Brief Analysis

Lentorykmentti 2 fighters had flown 3,486 sorties, claimed 193 aircraft shot down and another 70 damaged, producing ten aces, and lost 23 fighters and 14 pilots in action. By squadron these were split:

Squadron	Sorties	Victories	A/c lost	Pilots lost
LLv 24	2388	106*	10	7
LLv 26	1170	73**	12	7
LLv 28	288	14	1	-

* excluding 27 Fokker claims by LLv 26 pilots

** including 27 Fokker claims with LLv 24

Winter War aces of *Lentolaivue 24:*

Rank	Name	Flight	Victories	Plane
1Lt	Sarvanto, Jorma	4	13	FR-97
WO	Pyötsiä, Viktor	3	7½	FR-110
1Lt	Karhunen, Jorma	1	6½	FR-112
1Lt	Huhanantti, Tatu †	3	6	FR-108
1Lt	Nieminen, Urho*	5	6	FR-111
MSgt	Virta, Kelpo	2	6	FR-84
1Lt	Sovelius, Per	4	5½	FR-92
1Lt	Puhakka, Olli*	1	5	FR-117
2Lt	Vuorimaa, Toivo	1	5	FR-93
WO	Turkka, Yrjö	4	5	FR-83
MSgt	Nissinen, Lauri	3	5	FR-98
SSgt	Tilli, Pentti* †	3	5	FR-103

* member of LLv 26, † killed in action

The victories have been rounded up to nearest half. They contain both confirmed and damaged cases, which were upgraded to confirmed by Soviet loss records.

Preparing for the Next War

In mid-April 1940, just a month after the Winter War, LLv 24 exchanged its Fokkers for Brewster Model 239s, as was originally planned. August 1940 saw a move from Helsinki Malmi to a new base at Vesivehmaa near the city of Lahti, where intensive type training began for the pilots, many over 100 combat mission veterans from the Winter War.

Germany had obtained permission from Finland to transport personnel, material and ammunition to troops occupying northern Norway. In return Finland could acquire military equipment including aircraft, which the Germans had taken as war booty from occupied countries. In late May 1941 the Germans informed the Finnish High Command that they would commence Operation Barbarossa, the offensive on the Soviet Union, on 22 June 1941. Finland made a full-scale mobilization four days prior to this date.

The Finnish fighter arm was concentrated into two aviation regiments. *Lentorykmentti* 2 was in the front-line and possessed *Lentolaivue* 24 with Brewsters, *Lentolaivue* 26 with FIATs and *Lentolaivue* 28 with Moranes. *Lentorykmentti* 3 had *Lentolaivue* 30 with Fokkers and Hurricanes and *Lentolaivue* 32 with Fokkers. The latter was deployed in homeland defence.

The order of battle of *Lentolaivue* 24 was then:

- Squadron commander Major Gustaf Magnusson, Vesivehmaa
- 1st Flight, Captain Eino Luukkanen, Vesivehmaa (9 BW)
- 2nd Flight, Captain Leo Ahola, Selänpää (8 BW)
- 3rd Flight, 1st Lieutenant Jorma Karhunen, Vesivehmaa (8 BW)
- 4th Flight, 1st Lieutenant Per Sovelius, Vesivehmaa (8 BW)

Continuation War

Large numbers of German aircraft were based just before the offensive on airfields in southern Finland, carrying out reconnaissance and channel mining missions. The Soviet intelligence had found this out and the Russians assumed that these bases would also be used for major attacks against Leningrad. Therefore they decided to attack these airfields and drew up a plan for a six-day bombardment offensive. For this purpose the Red Air Forces (Leningrad Military District, parts of Baltic

Military District, Northern Fleet and Baltic Fleet) had at their disposal from the Arctic Sea to the Baltic Sea 2,503 warplanes, of which 933 were bombers and 1,327 fighters. In addition 202 long-range bombers were in the rear. The operational border between Germany and Finland ran at the Oulu – Kajaani – Belomorsk level. South of this line half of this force could be directed against Finland.

The air raids began early in the morning of 25 June 1941. During the course of the day the Russians flew 263 bomber and 224 fighter sorties attacking several locations in southern and south-western Finland, among them both airfields and purely civilian targets. After these bombardments the parliament considered the same day that Finland was in a state of war with the Soviet Union and the government declared war. The Continuation War began.

The first observations of large bomber formations, entering the airspace of southern Finland, were made over Turku at on 25 June 1941 at 6 o'clock in the morning.

Brewster 239 BW-366 of 3/LLv 24 in September 1940, based at Vesivehmaa. (Karolina Hołda)

The new fighters of LLv 24 were American Brewster 239s (de-navalized F2A-1). Here is BW-356 seen at the air depot at Tampere on 29 May 1940. It was flown to the squadron's new base, Vesivehmaa, on 13 August 1940. (Finnish Air Force)

The message went quickly to Selänpää, where 2/LLv 24 was stationed in case of a possible air raid. At 07.10 hours the first Brewster pair took off and was immediately engaged in a combat, Cpl Heimo Lampi being the first to do so:

"Five minutes after the take-off I observed a large enemy aircraft formation. I attacked the plane on the extreme right flank and hit it with the first burst. The aircraft went in a vertical dive into the forest. I fired then at the two right hand side aircraft of a three-plane patrol on the left flank. My last attack was against the left flank aircraft of the same patrol. The aircraft began to smoke and it descended down to the surface. I chased after it, when it slowed down so much that I had to pull away right from its side, when the rear gunner got the opportunity to fire at me from a very close range. I pulled up and turned again back behind its tail. With a short burst I got the port engine of the aircraft to catch fire, when the aircraft dived into the water in flames. I saw SSgt Kinnunen shoot down two aircraft in the same battle. My own aircraft was BW-354."

These were 27 bombers of 201 SBAP approaching Heinola at 1,500 metres. Five aircraft were shot down from this detachment, equally split between Lampi and Kinnunen. LLv 24 flew 77 sorties during the course of the day in its operating area, the Soviet air forces lost a total of ten bombers.

Four fighter squadrons claimed 26 downed Soviet bombers (23 later on admitted), which was not a bad start for the Continuation War. But the air surveillance and fighter control system had large gaps. 121 fighters were ready for interception, but only one fifth could be directed to deal with the enemy. The weak spots were detected and put into an excellent working order in due course.

The air force commander gave an order on 29 June 1941 according to which LeR 2 was to fly top cover for a 100,000 man strong Karelian Army, which was formed to take back the area lost in the Winter War. The main offensive was planned to occur north of Lake Ladoga. The LeR 2 commander defined the new areas of operations: for LLv 24 Savonlinna-Elisenvaara-Sortavala-Savonranta, for LLv 26 Joensuu – Tohmajärvi

– Pälkjärvi – Soanlahti – Korpiselkä and for LLv 28 Savonranta – Sortavala – Suistamo – Korpiselkä.

This day only LLv 24 was engaged in air battles with the Russians. In the first encounter two flying boats were downed over the Gulf of Finland and in the second one Pe-2 bomber. 1Lt Jorma Sarvanto of the 2nd Flight reports:

"At 1230-1340 hours. Two aircraft flew over Kuusankoski heading towards Utti. The distance between the two was about 600 m. I radioed my wingman to attack the southern one while I turned after the northern one. Our own flak tracers and explosions surrounded me. The enemy was flying at about the same speed as I. I aimed carefully and from 400 m got the port engine to smoke. Now I got closer and shot the plane in flames from below and behind. Then I followed the other one at full boost out to the sea up to Seiskari, but it just opened the distance. The downed machine opened fire at about 800 m distance when I turned after it. I was doing over 500 km/h. One man bailed out. My plane was BW-357."

On the last day of June LLv 24 flew 69 sorties during the course of the day, three times being involved in combat and sending down seven aircraft. WO Veikko Rimminen of the 2nd Flight reported:

"At 0955-1055 hours. We were patrolling with Cpl Peltola at 2,000 metres north of Hamina, when we discovered two SB bombers heading to the south-east as informed by the radio announcement. South-west of Virolahti the enemy anti-aircraft gunnery disturbed the pursuit, so we caught the SBs out in the sea north of Lavansaari Island. I shot with a few bursts one into flames and it crashed into the sea at 10.35 hours. One man bailed out. My plane was BW-367."

The Soviet bombing offensive lasted for six days, during which they recorded having attacked on 39 occasions with a total of 992 aircraft Finnish (and German) air bases, destroying 130 aircraft on the ground and in the air. The Germans had no losses, since they had already flown out. The Finnish losses were two slightly damaged aircraft. On the other hand the Finnish fighters claimed 34 Russian bombers as shot down during the same period.

A guard is standing in front of Brewster BW-352 of 2/LLv 24 at Selänpää in the evening of 25 June 1941. Earlier that day SSgt Eero Kinnunen flew this plane to claim 4½ air victories in two missions. (SA-kuva)

On 8 July 1941 enemy intelligence observed the Finnish troop concentrations on the south-east border and began air raids against these.

The Finns flew constant combat air patrols above the troop concentrations and LLv 24 shot down, in three encounters, two bombers and six fighters. SSgt Lauri Nissinen of the 3rd Flight got his first kills in this war thus:

"At 1230 hours. I was flying in 1Lt Kokko's swarm leading the top pair. Above Enso I observed two Chaikas and an R-5 at about 500 metres altitude below left. I attacked the R-5 from above and left opening fire at 100 metres. The tracers hit the fuselage behind the pilot, and all I noticed was that the gunner did not shoot. Then I had to start a turning fight with the other Chaika and was twice forced to break off by diving, after which I pulled steeply up in the opposite direction. We came towards each other several times and both were shooting. Last time I got a hit in the Chaika's engine, which started to pour smoke and the gear dropped down.

I could not follow the plane, because I was forced to evade another Chaika, which had slipped behind my tail. I dived away far enough and turned towards it. After meeting a couple of times the Chaika tried to break off along the surface, but I caught it easily. The enemy did not notice me since he kept flying straight ahead. Because I had only the starboard fuselage gun working, I opened fire just behind the Chaika's tail. It caught fire immediately and crashed in the woods at Enso on the west side of the river. Altitude when firing only 20 metres. The Chaika was more manoeuvrable since after 4-5 meetings from opposite directions it gets behind the tail of the Brewster. The BW is considerably faster. My plane was BW-353."

On 9 July 1941 Major Gustaf Magnusson led twelve Brewsters of LLv 24 on a combat air patrol at 4 am. 70 minutes later they engaged fifteen *Chaikas* of 65 ShAP over Lahdenpohja. In ten minutes the Russians lost eight aircraft and another four were damaged. WO Ilmari Juutilainen participated in the battle:

"I observed 1Lt Kokko's pair hit the I-153 formation. I attacked the gaggle followed by Cpl Huotari and fired from close

range at several aircraft. I observed later one I-153 at low level heading to Lahdenpohja, I dived after it and shot it from 50 m distance and 10 m height into the forest between Miinala and Lahdenpohja. The two aircraft which were left over from 1Lt Karhunen began heading towards Sorola island with me after them. I waited until the wingman calmed down to fly straight forward, dived behind it and shot half a dozen bullets with both fuselage guns (wing guns did not work during the whole battle), when the I-153 dived to port banking on its back to Lake Ladoga. When my engine began to run unevenly I gave up the chase of the third aircraft. 1Lt Karhunen saw the beginning of the latter case. My plane was BW-364."*

Offensive Begins

At this point the Soviet land forces had the 23rd Army on the Karelian Isthmus and the 7th Army north of Lake Ladoga, with front line responsibility continuing up to Uhtua. The air forces of the 23rd Army consisted of 5 SAD (7 and 153 IAP plus 65 and 235 ShAP. 65 ShAP was transferred in August 1941 to 7th Army air forces) and respectively the 7th Army air forces 55 SAD (72 SBAP, 155, 179, 197 and 415 IAP plus from August 1941 on 65 ShAP).

The offensive by the Karelian Army commenced on 10 July 1941 from Kitee-Ilomatsi area towards the north-western coast of Lake Ladoga. The CO of LeR 2 specified the operational areas and tasks: the sector of LLv 24 and 28 was Saarivaara – Korpijärvi – Kolosenjärvi – Mannervaara – Tohmajärvi – Pälkjärvi – Kakunvaara – Kaurila – Matkaselkä. Air superiority was to be held in this area in turns.

On 13 July 1941 six Brewsters of 2/LLv 24 were patrolling over Tolvajärvi and engaged a mixed formation, sending three aircraft down. SSgt Eero Kinnunen shot down his, as he recorded:

"At 1230-1300 hours. Our flight discovered south-east of Tolvajärvi three SBs and four I-16s flying top cover. I attacked the latter, when a turning battle built up. I saw then one I-16 dive to the surface, where it levelled and flew to the south-east. I dived

Men of 2/LLv 24 in front of Brewster BW-352 at Selänpää on 25 June 1941. From the left: Cpl Heimo Lampi, WO Yrjö Turkka, flight leader Capt Leo Ahola, SSgt Eero Kinnunen and mechanic Kauko Pyötsiä. On this day the unit shot down nine Tupolev SB bombers from 2, 201 and 202 SBAP. (SA-kuva)

then after it and shot from straight behind several bursts into it, when it banked to the port wing and crashed into the forest from 100 metres. The Russians used rolls and false manoeuvres (absurd) as evasive actions. My plane was BW-352."

On 19 July 1941 1Lt Jorma Karhunen's 3/LLv 24 took off with eight aircraft for an intercept in the Käkisalmi direction, where it engaged I-153 fighters. In the ensuing combat three were destroyed, one of them by MSgt Lauri Nissinen:

"At 1305-1310 hours. We circled the Käkisalmi airfield at about 4,000 metres. I was flying as the lead aircraft in the top cover for 1Lt Kokko's swarm. When 1Lt Kokko's pair dived to attack a Chaika circling north of the airfield, I descended down to about 2,000 metres. Then I noticed on Chaika circling in the Kaukola direction at about 500 metres. I took a shallow dive at low speed in that direction. The Chaika turned first towards me, but made then a shallow bank away. I opened fire from about 100 metres

Other 2/LLv 24 pilots behind Brewster BW-367 at Selänpää in late June 1941. From the left: Cpl Eino Peltola, 1Lt Jorma Sarvanto, WO Veikko Rimminen and Cpl Eino Myllymäki.

1/LLv 24 pilots behind Brewster BW-392 at Vesivehmaa in late June 1941. From the left: WO Viktor Pyötsiä, flight leader Capt Eino Luukkanen, SSgt Paavo Mannila and Cpl Curt Ginman.

Brewster BW-367 of 2/LLv 24 at Rantasalmi in July 1941, showing clearly the standard camouflage pattern. The regular pilot was WO Veikko Rimminen, who became later an ace.

behind, slightly from the side. I pulled the trigger to 20 metres and a few pieces came loose from the Chaika. *The rest of the burst I shot straight behind since the* Chaika *made no evasive moves at all. After pulling up I noticed that the* Chaika *went into a spin and hit the ground. Immediately after this I saw a* Chaika *shot by Cpl Huotari catch fire above me and the pilot bail out. My plane was BW-363."*

By 23 July 1941 the VII Army Corps of the Karelian Army reached Säämäjärvi and C-in-C Marshal Mannerheim called the advance to a halt. A week later the air force CO gave LeR 2 an order according which the new operations areas were for LLv 24 Jaakkima – Lumivaara – Laatokka – Käkisalmi – Kaukola – Kirvu – Ruokolahti. The squadron was subordinated directly under the air force commander.

Karelian Isthmus

On 31 July 1941 the offensive to take back the Karelian Isthmus commenced. In order to increase fighter power, the 3rd Flight of LLv 24 was placed under LeR 3.

On 1 August 1941 LLv 24 was engaged twice in combat over the Karelian Isthmus. The swarm of 1Lt Jorma Sarvanto downed near Immolanjärvi two *Chaikas* out of ten. 1Lt Jorma Karhunen's seven Brewsters destroyed six aircraft from an I-16 squadron in the Rautjärvi area. WO Ilmari Juutilainen claimed two fighters shot down:

"At 1600-1610 hours. In a battle, which started over the front between Immolanjärvi and Rautjärvi against eight I-16bis planes I shot down two I-16bis planes. One caught fire and the other crashed to the ground throwing pieces all over. The one that

Brewster BW-358 of 2/LLv 24 parked alongside LLv 26 FIATs on the dry beach of Pyhäselkä near Joensuu, in late July 1941. This fighter was frequently flown by Cpl Eino Myllymäki.

Brewster 239 BW-373 of 1/LLv 24 flown by 1Lt Olli Mustonen in July 1941, based at Mikkeli. (Karolina Hołda)

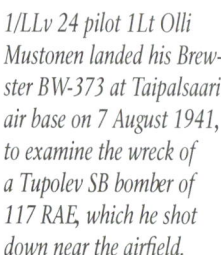

1/LLv 24 pilot 1Lt Olli Mustonen landed his Brewster BW-373 at Taipalsaari air base on 7 August 1941, to examine the wreck of a Tupolev SB bomber of 117 RAE, which he shot down near the airfield.

caught fire I shot at from behind 1Lt Kokko and the other just in front of 1Lt Karhunen's nose. I fired at a third one from a close range from behind and it disappeared in the smoke of the front in a dive. I did not see it again. BW is more manoeuvrable and climbs better. My plane was BW-353."

On 12 August 1941 Capt Jorma Karhunen led six 3/LLv 24 Brewsters in the Antrea area to attack a mixed formation of about twenty aircraft. The combat began at 1 pm and was over in half an hour. The Finns claimed nine *Chaikas* destroyed. WO Ilmari Juutilainen sent three down this way:

"When 1Lt Strömberg shouted in the radio – Chaikas below northbound – I observed them. I repeated the announcement and attacked the rearmost planes with Sgt Huotari, others joining a while later. I counted the enemy to be twenty-two. I managed to take the planes by surprise and after a long burst the first camouflaged I-153 started to smoke, banked calmly to the right and vent vertically down smoking heavily. Nobody parachuted. The second one I shot from above and behind and it lost several pieces going straight ahead for a moment and then went into a spin doing that as far as I could see. Nobody jumped. The third I shot from straight behind, got the oils on me and without any evasive action it went into a dive banking to the right. I followed to about 1,000 metres, but it continued in the same attitude crashing between Kirvu and Koljola. The pilot did not bail out. The fourth I shot

also from straight behind and it started a slow bank, I fired one more burst before passing it and after curving back I did not see it anywhere. In all I fired at ten planes, which were all camouflaged. My plane was BW-364."

18 August 1941 was particularly busy for LLv 24 with 75 accumulated sorties. Capt Eino Luukkanen's enforced 1st Flight fought with eight aircraft against six *Chaikas* over Paakkola, west of Lake Ladoga, being able to shoot down four aircraft.

On 29 August 1941 Viipuri, abandoned by the Russians, was occupied.

Back to Karelia

On 3 September the Karelian Army began the advance from Tuulosjoki towards the River Svir. The advance was rapid as the spearhead arrived at the River Svir on the morning of 7 September 1941. VI Army Corps continued east towards Lake Onega and VII Army Corps north towards Petrozavodsk.

On 6 September 1941 Capt Eino Luukkanen's swarm of 1/LLv 24 encountered, on an interception towards Aunus, a patrol of three SB bombers, shooting them all down.

The commander of LeR 2 ordered on 12 September 1941 the entire *Lentolaivue* 24 north-east of Lake Ladoga in the Olonets Isthmus. The squadron became also subordinated back to the regiment. The main task of the squadron was to protect the

VII Army Corps. The 1ˢᵗ Flight of Capt Eino Luukkanen was ordered to support the advance of the VI Army Corps and its supply units and crossing waterways at River Svir.

On 17 September 1941 the main parts of LLv 24 were now east of Lake Ladoga at Lunkula and on the nearby Mantsi Island. Capt Per Sovelius' 4ᵗʰ Flight attacked a 14-plane MiG formation near Petrozavodsk and shot down eight Russian aircraft, as told here by 1Lt Iikka Törrönen:

"At 1600-1710 hours. As the leader of the swarm flying top cover I attacked five MiG fighters. I fired at one obliquely from behind, the aircraft got numerous hits in the fuselage and went down in a spin. After this I fired at another obliquely from ahead. I got hits in the engine and the plane began to smoke heavily and crashed. After pulling up I observed a further four MiG fighters. My swarm was together again and I attacked the Russians. I fired at one obliquely from behind, the plane got hits again and down

Brewster BW-372 was assigned to 2/LLv 24 leader Capt Leo Ahola, seen here at Joensuu in late July 1941. The flight was tasked with bomber escort duties to the far rear in Karelia.

Four successful pilots of 3/LLv 24 at Rantasalmi on 10 July 1941. From the left: WO Ilmari Juutilainen, flight leader 1Lt Jorma Karhunen, 1Lt Pekka Kokko and MSgt Lauri Nissinen. In combats on two previous days these men downed nine Chaikas, some of 65 ShAP. (SA-kuva)

Pilots of 4/LLv 24 at Immola on 25 August 1941. From the left: Cpl Tapio Järvi, Cpl Aarne Korhonen, 2Lt Aulis Lumme, 1Lt Henrik Elfving, 1Lt Urho Sarjamo, flight leader Capt Per Sovelius, 1Lt Iikka Törrönen, SSgt Martti Alho and SSgt Jalo Dahl. (SA-kuva)

Right: *SSgt Martti Alho of 4/LLv 24 points at a bullet hole in the propeller blade of his Brewster BW-383, in August 1941. It was caused by a malfunction of the gun synchronization gear.*

Brewster BW-393 of 4/LLv 24 at Rantasalmi in August 1941. This fighter was assigned then to SSgt Jalo Dahl. It was to become the most successful individual plane in the Finnish Air Force, with 41 kills.

in a spin. I attacked a fourth MiG, but it went into a dive and I could not catch it any more. I returned to the base. My plane was BW-385."

On 19 September 1941 Capt Jorma Karhunen's six aircraft of 3/LLv 24 were engaged in a combat in the Pyhäjärvi area and shot down three SB bombers and one fighter. The latter was shot down by MSgt Lauri Nissinen:

"At 1125-1135 hours. We saw three SBs diving from the clouds to low level. I circled under the clouds over the aircraft to observe the national markings, when I noticed three I-18s on my left side. I attacked instantly the rightmost SB before the I-18s could make it behind my tail. I shot from a close range a burst into the port engine, when the SB caught fire and crashed in the forest. When all SBs were shot down we formed up. I saw then one aircraft coming towards me from below. The aircraft waddled at us. Since I could not determine the plane's nationality, I dived after and pulled by over speed to the side to see. The aircraft waddled then again. After identifying it as an I-18, I went behind it and started shooting, the aircraft smoked and leaked fuel from the port wing when my guns jammed. I stayed all the time behind it making cocking manoeuvres. I got two guns functioning and shot from a close range again. The aircraft crashed strongly smoking in the forest. Combat height 50-100 m. My plane was BW-384 with six bullet holes."

On 23 September 1941 Capt Jorma Karhunen of 3/LLv 24 attacked with eight aircraft three *Ratas*, which were harassing Finnish infantry which had advanced to Derevyannoye. All were sent down at 13.30 hours. The Finnish plan consisted then of a half hour circle at low level over the wilderness in complete radio silence. When returning to the target six *Ratas* were already firing at the Finnish troops. In the ensuing combat only one got away. 1Lt Pekka Kokko acted this way:

"*I was leading the swarm flying top cover, when three I-16s appeared in front of us. I attacked the lead aircraft. I fired a burst from straight behind from 100 to 20 metres. I saw hits behind the engine and the auxiliary tank under the starboard wing. The aircraft continued its dive straight ahead in a 35 degree angle. Cpl Mellin coming after me shot however one burst in the aircraft just a short while before it crashed in the forest, where it caught fire. On the same mission at 14.00 hours I observed another three-plane I-16 patrol slightly above us. By gaining altitude I got to attack straight behind, but the mgs did not work. I pulled up and got the guns to shoot again. From there I dived after one I-16, which was escaping from the battle straight to Petrozavodsk. I followed about 50 m behind and at 10 m altitude. I shot at it several times, but could not get it in flames. However, it got engine trouble and close to Petrozavodsk it went on its wing in the forest. My plane was BW-364.*"

Three days later the Russians again suffered heavy casualties. Capt Jorma Karhunen's 7-plane flight repeated the tactics. In the first encounter they downed, north of Petäjäselkä, all six *Chaikas* at 10.15 hours. After the return at 11.30 hours to the combat scene they downed another three out of eight. Karhunen reported:

"*When I carried our a recce task at Mikonselkä station, I saw three I-153s under the clouds south-east of Petäjäselkä. I went towards them meeting on the way an I-153 at the surface heading north. I shot the left wing plane down to the forest, where a fire ensued. Promptly after pulling to the left I saw a new I-153 coming on at low level from south-east about 4 km south of the impact spot of the previous one. I shot at it, the aircraft went to the forest at low speed cutting tree tops on a distance of about 100 m. It did not catch fire. In the second combat at 11.30 hours I did not get to fire while chasing two I-153s under and in the clouds towards River Svir. I met or saw two I-15s, two I-16s and*

two I-153s. During this second encounter they acted tactically in a very cunning manner taking advantage of the weather. Combat altitude 10-600 m. My plane was BW-366."

On 27 September 1941 Capt Eino Luukkanen's 8-plane 1/LLv 24 met over Derevyannoye eight fighters, three of these were destroyed. Luukkanen's summary was:

"*On a search mission at 1600-1750 hours were Capt Luukkanen, 1Lts Wind and Savonen, 2Lt Tervo, WO Pyötsiä and Sgts Vahvelainen, Ginman and Malin. Four I-15s and four monoplane fighters were met just under clouds at Derevyannoye. After the attack the enemy planes tried to pull in the clouds, but not all got in time to save themselves. The one shot up by Sgt Ginman caught fire in the air and the other two crashed to the ground. Spent ammunition count:*

Capt Luukkanen	52 heavy 16 light
1Lt Wind	13 " 54 "
WO Pyötsiä	144 " 76 "
2Lt Tervo	97 " 108 "
Sgt Ginman	47 " 25 "
Sgt Malin	18 " 15 "

When the course of the combat was analysed on the ground it was concluded that the victors were: Capt Luukkanen and 1Lt Wind together one I-15, WO Pyötsiä and 2Lt Tervo together one I-15 and Sgt Ginman one I-15."

The VII Army Corps of the Karelian Army occupied Petrozavodsk on 1 October 1941 and continued to advance to the north along the west coast of Lake Onega, aiming for Karhumäki in the north tip.

On 7 October 1941 Capt Per Sovelius' eight Brewsters of 4 /LLv 24 fought with fifteen fighters taking off from Suopohja airfield, scattering the whole formation and sending down five aircraft. Sovelius shot down one and damaged another:

"*At 0930-1115 hours. I-153 and I-16 aircraft were taking off from the airfield. I attacked one I-153 and when it tried to dodge I followed behind at low speed and shot. Finally the plane crashed in the forest wing first. I pulled up and continued to fight observing an I.16 at 200 m altitude. I fired and the plane flew through*

Brewster BW-394 of 4/LLv 24 taking off from Immola in mid-September 1941. The regular pilot was MSgt Lasse Heikinaro. This plane is a good example of the early use of two yellows in the Eastern Front markings. (SA-kuva)

the burst. I had to pull up so I did not see the result. I fired at another I-16. Flak was strong above the airfield. The aircraft probably took off to harass MS planes. My plane was BW-378."

On 15 October 1941 Capt Eino Luukkanen's six Brewsters carried out a search to the south side of Lake Onega, engaging a detachment of three bombers all fitted with skis, though there was no snow within sight. The Finns quickly downed all the bombers. WO Viktor Pyötsiä shot down his in this manner:

"At 1505-1508 hours. Above Osta we met three separate bombers. The aircraft turned immediately back after having observed us. I attacked the one last observed when it was banking to the south. I fired on several occasions, got at first the port engine to smoke and then both engines into flames. The aircraft went into a dive, which continued down to the ground. Nobody bailed out of the plane. The type was unknown, on skis, undercarriage down, speed as DB, fuselage from ahead like Ju 52. Fired back at first and threw leaflets and all sorts of stuff. My plane was BW-376."

On 26 October 1941 LLv 24 fought two combats during the day. Capt Jorma Karhunen flew with six aircraft from the 3rd Flight infantry top cover over the western River Svir, shooting down a lone Pe-2 bomber. 1Lt Iikka Törrönen's eight aircraft from the 4th Flight flew a search in the same area, meeting nine Yak-1 fighters, of which four were damaged. From Russian sources it has become clear that Törrönen's formation fought with a detachment consisting of 238 and 524 IAP flying LaGG-3s, of which two were shot down.

When Capt Eino Luukkanen's 1/LLv 24 was on a search mission on 7 November 1941 to the western River Svir they jumped on a Pe-2 escorted by three LaGG-3s. Luukkanen shot down one fighter:

"When leading a seven-aircraft search mission I observed three I-18s at the same altitude (1,500 m) south side of River Svir. I attacked with 1Lt Wind and 2Lt Suhonen, when I shot down one I-18 crashing in Lyugovitsa. The aircraft came down vertically. The whole battle lasted for 5 minutes (9.40-9.45). All aircraft were shot down, but a nearby flying bomber managed to evade due to its higher speed. I used 120 rounds of ammunition. The evasive manoeuvres of the enemy aircraft appeared to be very ineffective. My plane was BW-375."

1Lt Pekka Kokko's swarm of 3/LLv 24 reconnoitred in the Lake Seesjärvi direction on 14 November 1941 and made firing passes on Kesä airfield located on the south shore, setting fire to two bombers and three fighters.

The last point of the advance of the Karelian Army was the town of Poventsa, which was occupied on 6 December 1941. The C-in-C called a halt to the Finnish offensive. A stalemate for two and a half years followed.

The last sizeable air combat of the year was fought on 17 December 1941. Two swarms of LLv 24 encountered on a reconnaissance mission to the Maaselkä isthmus six fighters of 65 ShAP, escorted by three Hurricanes of 152 IAP. The Finns decreased the number of the Russians by five, of which Capt Jorma Karhunen claimed two:

"At 0925-0945 hours. I led a swarm on a railway reconnaissance mission. I saw below me three Hurricanes and six I-153s. I shot down immediately one Hurricane and in the ensuing "rocking chair" battle one I-153. During the combat the Brewsters had a better control of the situation due to better tactics, where they remained all the time above the enemy aircraft making dive attacks. The enemy was using their Hurricanes very elementarily.

The mixed formation flew at the same altitude with the same speed. Combat height 1000-50 m. My plane was BW-366."

Brief Analysis

1941 aces of *Lentolaivue* 24:

Rank	Name	Fligh	Victories	Plane	Total victories
WO	Juutilainen, Ilmari	3	13	BW-364	13 + 2
MSgt	Nissinen, Lauri	3	13	BW-384	13 + 5
1Lt	Kokko, Pekka	3	10	BW-379	10 + 3½
Capt	Karhunen, Jorma	3	8½	BW-366	8½ + 6½
MSgt	Kinnunen, Eero	2	6½	BW-352	6½ + 3½
Sgt	Katajainen, Nils	3	6	BW-368	
Capt	Luukkanen, Eino	1	5½	BW-375	5½ + 2½
WO	Pyötsiä, Viktor	3	5½	BW-376	5½ + 7½
Capt	Sovelius, Per	4	5	BW-379	5 + 5½
1Lt	Törrönen, Iikka	4	5	BW-385	5 + 1

This table contains the victories of named pilots in 1941. They are rounded to the nearest half.

The right hand column shows the victories in the squadron by the end of 1941. The Winter War victories are after the + sign.

1942

At the beginning of 1942 the fronts had been in a stalemate for almost a month. LLv 24 had thirty serviceable Brewsters. Soon attention was drawn further north, to repel the flow of Lend-Lease Hurricanes from Murmansk on the Arctic Sea. On 8 January 1942 the 2nd Flight was posted 500 km north to Tiiksjärvi to counter this threat.

On 9 January 1942 four Brewsters led by Capt Per Sovelius of 4/LLv 24 headed toward the Maaselkä isthmus. On the east coast of Seesjärvi lake the swarm bounced fifteen R-5 biplanes, which had an escort of six fighters. In the ensuing combat the Finnish pilots sent down four R-5s and one I-153 fighter.

Next day 3/LLv 24 headed for reconnaissance with 1Lt Osmo Kauppinen's seven Brewsters to the Meri-Maaselkä area and they met six MiGs at Voljärvi, shooting down four. Three Brewsters of 1Lt Iikka Törrönen's 4/LLv 24 recce outfit entered combat over Haapaselkä with a dozen MiGs and claimed one shot down and two damaged.

On 24 January 1942 1Lt Lauri Ohukainen's (Pekuri from July 1942) five Brewsters of 2/LLv 24 entered a combat over Ontajärvi with ten biplanes, which were escorting a pair of R-5s. The Finns claimed four aircraft shot down. In addition 1Lt V.A. Knizhnik of 65 ShAP collided in his *Chaika* with the Brewster flown by Sgt Paavo Koskela. Both pilots claimed an aerial victory. The I-153 made a forced landing, but Koskela flew his BW-372 back to Tiiksjärvi.

On 6 February 1942 the 3rd and 4th Flights of LLv 24 sent eight Brewsters in the morning under Capt Per Sovelius to reconnoitre in the Petrovskiy-Jam direction. The detachment encountered seven bombers escorted by twelve fighters. In the ensuing combat the Finns shot down two SBs and two MiGs. WO Ilmari Juutilainen claimed both of the bombers:

"At 0905-1020 hours, altitude 3000-100 m. When I noticed the bombers I radioed the boys about them. I observed during the battle three SBs flying towards the railway. I dived after them and shot the port wing plane into flames. It fell just next to the railway. I took the second wing plane into my sight when I observed a fighter closing in. In spite of that I shot the bomber in the starboard engine, which poured out smoke and oil and the plane flipped one turn to the right and then went into the forest. This one close to the railway too. I managed to shoot at the MiG from ahead and below when it began to emit black smoke. The MiG banked eastwards losing altitude. Then I got 5-6 MiGs in my neck and got rid of them after 5 minutes fight. The MiGs picked on me one by one and I had difficulties in getting above them. The Russian fighters were staggered at several altitudes so all could not at once be observed. My plane BW-364."

On 13 February 1942 MSgt Lauri Nissinen's three Brewsters of 2/LLv 24 reconnoitred the Segesha-Belomorsk area. On the return flight they engaged three Russian Hurricane fighters and shot down all of them.

On 26 February 1942 Capt Jorma Karhunen's seven planes of 3/LLv 24 fought with 17 MiGs over the north-eastern corner of Seesjärvi lake, claiming seven shot down for the cost on one Brewster. Karhunen reported:

"At 1015-1040 hours. We were on a search mission and engaged 15-17 MiG-1s and MiG-3s at 2,500 metres over Liistepohja. The first phase of the battle was close to the frontline and the second phase west of Juka and Nopsa stations on the Murmansk railway. I shot one MiG-1 straight from behind, it made two fast rolls. I continued the firing, it pulled the nose up a little and went into a spiral, I followed it down to 800 metres when I continued the fighting with other planes. In a turning battle I shot one MiG-3 in the engine, and the plane started to make

Brewster BW-351 of 2/LLv 24 after a forced landing at Karhumäki on 10 January 1942. 1Lt Lauri Ohukainen (later Pekuri) did this successfully after combat damage and the plane was later flown off.

Brewster BW-358 of 2/LLv 24 at Tiiksjärvi. It was there for only two weeks, as on 24 January 1942 it was shot down near Belomorsk, killing Sgt Eino Myllymäki.

a forced landing with the propeller wind milling, location 4 kilometres north-east of Nopsa station. The MiGs were staggered in about 500 metres in 4-6 plane divisions flying in pairs. They did not exploit the advantage given by the altitude. In all I fired at five aircraft. My plane BW-366."

The second double scorer was MSgt Martti Alho:

"At 0845-1100 hours, altitude 1500-5 m. I dived on a MiG pair west of Liistepohja, I fired at the other MiG from above and behind in a left bank, when it swiftly threw to the right, made a short dive and began to glide westward. Judging by the tracers the mg burst hit the engine and wing roots. After this I fired at three other aircraft. When the Russians withdrew under the flak cover of Juka station and other MiGs there, I followed one MiG

in a dive. I shot from straight behind and after this it levelled on the deck and flew directly towards Juka station. I shot from 100 metres when the MiG pulled slightly up, turned to the starboard wing and crashed in the forest next to a swamp south-west of Kärkijärvi. Ammunition expenditure 680 rounds. The flak shot all the time during the battle. My plane BW-383."

On 4 March 1942 MSgt Lauri Nissinen's swarm of 2/LLv 24 took off for a strafing run at Suikujärvi. The aircraft observed on the shore in Lehto village were to be destroyed and reported thus:

"At 0745-0800 hours, altitude 100-10 m. BW-384 MSgt Nissinen, BW-381 2Lt Pokela, BW-352 Sgt Peltola and BW-372 Sgt Lehto. At 7.15 I received an order from Capt Ahola to destroy with four planes two DBs and one I-16 in the location mentioned.

Brewster BW-356 of 2/LLv 24 at Tiiksjärvi in February 1942. At this point it was assigned to 1Lt Lauri Ohukainen.

Pilots of 2/LLv 24 at Tiiksjärvi in April 1942. From the left: 2Lt Väinö Pokela, 1Lt Uolevi Alvesalo, 2Lt Lauri Nissinen, SSgt Heimo Lampi and Sgt Urho Lehto.

During my first firing dive, mg fire from a nearby island scored five hits in my plane. In the second dive I silenced the mg post after which could spend all our ammunition without interference. The aircraft did not catch fire, but smoked occasionally. Cartridge expenditure 1,500 heavy and 1,000 small rounds. On the scene the clouds were at 150-200 m."

On 9 March 1942 Capt Jorma Karhunen's eight Brewsters of 3/LLv 24 fought another battle east of Seesjärvi lake. The opposition was ten MiGs escorting one SB bomber of 80 BAP. The Russians lost five aircraft, but one Brewster did not return to its base. Sgt Nils Katajainen wrote in his combat report:

"At 1420-1450 hours, altitude 50-2000 m. Over Liistepohja I observed 1 SB and 1 MiG-3. I began to chase the SB and fired at it four times. After the last shots it landed on the ice with both engines smoking. When it was on the ice I shot it into flames. In the fight against the MiGs I got in the opposite direction of one MiG and shot a short burst towards it. A thick stream of smoked puffed out of the MiG. I made a fast vertical turn but did not see the MiG anywhere. The altitude was 400 m. My plane BW-368."

By March the Russians flying units had received considerable reinforcements in the Murmansk railway area. On 10 March 1942 the Hurricanes of 152 IAP took by surprise a Fokker reconnaissance pair of 1/LLv 14 near Lake Tungutjärvi and shot both planes down. After this, the reconnaissance missions were avoided in the immediate vicinity of the railway.

Suursaari Occupation

On 19 March 1942 he Finns decided to take Suursaari (in the middle of the Gulf of Finland) back, when the advance could be made on the ice. Detachment Pajari was formed for this purpose. LeR 3, LeR 4, LLv 6, 3/LLv 24 and an ambulance/liaison flight were ordered to provide air support. Two days later Maj.Gen Aaro Pajari, commanding the 3,500 strong invasion force, issued the order for the attack:

"For the invasion LeR 3 was ordered with tasks

1) to protect the march of Detachment Pajari to Haapasaari and from there onwards,

2) to support the attack on Suursaari,

3) to hamper the retreat of the enemy from Suursaari by causing losses and

4) to carry out reconnaissance on every mission."

On 27 March 1942 the invasion troops began the approach to Suursaari. 57 aircraft were assembled for air cover, the main part from LeR 3: sixteen Fokkers of LLv 30 and thirteen Curtiss from LLv 32. Additionally LLv 6 provided five SB bombers and six I-153 fighters, LLv 24 six Brewsters and LLv 42 eleven Blenheims.

Next morning 1Lt Osmo Kauppinen's five Brewsters of 3/LLv 24 fought against ten *Chaikas*, sending half of them down. O Ilmari Juutilainen claimed two:

"At 0745-0815 hours, altitude 2000-10 m. I flew in the top patrol with Sgt Huotari. Over the Suurkylä shore at Suursaari I observed beneath me enemy fighters and we went to attack. During the combat I followed one I-153 shooting at it several times from 50 m, until it suddenly flipped from 200 metres altitude to the ice. Then I chased with Sgt Huotari two I-153s to Lavansaari, where I shot the aircraft burning on the west shore. The Russians were shooting rocket projectiles, which when exploding created a black

4/LLv 24 pilots at Kontupohja on 17 March 1942. From the left: MSgt Sakari Ikonen, 1Lt Aulis Lumme, 2Lt Erik Teromaa, 1Lt Hans Wind, Sgt Aarne Korhonen, 1Lt Urho Sarjamo and Sgt Martti Immonen. Only the flight leader, 1Lt Iikka Törrönen, is missing from the picture. (SA-kuva)

cloud resembling an explosion of heavy flak, though a bit bigger. They exploded 100-200 m ahead of my plane. I-153s had four of the them under each wing. My plane BW-364."

Hurricane Party

On 30 March 1942 1Lt Lauri Ohukainen took off with eight Brewsters of 2/LLv 24 to reconnoitre the Seesjärvi-Ontajärvi isthmus. East of Rukajärvi the detachment took by surprise one dozen Hurricanes flying in two groups and shot down eight. MSgt Heimo Lampi wrote in his combat report:

"At 1550-1610 hours, altitude 500 m. I attacked from ahead and above against a 6-7 fighter formation. I fired at two aircraft first, but did not see any results. I fired then at one Hurricane flying ahead and below me when this plane turned at about 300 metres on its back and crashed steeply in the forest. I jumped then on a new Hurricane formation arriving from south-east and shot at a couple of planes, but they dodged. The Hurricanes then turned to flee and I gave chase to four of them with Sgt Koskela. I got to fire at one in level flight from 80 metres, when it crashed in the forest smoking heavily. The combat involved altogether some 12 Hurricanes. At the beginning they fought hard, but once they began fleeing they did not even bother to dodge and only tried to reach their base as quickly as possible. My plane was BW-354."

The Finnish air base opposite of Segesha was at Tiiksjärvi. Perhaps the Finnish bombardment of Segesha the day before speeded up matters, because the Soviets carried out a long planned air raid on Tiiksjärvi.

On 6 April 1942 the Finnish air surveillance observed a formation of 26 aircraft approaching and an eight plane Brewster flight of LLv 24 on a recce mission was radioed for help. Just before the bombardment the Finns attacked the formation and reported having shot down two DB-3 bombers and twelve Hurricanes, without any losses of their own. The leader of the Brewster unit, 1Lt Lauri Ohukainen, destroyed three and reported:

"At 1525-1550 hours, altitude 3000-20 m. When I returned from a recce mission I received an air surveillance message that

7 bombers and 18 fighters were approaching from SE. We engaged the planes about 20 km before the base. I attacked the fighter while two BWs took the bombers. In the combat I shot the first aircraft down about 5 km SE of the base, and it burned on the ground. I hit the second in the engine and cooler and left the plane south of Ontrosenvaara smoking heavily at 800 m altitude (found in that spot). In the chase on the deck I shot the third into the forest at Rukajärvi. The pilot was wounded and taken prisoner. I shot at 6-7 aircraft in all. The Hurricanes had 12 gun wings. Appearance generally helpless. My plane was BW-372."

2Lt Lauri Nissinen sent down three, writing:

"At 1525-1550 hours, altitude 3000-10 m. I flew as the lead plane of the top swarm. We were returning from a Belomorsk reconnaissance, when we were alerted by the radio of about 25 Russian aircraft. We started to climb while heading towards them. A big dogfight started with the fighters. At first I fired at several planes, but was too busy to follow any. After shooting at one Hurricane it went in a half roll to a dive. I followed and fired a short burst at a close range, when the machine fell vertically and smoking into the forest. When the battle had lasted ten minutes, the remaining Russians evaded us. At Rukajärvi I caught one at 500 metres. I got into a position to shoot from 50 metres slanted behind, and just after pulling the trigger the plane's fuel tank exploded. I did not have time to pull aside to dodge the fragments, and one piece snapped off the other exhaust pipe. I continued the chase of the two remaining Hurricanes with 1Lt Ohukainen and Sgt Korhonen. When the other tried to bank towards Ohukainen, I slipped behind its tail, then the Russian took evasive action. I got it to smoke first and after a short chase the Russian crashed into the forest. In spite of numerical superiority the Russians only tried to evade. Appeared to be very helpless. No hits in my plane. Exhaust pipe cut off. My plane was BW-384."

The Russians saw it differently. The Hurricane pilots of 609 and 767 IAP reported four aircraft destroyed at Tiiksjärvi air base and seven Brewsters shot down in the aerial combat.

2/LLv 24 pilots at Tiiksjärvi on 12 April 1942. From the left by rank: flight leader Capt Leo Ahola, 1Lt Lauri Ohukainen, 1Lt Uolevi Alvesalo, 2Lt Lauri Nissinen, 2Lt Väinö Pokela, 2Lt Martti Salovaara, MSgt Eero Kinnunen, Sgt Aarne Korhonen, Sgt Martti Lehtovaara, Sgt Jouko Lilja, Sgt Urho Lehto and Sgt Sulo Lehtiö. (SA-kuva)

Two very successful fighter pilots at the tail of 3/LeLv 24 Brewster BW-364 at Hirvas in May 1942. At left WO Ilmari Juutilainen (94 victories and twice Mannerheim Cross) and on the right Sgt Nils Katajainen (34½ victories and Mannerheim Cross).

80 BAP lost two SB bombers. 609 IAP reported having lost two Hurricanes and 767 IAP four.

Warrant Officer Ilmari Juutilainen of 3/LLv 24 had reached the score of twenty aerial victories, the last two during the Suursaari invasion. On 26 April 1942 he was awarded the Mannerheim Cross, as the first member of *Lentolaivue* 24 to be so decorated.

When the ice melted in the Gulf of Finland in May 1942, the Red Banner Baltic Fleet commenced sending out submarines from Kronstadt, a huge naval and air base outside Leningrad, where the Russians had retreated during the previous autumn. The aim of the operation was to harass the German and Finnish shipping in the Baltic Sea as much as possible. By the early summer of 1942 the Baltic Fleet air forces had received substantial reinforcements in order to fly top cover for maritime operations and especially protect the departures and arrivals of the submarines at Kronstadt.

On 4 June 1942 1Lt Joel Savonen's swarm of 1/LeLv 24 and 1Lt Hans Wind's swarm of 4/LeLv 24 few top cover in the morning for the German leader Adolf Hitler's Fw 200 Condor aircraft from Helsinki to Immola and back in the evening. Hitler came to congratulate Marshal Carl Gustaf Emil Mannerheim on his 75th birthday.

On 8 June 1942 1Lt Lauri Ohukainen's five Brewsters of 2/LeLv 24 took off from Tiiksjärvi to a chase straight to the east. At Kesä air base by the Murmansk railway three Hurricanes were caught, but another ten appeared on the scene. In the ensuing combat five Hurricanes were downed, but also one Brewster received hits and it was destroyed on the return flight when making a forced landing.

A great air battle was fought on 25 June 1942 north-east of lake Segozero. The swarms of both 3/LeLv 24 and 4/LeLv 24 struggled for 15 minutes against Hurricanes. The Brewsters of the 3rd Flight downed four and those of the 2nd three Soviet fighters, but lost two Brewsters. 1Lt Lauri Ohukainen was one of the shot down pilots:

"*At 1335-1440 hours, altitude 6500-10 m. I flew an alert mission with four planes. West of Segesha we met the swarm of WO Juutilainen. While Juutilainen attacked a Hurricane, which took off from the air base, I remained with Sgt Anttila to fly top cover at about 5,000 m altitude. When the battle shifted to the south-east I followed at about 3,000 m altitude. Suddenly we were attacked by one MiG and four Hurricanes. I observed one Hurricane fire at Anttila and score hits on him, before I could assist him. I shot at the Hurricane at a close range, when it went in a vertical dive and crashed in a swamp beneath us. Finally I had four enemies attacking at me while Anttila descended to the west alone. After a short fight I managed to break off. About 12 km north of lake Kalitsin two Hurricanes took me by surprise. At 4,000 m altitude one of them scored hits on my engine and rear armour plate, setting the port wing fuel tank in fire. When I dodged I managed to shoot at the attacker from very close range, when it caught fire and disappeared southward smoking. My engine stopped and when I was trying to make a forced landing on a small lake about 15 km north of lake Kalitsin, the other Hurricane fired at my starboard tank setting the whole plane on fire. From 10 m altitude and 250 km/h speed I pushed the plane into the lake, where it flipped over. I dived out of the plane and swam ashore. The plane sank in half a minute. After a walk of 20 km I reached a Finnish outpost south of lake Jolmozero. The infantry observed that a Hurricane had crashed and started a forest fire in the mentioned location. About 15 Hurricanes and MiGs participated in the battle. The HCs appeared much faster than the BW at altitude and also relatively manoeuvrable. My plane BW-372 became a total write-off."*

The Brewsters of LeLv 24 had claimed 45 Hurricanes in six months in the Rukajärvi direction south-west of the White Sea. In its sector of operations a quiet period of a few weeks followed. Part of the Soviet air forces based in the southern part of the Karelian Front were transferred to the north, to protect the Allied ship convoys, which were arriving at the ports of Murmansk and Archangel.

On 2 July 1942 special arrangements were made to catch a fast Pe-2 reconnaissance plane. The leader of 3/LeLv 24 Capt Jorma Karhunen scrambled towards the arriving Pe-2 over Lake Onega and shot it down. This was the 100th aerial victory for the flight.

2Lt Lauri Nissinen of 2/LeLv 24 was awarded the Mannerheim Cross on 5 July 1942 , after gaining 20 kills in the present war, as the second such pilot of *Lentolaivue* 24.

Gulf of Finland

A noteworthy Soviet operation was the invasion attempt of Someri island in the second week of July. The Brewsters of 1/LeLv 24 fought on 8 July 1942 the first battles over the Gulf of Finland. 1Lt Joel Savonen's swarm reported having shot down two and damaged several on two interceptions. The Russian introduced new types in this area. In aerial com-

WO Ilmari Juutilainen (centre) demonstrates his latest combat to Sgt Emil Vesa (left) and SSgt Jouko Huotari (right) at Hirvas on 27 June 1942. Behind is Juutilainen's Brewster BW-364 of 3/LeLv 24. (SA-kuva)

The same three 3/LeLv 24 pilots around Juutilainen's machine, BW-364, at Hirvas on 27 June 1942. This plane went on to score 37 kills, 28 by Juutilainen. (SA-kuva)

1/LeLv 24 pilots at Suulajärvi. From the left: SSgt Matti Pellinen, 1Lt Kai Metsola, flight leader Capt Eino Luukkanen and MSgt Paavo Mannila. Behind at left is BW-374, which was damaged in a landing on 19 July 1942.

2/LeLv 24 leader Capt Pauli Ervi standing in front of Brewster BW-352 at Tiiksjärvi in August 1942. This plane was assigned to MSgt Eero Kinnunen, bearing his victory markings on the fin, which also carries the flight emblem.

bats the Baltic Fleet air forces recorded the loss of two Yak-1s and one LaGG-3 from 21 IAP, in addition to one Il-2 from 57 ShAP.

By August the Russians had increased considerably their aerial activity over the Gulf of Finland, related to the submarine launches from and returns to Kronstadt's massive naval and air base. To repel this, the main part of *Lentolaivue* 24 was transferred to Römpötti on the Karelian Isthmus and subordinated to *Lentorykmentti* 3.

Though the role of LeR 3 did not essentially change, the tasks of the squadrons were re-specified. Only interception mission were to be flown, as far west over the Gulf of Finland that the return of enemy aircraft reaching the Kotka line could be cut. LeLv 24 was to prevent enemy air operations west of the Vuoksi-Pölläkkälä-Valkjärvi-Lempaalanjärvi line and cut the return route of enemy aircraft arriving over the Finnish coast,

in addition to flying top cover for their own transports and bombers on the western isthmus. The regiment's line of operations with the Germans was the south coast of the Gulf of Finland, east of Karavaldai.

The order of battle of *Lentolaivue* 24 was then:
- Squadron commander Lieutenant Colonel Gustaf Magnusson, Römpötti
- 1ˢᵗ Flight, Captain Eino Luukkanen, Römpötti (6 BW)
- 2ⁿᵈ Flight, Captain Pauli Ervi, Tiiksjärvi (6 BW)
- 3ʳᵈ Flight, Captain Jorma Karhunen, Römpötti (5 BW)
- 4ᵗʰ Flight, 1ˢᵗ Lieutenant Iikka Törrönen, Römpötti (7 BW)

The battles over the Gulf of Finland commenced in a true sense on 6 August 1942. The Finnish aim was to cut the Soviet flying activities over the sea to the west in the Koivisto – Seiskari – Suursaari area. The task did not succeed, since the Russians flew under the cover of Oranienbaum (encircled by the Germans) via the south side of Seiskari, to where it was impossible to fly in time after air surveillance messages.

Capt Eino Luukkanen's five Brewsters of 1/LeLv 24 were alerted to Seiskari, where four I-16 fighters patrolled, two were shot in the sea. In the Rukajärvi direction WO Yrjö Turkka's three Brewsters of 2/LeLv 24 escorted a Fokker C.V reconnaissance plane. Two Tomahawks appeared over Tiiksjärvi airfield, dropping their bombs. One of them was caught and sent down.

LeLv 24 had a good observation and control post at Ino, from where you could see the enemy's take-offs and landings both at Kronstadt and Oranienbaum. New tactics were now employed and the long endurance Brewsters were sent into the air to wait for the return of the Soviet formations.

On 12 August 1942 1Lt Aulis Lumme with Brewsters of 4/LeLv 24 fought against sixteen I-16 fighters and ten Il-2 ground attack aircraft in the vicinity of Seiskari. The Finns destroyed two aircraft.

On a search mission two days later a combat broke out with Capt Jorma Karhunen's six Brewsters of 3/LeLv 24 and a Hurricane squadron of 3 GIAP, KBF. Six Russians planes were shot down. Karhunen's combat report stated:

"At 1020-1200 hours, altitude 100-500 m. I led the flight in an interception mission. I met an enemy fighter formation, eleven Hurricanes on the way from Seiskari to Kronstadt. We made an attack with six Brewsters from above and aside. I shot from straight behind at one HC, which instantly puffed black smoke. It went into the sea after the second burst. When I fired at another HC in a turning battle, it threw out its coolant liquids. While a fierce battle went on with other Hurricanes I could not fire any more at it. After 1-2 minutes I saw the plane flying on the deck in the combat area, trailing a thickening stream of vapour. After that I did not see the plane any more. My plane was BW-388."

A couple of hours later 1Lt Hans Wind's five Brewsters of 1/LeLv 24 were fighting against thirteen Hurricanes, claiming three shot in the sea. Wind scored two and reported the encounter:

"At 1230-1345 hours, altitude 100 m. While on an interception mission with five aircraft to Tolbukhin lighthouse we met thirteen Hurricanes. I attacked a lower flying swarm, of which the port wing plane fell from 500 m altitude into the sea after two short bursts. The swarm leader turned towards me and I was forced to fire at it in the opposite directions and began a turning battle right down to the deck. After 3-4 minutes of fighting I managed to shoot at it from straight behind from 20-30 m distance. The oil came out of the HC and it went in a steep dive into the forest at Oranienbaum. The oil from the HC covered my windscreen and I had to return to the base since I could fly only with the hood open. Ammunition spent 280 large and 125 small. The Hurricanes flew with open cockpits, pilots wearing goggles. My plane BW-393."

On 16 August 1942 Capt Jorma Karhunen's 3/LeLv 24 encountered a large enemy formation and reported having shot down eleven I-16s. Karhunen claimed three aircraft:

"At 1745-1800 hours, altitude 50-500 m. I led a six-plane BW-flight on an intercept mission. South of Seiskari we engaged an enemy formation, 8 SBs, 3 MiGs and 15 I-16s, which were flying at 200 m altitude. We attacked the escort fighters, the I-16s chose to fight while the others fled. During the first dive I shot the plane flying at the extreme port wing which caught fire. The aircraft went into a bank into the sea. Another I-16 went to the deck smoking and then flipped over. The third had just dodged after another BW had fired at it without hitting, I continued the attack instantly, the plane turned first on its back, received several hits spiralled finally wing first into the sea. I made in all 12 attacks. The escort fighters flew at the same altitude as the SBs. The I-16s fought bravely, but they did not realize to pull up from the flanks and hit back since they had numerical superiority. My plane BW-388."

SSgt Nils Katajainen was on the same mission and he told shortly sending down two fighters:

"At 1745-1800 hours, altitude 50-500 m. During the battle I fired at one I-16 in a bank, the first burst hit and the I-16 began to dive to the deck when smoking, The other I-16 I shot also in a bank and it went into the sea in a shallow dive. My plane BW-373."

The greatest air combat of the Continuation War up to now was fought on 18 August 1942. First came an announcement

Brewster BW-387 of 2/LeLv 24 was repaired at the field air depot at Onttola. On 15 August 1942 1Lt Lauri Pekuri flew it back to Tiiksjärvi. It was assigned to Sgt Aarne Korhonen, still carrying colours of 4/LeLv 24. (Finnish Air Force)

Brewster BW-356 of 2/LeLv 24 parked at Tiiksjärvi in Eastern Karelia during September 1942. The regular pilot at this point was Sgt Martti Lehtovaara, the little brother of Morane ace and future Mannerheim Cross holder Urho Lehtovaara.

that ten I-16s were going eastwards at Tytärsaari. 1Lt Hans Wind scrambled with eight Brewsters and flew towards Seiskari to wait for the Russians to return. There the Finns were attacked by a numerically superior enemy formation. Capt Jorma Karhunen's and 1Lt Aulis Lumme's swarms were immediately dispatched to assist. The Russian were estimated to be about 60 aircraft, against sixteen Brewsters of LeLv 24. The final result was sixteen Russians and one Brewster shot down.

Wind claimed three aircraft over Kreivinlahti in this way:

"At 2000-2100 hours, altitude 50-2,000 m. I met four Hurricanes with eight Brewsters, after a short chase I shot one into flames with two bursts. It fell into the forest in flames. Then about 60 I-16s appeared on the scene. Early in the battle I shot one of these smoking, at the same time another I-16 managed to shoot a cannon shell through my port wing. I pulled up and saw the I-16 at which I had fired catch fire on the deck. Right after this I ended up firing in opposite directions at three I-16s, but I did not see the results. At the end of the combat I got conveniently behind one I-16. The plane caught fire after the first burst and went into the sea. A cannon hit about 5 cm outside the wing tank, damaged slightly the aileron rod. My plane was BW-393."

Karhunen told about his struggle as follows:

"At 2000-2120 hours, altitude 50-500 m. After receiving information of the combat I led my swarm on the scene. I observed about twenty I-16s flying top cover for about ten patrol motor boats. In the ensuing combat I managed to shoot one I-16 from straight behind. The plane developed a black smoke trail and flames going in a dive bank into the sea. I fired at the second one from opposite directions. I had just attacked one I-16 flying on the deck, when during the pull-up five I-16s came towards me from ahead and above. When one fired from ahead I returned the fire and hit straight in the engine. The plane made a swift flip over at about 75 n distance and crashed in the sea. Harassed by four I-16s I had to pull to the north-east of the combat scene. Suddenly I saw on the deck a Pe-2 in front of me. I shot it into flames.

It went into the sea and sank. The enemy was flying in several separate formations, totally about 60 planes staggered in altitude. I-16s very active, a bullet scratch in the starboard aileron. My plane was BW-388."

Juutilainen got his third triple victory like this:

"At 2000-2120 hours, altitude 500-50 m. We received a message that a combat was going on east of Seiskari. The flight took off with four planes directly to Kreivinlahti, where the enemy was met fighting with Brewsters. Right at the beginning I shot one I-16 smoking into the sea next to patrol boats cruising below. The second I shot from ahead banking straight to the sea south-west of Kronstadt. And the third in fire to the sea, about 500 metres southwards of the previous one. I could not determine the final number of the enemy planes, but there were many. The 14–15 patrol boats below us, Tolbukhin lighthouse, Kronstadt and Oranienbaum, were firing all the time heavily with flak. My plane was BW-364."

The Russian were very quiet about this combat and the Baltic Fleet chronicles revealed that only that 21 IAP had lost one Yak-1 and one LaGG-3 in addition to 73 AP's loss of a Pe-2.

On the last day of August Capt Eino Luukkanen led five Brewsters of 1/LeLv 24 to a search in Seiskari-Lavansaari area. In a clash with eight *Chaikas*, half of theSoviet planes were shot down.

The Brewster pilots had one trump card. LeLv 24 had flown in other parts of the front developing their skills and arrived as a complete surprise to the Gulf of Finland with an aircraft which was not mentioned in Soviet recognition manuals. The Brewsters were identified as FIATs or Capronis. Some 39 Soviet aircraft were claimed in one week and 50 during the whole of August. Judging by the following passivity of the Russians, the Brewster pilots of LeLv 24 had done a good job over the Gulf of Finland.

The victory score of 3/LeLv 24 leader Capt Jorma Karhunen rose to 20 in this war, for which achievement he received the Mannerheim Cross on 8 September 1942, as the third recipient from *Lentolaivue 24*.

On 20 September 1942 Capt Jorma Karhunen's seven Brewsters of 3/LeLv 24 were sent for search the Kronstadt – Tolbukhin – Seiskari area. Near the Estonian coast they jumped ten fighters, of which three were sent down. WO Ilmari Juutilainen claimed two:

"*At 1300-1440 hours, altitude 1000-4000 m. I attacked after Capt Karhunen about ten MiGs and Spitfires. In the ensuing combat I managed to shoot at one Spitfire in a turn, when it went in an inverted dive smoking into the sea about 4 km SE of Peninsaari. I was able to shoot at the MiG from below and behind, when it crashed burning in the sea. I fired at another MiG, but without results. The Spitfires were fast, had a good rate of climb and manoeuvrability. My engine did not run properly during the fight. My plane was BW-364.*"

On 4 October 1942 two Pe-2 bombers escorted by 30 fighters attacked Tiiksjärvi. 1Lt Väinö Pokela's swarm of 2/LeLv 24 scrambled and the Brewsters sent down three fighters. These were the squadron's last victories in the Rukajärvi area. In addition the flak of the base claimed two bombers.

On 22 October 1942 Capt Eino Luukkanen's pair of 1/LeLv 24 took of for an intercept in the Seiskari-Retusaari area, where one bomber and escort fighters were met. Luukkanen achieved his first double victory and reported it:

"*At 0930-0945 hours, altitude 400-50 m. When I was firing at the Pe-2 I saw above me four fighters about 20 km away south-east. I pulled up just below the clouds (500 m) and shot at one I-16 from behind, when it caught fire and fell in the sea. After this I fired at another head-on with visible results. I had to shoot at the third also from ahead, and after turning away I saw the plane crash in the shore forest cutting trees on the way. I noticed having spent all my ammunition so I broke off the fight and flew at 10 metres to Kreivinlahti, when I all the sudden observed another four I-16s above and about 200 metres higher than me. Since I had no ammunition left, I tried and succeeded in hiding by flying low. I had used 700 small and 660 heavy rounds, of which about half to the Pe-2, which dropped one torpedo. My plane BW-393.*"

Three days later 1/LeLv 24 sent three Brewsters of 1Lt Joel Savonen for interception in the Lavansaari-Seiskari area, where four Hurricanes escorted two Pe-2 bombers. Capt Jorma Karhunen's five Brewsters of 3/LeLv 24 scrambled to the area and in the aerial combat the Russians lost all their fighters belonging to 3 GIAP, KBF. 1Lt Väinö Suhonen fought this way:

"*At 1400-1410 hours, altitude 0-500 m. Our three-plane patrol met at Tolbukhin-Kronstadt area an enemy formation consisting of 2 bombers and 4 HC fighters. In the ensuing combat I fired at one HC from above and behind and it crashed in the sea after a short burst. I fired at another HC from the side and a third from above and behind, the latter going in a dive. When I was dodging at the same time a stream of bullets shot from aside by another enemy, I could not see the result. My plane BW-376.*"

On 26 October 1942, just before noon, Capt Karhunen's six Brewsters of 3/LeLv 24 took by surprise at the Oranienbaum coast two DB-3 bombers, which were escorted by nine fighters. Five of the Brewsters tied the escort fighters in combat and MSgt Eero Kinnunen claimed both bombers of 1 GMTAP, KBF shot down:

"*At 1100-1200 hours, altitude 10 m. We met two DBs and six HCs. I attacked one of them and shot at it from above and*

behind 100-50 m distance, when it caught fire. I moved right away to shoot at the other from slightly aside and behind, but I had to dodge as I was about to collide with it. I made another attack from straight behind, when it began to smoke heavily and the starboard gear fell out. After continuing flying for a while, it also caught fire and went burning into the sea, like the one before. The rear gunners scored one hit in a propeller blade. My plane was BW-351.*"

Half an hour later 1Lt Hans Wind's detachment of 1/LeLv 24 fought against fifteen fighters in the same area, sending four *Ratas* into the Gulf of Finland. Wind wrote in his combat report:

"*At 1215-1400 hours, when I was on an alert mission with six aircraft we met north of Kreivinlahti 4 I-16s heading towards Seiskari. We attacked these. The battle shifted gradually to the Shepelevskiy lighthouse. At this point at least another eleven I-16s arrived on the scene. After a short curve battle I shot one of these I-16s into flames at about 20 m altitude. The aircraft went straight into the forest on the west shore of lake Harjavallanjärvi. Of the eight lowest harassing I-16s I shot one in the central fuselage a long burst. This plane went to its wing at 50 m altitude and cut trees going in the forest. Crash point about a half km southeast of Shepelevskiy lighthouse. I fired at another three I-16s, did not observe any effect. I-16s very active fighting at the deck. I-16 much more manoeuvrable than BW. An incendiary bullet hit the wing tip, burning the paint slightly, which caused the wing to smoke heavily for half a minute. Another hit came from frontal shooting to the port stabilizer. My plane was BW-377.*"

On 30 October 1942 Capt Eino Luukkanen's detachment of 1/LeLv 24 was met over Oranienbaum by eight I-16s and during the battle they met one escorted Pe-2. Twenty minutes later there were three fewer Russians, but also one Brewster had been shot down. This was also Luukkanen's last mission with the Brewster:

"*At 1130-1150 hours, altitude 500-3500 m. When I was flying the lead aircraft of the swarm I observed eight I-16s at 1,000 metres while my pair was at 400 m altitude. The top pair being at about 1,000 m attacked immediately the I-16s, but because we could not gain more altitude than the I-16s, we and the top pair broke off a bit to the side. After this we climbed up to 3,500 metres altitude, but the I-16s came behind and managed also to reach 3,400 metres. The I-16s were then above Harjavallanjärvi lake and my swarm attacked the eight I-16s from the south (from the sun) as they were flying in a large circle (diameter about 1 km) about 100 m below us. Before the attack Sgt Tolonen was with us. During the battle I fired at three I-16s, of which one went in a vertical dive smoking from the fuselage. After this I became harassed myself, so I went vertically down to about 1,000 metres. I saw below me two Spitfires and a Pe-2, attacking one of the Spitfires from obliquely behind, when it puffed into smoke on my first burst crashing in the forest. Sgt Tolonen did not return from this mission and his disappearance was not observed by anyone. In addition to the eight I-16s, there were four more below them. I saw three I-16s crash down smoking and burning further on the ground. My plane was BW-393.*"

The airspace over the eastern Gulf of Finland stayed quiet until 22 November 1942. In the morning 1Lt Aulis Lumme with six Brewsters of 4/LeLv 24 fought with an equally strong Yakovlev formation in Kronstadt area, shooting down a single Il-2 ground attacker appearing on the scene and three fighters. 1Lt Erik Teromaa claimed two aircraft:

3/LeLv 24 pilots at Römpötti on 10 September 1942. From the left: Sgt Erkki Pakarinen, 2Lt Jalo Ahlsten, 1Lt Martti Salovaara, flight leader Capt Jorma Karhunen, WO Ilmari Juutilainen, SSgt Jouko Huotari and Sgt Erik Lyly. Behind is Karhunen's Brewster BW-366. (SA-kuva)

"At 0915-0940 hours, altitude 300-2600 m. I fired at the Il-2 from above, when it began to smoke heavily and the engine probably quit since it made a forced landing. At the same time I got a Tomahawk shooting several holes in my plane, I managed to break off by pulling into the clouds. Later I observed below me a Tomahawk, which I fired at while curving, when the aircraft dropped its wing and dived vertically into the sea. Tomahawk is faster and more agile than BW. These, unlike the other Russians, were very active. The Tomahawk which surprised me remained remarkably easily behind my tail. My plane BW-367."

After re-arming and re-fuelling, Lumme's detachment hurried back, together with Capt Jorma Karhunen's six aircraft of 3/LeLv 24, engaging in the same area a bomber escorted by several fighters. The Il-4 and three fighters were destroyed.

Next day LeLv 24 fought three separate air battles almost simultaneously in the Lavansaari-Kronstadt-Kreivinlahti area. The first to struggle was the 4th Flight with 1Lt Aulis Lumme's five Brewsters, against four bombers escorted by six fighters. Two Pe-2s, one Tomahawk and one Spitfire were claimed shot down. Next in turn was the 1st Flight, when 1 Lt Hans Wind's swarm attacked a similar formation sending two Pe-2s and one I-16 down. Wind's combat report told:

"At 1105-1225 hours, I shot the I-16 from behind, when it crashed on the shore (both wing broken, plane upside down). After this I engaged three Pe-2s and one Er-2, of which I shot one Pe-2 into the sea. The aircraft sank at once. I shot four bursts in another Pe-2, which went in a shallow glide towards Kronstadt. Due to fierce flak I did not see the final result. I shot from about 300 m the Er-2 into the fuselage without effect. Er-2 as fast as the

Pe-2, appeared to be larger than the Pe-2, the wings were strongly bent. My plane was BW-393."

The last to arrive on the scene was Capt Jorma Karhunen's swarm of the 3rd Flight, meeting six fighters and destroying one Tomahawk. The Russians in this area did not possess Spitfires and according their loss statistics these planes were Yak-1s or Yak-7s.

Brief Analysis

1942 aces of *Lentolaivue 24*:

Rank	Name	Flight	Victories	Plane	Total victories
WO	Juutilainen, Ilmari*	3	20	BW-364	33 + 2
Capt	Karhunen, Jorma*	3	13	BW-366	21 + 6½
1Lt	Wind, Hans	1	12½	BW-378	15
1Lt	Ohukainen, Lauri**	2	10½	BW-372	10½
Capt	Luukkanen, Eino	1	9	BW-393	14½ + 2½
MSgt	Kinnunen, Eero	2	9	BW-352	15½ + 3½
SSgt	Peltola, Eino	2	7	BW-379	7
2Lt	Nissinen, Lauri*	2	6½	BW-384	19½ + 5
1Lt	Teromaa, Erik	4	6	BW-367	8
SSgt	Huotari, Jouko	3	6	BW-355	9
SSgt	Katajainen, Nils	3	6	BW-368	14
WO	Turkka, Yrjö	2	5	BW-357	9 + 5
MSgt	Alho, Martti	4	5	BW-383	9 + 1½

* Mannerheim Cross, ** from July 1942 Pekuri

This table contains victories by named pilots only in 1942. They are rounded to the nearest half. The right hand column shows victories in LeLv 24 by end 1942, after + sign are Winter War victories

1943

The first combat of 1943 took place on 24 January, when five Brewsters of 1/LeLv 24 led by Capt Jorma Sarvanto, on a search mission to the Gulf of Finland, engaged a ten-plane and a twenty-plane formation both having assault planes with a fighter escort. In the attacks one I-16 fighter was downed over Kronstadt.

Lentolaivue 24 was not spared from the attrition of Brewsters during eighteen months of combat, causing a reorganization in the squadron, which was carried out on 11 February 1943. The number of flights was reduced to three.

The order of battle of *Lentolaivue* 24 was then:
- Squadron commander Lieutenant Colonel Gustaf Magnusson, Suulajärvi
- 1st Flight, Captain Jorma Sarvanto, Suulajärvi (8 BW)
- 2nd Flight, 1st Lieutenant Iikka Törrönen, Suulajärvi (8 BW)
- 3rd Flight, Captain Jorma Karhunen, Suulajärvi (8 BW)

February 23 1943 was the busiest day of the month for LeLv 24 with 32 sorties. The first large air combat of the year was fought on the south side of Lavansaari, when six Brewsters of 3/LeLv 24 leader Capt Jorma Karhunen attacked four Pe-2 bombers, which were escorted by twelve *Ratas*. The Finns were not able to make contact with the bombers, but half of the fighters were shot down, as told here by WO Eero Kinnunen:

"*At 1115-1230 hours. I flew in Capt Karhunen's flight as the leader of the top cover patrol. We met south of Peninsaari twelve I-16s and four Pe-2s heading west. I attacked from above and behind one I-16, which I got to pour smoke. Sgt Kauppinen shot at it after me. Then I observed three I-16s coming from the Peninsaari direction, I attacked these from the rear and above. I first*

shot at the starboard wingman, which caught fire and crashed through the ice. The pilot bailed out and appeared to land luckily on the ice. The I shot in three separate dives at the port wingman, which got enough from the third burst and went strongly smoking in the open sea. In the meantime the lead aircraft had climbed to the same altitude as me and I had no other alternatives than shooting in the opposite direction at it. On the second time neither of us managed to get to a firing position and on the third time it however got behind my tail. I was able to break off in a dive and bank. During the battle the fighters were dispersed over a wide area. My plane was BW-352.*"

On 2 March 1943 the leader of 3/LeLv 24, Capt Jorma Karhunen, met with five Brewsters six *Chaikas* around Seiskari and dispatched three of them. Some time later the five plane patrol of Capt Jorma Sarvanto, boss of 1st Flight, found a solitary I-16 and shot it down.

On 10 March 1943 seven Pe-2s and ten LaGG-3s were approaching Kotka when their flight was met by eight fighters of 2/LeLv 24 under 1Lt Iikka Törrönen, in the Haapasaari area. The Russian turned around to evade and the Finns chased the formation across the Gulf of Finland up to Oranienbaum, shooting down one bomber and six fighters during the chase. Near Oranienbaum the fierce flak forced the Brewsters to break off. 2Lt Eero Riihikallio describes the melee this way:

"*At 1515-1625 hours. In combat with seven Pe-2s and at least ten MiG-3s I saw one MiG-3 attacking one of our Brewsters. When the Brewster dodged, the MiG-3 pulled up to the side ending up straight in front of me, when I fired from a short distance several short bursts, when it fell on the starboard wing and crashed on the ice about 2 km north-west of Seiskari becoming*"

Brewster BW-355 of 3/LeLv 24 paying a visit to sister squadron LeLv 26's base Kilpasilta on 11 February 1943, on the way to an overhaul at a field air depot at Immola. It was flown on this occasion by 1Lt Heikki Herrala.

completely destroyed. After this I fired at another two MiG-3s. My plane BW-387."

During the winter the Germans had twice shown great innovation. In the first place a double anti-submarine net was laid across the Gulf of Finland between Porkkala and Naissaari in Estonia, keeping effectively all Soviet underwater shipping on the east side of it. Secondly they laid further east between Kotka and Narvi two large minefields, which were overhauled from the islands and by vessels, which were based at the northern servicing point Kotka harbour. For over a year these supply points and harbours were the main target of the Red Banner Baltic Fleet air forces. The naval flying units began also exchanging the nearly obsolete equipment for modern aircraft, getting La-5 and Yak-7 fighters while the numbers of Boston, Pe-2 and Il-2 planes kept growing.

In April a significant change occurred over the Gulf of Finland. The Russian activities accelerated and both the equipment and tactics changed suddenly to first class. In addition the Russian had built a new airfield at Seiskari and extended the Lavansaari airfield so that Pe-2s and new generation fast Soviet fighters were able to land there. The fighter escort of Il-2 and Pe-2 aircraft was ingeniously arranged and thus LeLv 24 only got rare opportunities to shoot at them. The bombers had direct cover, usually manoeuvrable I-153, Yak-1 or LaGG-3 planes, of which some were staggered at different altitudes, in addition to a tactical cover of La-5, Yak-1 and Yak-7 fighters up to 6,000 m altitude in separate formations. Finally from Krasnaya Gorka and Kronstadt fighters were sent, which flew up to Seivästö to tie up the Brewsters in combat.

On 14 April 1943 fierce air battles for LeLv 24 broke out. 1Lt Hans Wind led four Brewsters of the 1st Flight over the eastern Gulf of Finland to combat thirty LaGG-3 and LaGG-5 fighters, which were escorting a bomber detachment arriving from the west. Five Soviet fighters were shot down into the sea, two of them by 1Lt Kim Lindberg:

"At 0645-0815 hours. When we observed the enemy planes straight below, I attacked the nearest one from behind. At about

200 m altitude I fired at a plane, which was a LaGG-5. The plane crashed vertically into the sea soon after this. The firing distance was about 70 m. After this I shot at a Spitfire, which was flying at 150 m. Thick white smoke puffed instantly out of the plane. It was heading north and continued to do so after the shooting. This occurred about 12 km south of Seiskari. I could not see the fate of this plane as I was engaged in combat with other Spitfires. In spite of later searches the plane was not found at Seiskari. My plane was BW-373."

Four days later a large air battle was fought over the Gulf of Finland. At 17.00 hours 1Lt Aulis Lumme took off with seven 2/LeLv 24 Brewsters and 1Lt Joel Savonen followed five minutes later with six 1/LeLv 24 machines to intercept eight Il-2s escorted by 50 fighters near Kronstadt. In an air battle which lasted one hour, the Finns shot down two assault planes and eighteen fighters, without losses. Lumme claimed two aircraft in the following manner:

"At 1700-1820 hours. Scramble with seven aircraft. We encountered over Seiskari eight Il-2s heading east with at least 25 fighters escorting (8 La-5s, 9 LaGG-3s and 8 Yaks). A fierce combat continued uninterrupted from Seiskari to Kronstadt for 45 minutes. The Il-2s flew at 5 m altitude and the escorting fighters stayed properly nearby. I fired at one Il-2 from behind and above and from the side from a very close range. The plane banked to left, hit the sea and sank instantly. After a short curving battle I managed to shoot at one LaGG-3 from behind at very close range, when black smoke puffed out of the plane and it went in a shallow bank into the sea. Additionally I fired at another twenty fighters at least, sometimes effectively. The score (11 aircraft) of our flight would not have been so high unless the escort duty of the slow Il-2s by the fast enemy fighters had not forced combat with our slow Brewsters. My plane was BW-370."

WO Viktor Pyötsiä also shot down two fighters:

"At 1705-1835 hours. The flight with six aircraft west of Harjavalta at 1500 m. On the deck came towards us 2/1, which we attacked. More Russians came from the clouds and in the ensuing combat I managed to make two Yaks smoke badly, but I was un-

Pilots of LeLv 24 at Suulajärvi on 4 May 1943. From the left: Sgt Viljo Kauppinen, 1Lt Hans Wind, squadron CO LtCol Gustaf Magnusson, 3rd Flight leader Capt Jorma Karhunen, SSgt Nils Katajainen, SSgt Leo Ahokas and air traffic controller SSgt "Skipper" Kauppinen.

able to follow the consequences (pilot bailed out from one), since there were plenty of Russians. The battle went on and north of Yhinmäki I put one La-5 into smoke shooting after a long battle in a dive from near behind. It was seen to go into the sea, but I had to dodge another La-5, which attacked me. The battle lasted 40 min, during which time I fired at several planes from close range without visual effect. Yak-1s fast and agile. La-5s good to climb, pretty agile and fast of course. My plane was BW-382."

Loyal to its habit of being quiet about big losses, the Baltic Fleet chronicle mentions only that strong flak over the target shot down one Il-2, and two fighters did not return to their bases.

On 21 April 1943 LeLv 24 flew 37 sorties. In the morning all three Brewster flights intercepted thirty-five Yak-1, LaGG-3 and LaGG-5 fighters of the Baltic Fleet air forces between Seiskari and Kronstadt. Capt Jorma Karhunen's six and Capt Iikka Törrönen's six aircraft took off to intercept the enemy formation. After the combat commenced Capt Jorma Sarvanto came to assist with five planes. The Russians shot down two Brewsters, but the score of the Finnish grew by nineteen aircraft. Törrönen described the destruction of two planes this way:

"At 0800–0930 hours. I took off with my flight of six Brewsters for a scramble to the Gulf of Finland. At Peninsaari I observed five Il-2s and their escort of six LaGG-3 and Yak-1 fighters. After noticing us the enemy aircraft went into a dive and by Seiskari they were already down on the deck. We chased the planes, but did not catch them until on the east side of Seiskari, where Capt Karhunen was already waiting for them. With the escort fighters ensued a fierce combat, where more enemy fighters appeared their number going up to twenty to thirty by the end of the air battle. I fired at several enemy fighters. Finally one LaGG-3 began to smoke after shooting and crashed in the forest near Lipeva village. After this I was forced into another combat over Tolbukhin-Kronstadt area. Three La-5s attacked me and 2Lt Riihikallio. I shot at an La-5, which had appeared behind the tail of 2Lt Riihikallio, so that it began to pour smoke and it crashed on the shore of the north-west tip of Kronstadt island. The enemy fighters carried out well their escort duty. La-5 proved to be rather agile and completely superior in both speed and ability to climb compared to the Brewster. My plane was BW-380."

The input of 1Lt Hans Wind was the following:

"At 0815-0930 hours. I shot at one Yak-1 from straight behind, pulled off after which WO Kinnunen shot at the same plane. After this I shot it from a close range. It began to smoke and dived in about 45 deg angle to the sea. Soon after this I got behind another Yak-1. I shot, the plane began to smoke instantly going vertically into the sea. The battle shifted over the Oranienbaum encirclement, where after a short curve battle with two Yak-1s I got to fire at one from 20 m at 600 m altitude. The plane went vertically into the forest. Combat altitude 10-6,000 m. My plane was BW-393."

On 2 May 1943 the Red Banner Baltic Fleet air forces attacked Kotka with 30 aircraft and they were engaged with 18 Brewsters on the south side of the town. An hour long battle waged across the Gulf of Finland and 2/LeLv 24 sent down four LaGG-3s, but not without losses, since the 2nd Flight leader Capt Iikka Törrönen did not return. 1Lt Aulis Lumme took over the command of the flight. 2Lt Olavi Puro described the melee:

"At 1000-1130 hours. We met an enemy bomber and fighter formation between Someri and Haapasaari heading south-east. I attacked the LaGG-3 fighters, which were escorting Il-2s, going towards Lavansaari. I managed to get behind one LaGG-3 and

fired at it for about 3 km distance from behind at 100-50 m range. We descended all the time and finally at about 150 m altitude the burst hit the cockpit and it went straight into the sea about 10 km SW from Someri. I chased with 1Lt Sarjamo a further three enemy fighters to east side of Lavansaari, where eight other LaGG-3s came from the south. They attacked us from the sun and in the ensuing curve battle I managed give a good burst at one in a bank scoring hits between the engine and the canopy of the LaGG-3, when plating sheets flew off and flames came out of the plane. Smoking strongly it went down halfway between Lavansaari and Peninsaari. I fired at another six enemy fighters at least, when one landed immediately to Lavansaari airfield once I had fired from the opposite direction. Low on ammunition the Russians finally broke off and headed east. Brewster pulls tighter turns than the LaGG-3. The speeds about the same at the deck. The Russians very aggressive and enterprising. My plane BW-387."

On 4 May 1943 a twelve Brewster detachment of LeLv 24 attacked near Seiskari five Il-2s, which were escorted by ten *Chaikas* with a dozen LaGG-3s flying as top cover. The Russians lost nine aircraft but the fighters claimed one Finnish plane shot down. 1Lt Hans Wind entered four air victory claims in the records, all seen to ditch:

"At 1040.1220 hours. I flew as the wingman of Capt Karhunen. I observed at Peninsaari four I-153s and five Il-2s. I attacked behind one I-153 and fired at it on the deck. The plane went on its starboard wing into the sea. The Il-2s continued towards Kopornoye bay. I managed to slip behind them when the other Brewsters tied up the fighters. I shot from the side in the port wing root of the first Il-2. The plane caught fire instantly and crashed into the sea. I continued and shot at the second Il-2 with the same results. At Shepelevskiy I shot at the third Il-2. It began to smoke and went into the sea near Tolbukhin. Side-slip was the only evasive manoeuvre of the Il-2s. Caught fire easily when shot in the junction of the wing and fuselage. My plane was BW-393."

On 9 May 1943 the Russians bombed fortifications at Suursaari. *Lentolaivue* 24 recorded its 500th air victory, when on a return leg fifteen Brewsters fought on two occasions against a thirty-plane escort detachment. In the first clash 1Lt Hans Wind's six Brewsters of the 3rd Flight shot down one La-5 and a Yak-7. In the second encounter five 1st Flight Brewsters of Capt Jorma Sarvanto and four 2nd Flight planes of 1Lt Aulis Lumme claimed one Yakovlev. At this stage it was stated that the aircraft thought to be Tomahawks were Yak-7s. Once more a misidentification was made, destroyed aircraft had also been claimed as Spitfires.

During the spring the Baltic Fleet air forces gained a new air base on Seiskari Island and the operations were very intense right from the beginning. On 20 May 1943 LeLv 24 took of for 43 sorties and their 15 Brewsters fought three air battles around Seiskari between 9.00 and 10.30 hours. First the 2nd Flight sent down both of a LaGG-3 pair, then the 1st Flight destroyed one Yak out of an 8-plane escort detachment and finally the 3rd Flight shot down two Yaks from a thirty-plane formation. Three hours later LeLv 24 returned with 14 aircraft to Seiskari and the same carousel continued with about 40 planes in mixed formations. Now the score was seven Lavochkins and one Yakovlev. 1Lt Hans Wind described this combat tersely:

"I shot one La-5 in a climb and bank, a thick smoke puffed out of the it and it went vertically into the sea. I shot up the Yak-1 from close behind. It fell in the sea next to two Russian patrol motor boats. My plane was BW-393." 1Lt Antti Saikko-

Three aces of LeLv 24 at the tail of Karhunen's BW-366 at Suulajärvi on 7 May 1943. From the left: 1Lt Hans Wind (74½ victories), WO Yrjö Turkka (17½ victories) and Capt Jorma Karhunen (31½ victories). (SA-kuva)

1/LeLv 24 pilots at Suulajärvi in April 1943. From the left: Unto Viinikka, Mikko Pasila, Martti Lehtovaara, Åke Roos, Kim Lindberg, Jorma Sarvanto, Hans Wind, Kaarlo Saukkonen, Aimo Vahvelainen, Joel Savonen, Jouko Lilja, Viktor Pyötsiä, Eero Hällfors, Matti Pellinen, Kalevi Anttila and Kai Metsola. (SA-kuva)

nen witnessed both cases: "*I saw aircraft remnants in the said locations.*" LeR 3 signed the matter: "*Do they stay afloat?*"

During six weeks the ageing Brewsters claimed 81 enemy aircraft shot down, losing three of its own. The excellent results for the Finns were based on staggering the flights at different altitudes and taking full advantage of the pendulum tactics, which came as quite a surprise to the Russians flying faster and better-climbing aircraft.

On 30 May 1943 LtCol Gustaf Magnusson, who had led LeLv 24, was appointed to command LeR 3 and his previous post was occupied by Capt Jorma Karhunen. 1Lt Hans Wind took over the leadership of the 3rd Flight.

On 5 June 1943 LeLv 24 took of for 31 sorties during the day. In the afternoon all three flights fought in the vicinity of Kronstadt, with sixteen aircraft, against two Russian mixed

formation both consisting of four Pe-2s, 7–8 Il-2s and 10–15 fighters. The Brewsters shot down without losses a Petlyakov, an Ilyushin, a Lavochkin and three Yakovlevs. Sgt Onni Avikainen claimed one fighter and one bomber:

"*At 1430-1550 hours. I observed south of Tolbukhin ten fighters and four Pe-2s. I made the attack from above and behind on one Yak-1 and fired a burst after which I pulled up to make another attack. When I began this I observed that the plane I had fired made a forced landing in the sea. Then I observed four Pe-2s north of Tolbukhin. I made an attack against the starboard wing plane from above and behind and kept firing down to 30 metres. Just when I managed to pull alongside the aircraft in question exploded. My plane was BW-370.*"

On 16 July 1943 LeLv 24 was reinforced when the six-plane Messerschmitt detachment led by 1Lt Lauri Pekuri arrived at

Suulajärvi and was subordinated into the squadron. The main task was to fly top cover for the two-year combat veteran Brewsters, still possessing reasonable combat capabilities.

On 31 July 1943 *Lentolaivue* 24 gained its fourth Mannerheim Cross Knight holder, when the commander of 3/LeLv 24 1Lt Hans Wind was awarded the decoration. He had gained 33½ aerial victories at that point.

The Soviets concentrated their bombing missions on Tytärsaari, held by the Germans, and the German ships on the Estonian coast. Even though LeLv 24 tried hard to sever the return route of the formations to Oranienbaum, it was a very difficult task. The Soviets had learned to stagger their fighter cover formations, and as bombers left or returned, fighter cover was in place between Seiskari and Oranienbaum. The bombers and ground attack planes were thus out of reach for the Brewsters.

On 20 August 1943 a large aerial battle was fought over the Gulf of Finland, with 16 Brewsters and three Messerschmitts battling 15 LaGG-3s and La-5s south of Seivästö. LeLv 24 shot down six Soviet planes and heavily damaged one more. SSgt Emil Vesa and his patrol shot down three:

"*At 1630-1800 hours. I flew top cover for the flight led by 1Lt Wind, assisted by Sgt Kauppinen. I attacked two LaGG-3s from a curve and scored many hits on one while Sgt Kauppinen shot at it too. The plane went smoking as it curved round Tolbukhin, towards Kronstadt, and after a crash landing caught fire on the ground. I pulled up and again went into a curve battle with a couple LaGGs, one of which I hit from a diving turn and saw it crash on Yhinmäki beach, smoking. I fought with several more planes, no results. My plane was BW-357.*"

On 31 August 1943 1Lt Urho Sarjamo of 2/LeLv 24 had his six Brewsters fighting two Il-2s and their fighter cover which had many planes, shooting down one Il-2. At the same time, 1Lt Hans Wind of 3/LeLv 24 flew his five planes to meet two La-5s at Oranienbaum, shooting both down. Finally the new leader of 1/LeLv 24, 1Lt Lauri Nissinen, had his five Brewsters wrestle with four Yakovlevs, shooting down two in exchange for one of their own. Nissinen wrote:

"*At 1645-1700 hours. I flew as lead plane of five. As we flew at some 2000 m south of Koivisto westward, we saw four LaGG-3s down and right, and we attacked. Early in the curve fight I shot at one LaGG and got it to smoke, and then it dove. The smoke stopped and the LaGG came away from the dive, and I fired a second time. This time it went into the sea leaving a large slick of oil. As I was shooting the second time, a LaGG got behind me, but I got behind it with a few quick turns. I got smoke out of it, and it tried to dive away. It pulled some 300 m distance, but in level flight I slowly gained on it and kept shooting at it close to the shore of the island, and turned away. As I left it it was all covered in smoke, only wingtips were visible. A moment later the Seivästö observation post "Seppo" reported it crashed on the Oranienbaum encirclement shore. Altitude of fight 10–1000 m. My plane was BW-373.*"

The last great aerial battles before the winter were fought above the eastern Gulf of Finland on 23 September 1943. At 13.00 four Brewsters of 3/LeLv 24, led by 1Lt Martti Salovaara, and two MT pairs of 1/LeLv 34, fought twenty Soviet fighters close to the Shepelevskiy lighthouse and reported three Yakovlevs and five Lavochkins shot down.

Two and a half hours later, 1Lt Hans Wind took eight Brewsters to attack fifteen Soviet planes returning to Seiskari. The Finns claimed one Il-2 and six Lavochkins. 1Lt Hans Wind gave a short report:

1Lt Joel Savonen of 1/LeLv 24 demonstrates his latest combat at Suulajärvi to Sgt Sulo Lehtiö (left) and Sgt Paavo Koskela and 1Lt Kai Metsola (right), in early May 1943.

"At 1530-1630 hours. I led the flight when we met 8 LaGGs and 6 Il-2s. First I shot at an Il-2 which was badly damaged and tried to reach Seiskari airfield, but crashed on the south end of the runway and burned. Then I shot down an La-5 which landed in the sea and sank almost immediately. My plane was BW-393."

On 4 November 1943 five Brewsters of 3/LeLv 24, led by 1Lt Antti Saikkonen, and five Brewsters from the 1st Flight, led by 1Lt Joel Savonen, fought a battle around Seiskari – Shepelevski lighthouse. Protected by three MTs of 1/LeLv 34, they battled a mixed formation of seven and nine planes, shooting down three Yakovlevs.

On 11 November 1943 1Lt Vilppu Perkko of 3/LeLv 24 led five Brewsters in a chase of four Il-2s and four Yaks to Oranienbaum, shooting down two of the fighters. One of them was piloted by 1Lt V.I. Borodin of 13 IAP, KBF, who died when he crashed in the sea. Before succumbing he hit the left wing of Perkko's Brewster, which caught fire. Perkko took to his parachute and was fished out into captivity by a Soviet patrol boat. The Soviet listed this as a "taran" or ramming victory, but the planes merely shot each other down without contact.

December 12 1943 saw the last clash of the year. 2/LeLv 24 sent Urho Sarjamo's Brewster patrol to chase an Li-2 bomber that flew over Suulajärvi, and managed to shoot it down over the Isthmus. Sarjamo reported:

"At 12.15 a DC plane passed the airfield very low (at 20 m). I took off in chase and caught it south of Suulajärvi. After I got to it, the rear gunner gave me a long burst, after which I attacked it from rear and above. First I silenced the gunner, and then I shot the right engine into flames. As I fired, the plane did an 180 degree turn. I stopped firing and followed it as it crashed in a bog; I saw three men emerge. It seemed like the fire would die out, but

1/LeLv 24 leader Capt Jorma Sarvanto, a 17 victory ace, at Suulajärvi in May 1943, after returning from his 250th mission. Behind is his assigned plane, BW-373.

Three aces of 3/LeLv 24 Suulajärvi on the tail of BW-351. From the left: Sgt Viljo Kauppinen (7½ victories), SSgt Nils Katajainen (34½ victories) and SSgt Leo Ahokas (12 victories), who points at his fourth kill, which came on 4 May 1943.

Brewster 239 BW-393 of 3/LeLv 24, flown by the flight leader 1Lt Hans Wind in September 1943, based at Suulajärvi. (Karolina Hołda)

when I went back to see the plane after 20 minutes, it was entirely in flames. Very easy to light it. It flew so low that I could only shoot from rear and above. My plane was BW-386."

The plane was apparently disoriented and later found to belong to a long-distance bomber unit whose flights were based on airfields north of Leningrad.

Brief Analysis

1943 aces of *Lentolaivue* 24:

Rank	Name	Flight	Victories	Plane	Total victories
1Lt	Wind, Hans*	3	23½	BW-393	38½
SSgt	Vesa, Emil	3	7½	BW-357	9½
1Lt	Sarjamo, Urho	2	6	BW-386	6½
2Lt	Puro, Olavi	2	5½	BW-387	5½
2Lt	Riihikallio, Eero	2	5½	BW-374	6½
MSgt	Alho, Martti †	2	5½	BW-383	13½ + 1½
SSgt	Katajainen, Nils	3	5½	BW-353	19½
1Lt	Lumme, Aulis	2	5	BW-370	11½
2Lt	Saarinen, Jorma	2	5	BW-380	5

* Mannerheim Cross, † killed in flying accident

This table contains victories by named pilots only in 1943. They are rounded to nearest half.

The right hand column shows victories in LeLv 24 by end 1943, after + sign are Winter War victories

1944

Mid winter was a quiet period of flying on both sides. At the beginning of 1944 *Lentolaivue* 24 still possessed seventeen serviceable Brewsters.

On 14 February 1944 the front-line squadrons received a task prefix, the name, *Lentolaivue* 24 becoming *Hävittäjälentolaivue* 24 (Fighter Aviation Squadron), HLeLv 24.

Initiated in mid-January, the main offensive of the Red Army against the German North Army Group broke the siege of Leningrad by the end of the month. In this new situation, the commander of LeR 3 redefined on 20 February 1944 the areas and tasks of the squadron, thus: HLeLv 24 has Virolahti-Seiskari as its right boundary and Pölläkkälä – Lempaalanjärvi as its left one. Crossing the line from Seiskari to Seivästö and Kronstadt, and finally the front line, was forbidden.

9 March 1944 saw enemy contact in the air again, when WO Viktor Pyötsiä's Brewster pair of 1/HLeLv 24 took off for a night intercept mission to the Karelian Isthmus, engaging a solitary Li-2 bomber and downing it.

On 2 April 1944 the last aerial victory of Brewsters of HLeLv 24 occurred, by 2nd Flight's Capt Jouko Myllymäki's detachment of six Brewsters. They sent down near Ino one LaGG-3 belonging to 11 GIAP, KBF.

Messerschmitts

The plane situation of HLeLv 24 changed in April 1944. On 4 April the 1st Flight was issued Messerschmitt Bf 109 G-2 fighters and it passed its remaining Brewsters to the 2nd and 3rd Flights. The 2nd Flight got its first Messerschmitts on 15 April, and on the last day of April Messerschmitts found their way to the 3rd Flight as well.

On 14 April 1944 HLeLv 24 scored for the first time on an intercept with Messerschmitts. 1Lt Lauri Nissinen, the 1st Flight leader, reported:

"*Air wardens reported Russians in the air over the Gulf of Finland. From my base at Suulajärvi I saw a heavy contrail in the south-west, somewhere between Lavansaari and Seiskari. I was scrambled and I moved in on the contrail. At 500 meters I noticed flak reaching out to the plane I chased. I assume it was the Saarenpää batteries firing. The plane changed course due northwest right after that and the contrail ended. I followed the plane which was at 7,500 m some 10 km ahead (I was at the same altitude). Slowly I gained on it. I was in radio contact with ground control all the time and they relayed more information to me. Over the Gulf of Viipuri the plane ducked and I lost it. I descended somewhat and shortly before Virolahti I noticed a plane 2 km ahead, 500 m down, and to my left. I flew to 500 m from him and looked for national insignia. I saw nothing and I fired our friendly plane sign, to no avail. I had never seen the type and assumed it to be a new Russian one. It looked dark and had no insignia. From previous battles with Russians I remembered that they have no insignia on wings. I decided to attack. From 300 m I fired both guns and cannon, and hit the engine. I went to 50 m and fired again, setting the right engine on fire. The plane went in burning at 1610 hours. My plane was MT-225.*"

The dispatched plane was a German Junkers Ju 188 F reconnaissance plane of 3.(F)/22. As the remains were searched through at Virolahti it was found that in the dirty winter camouflage the national insignia and Eastern Front markings were not visible. The Germans were given strict orders for reporting in advance to the Helsinki flight control. LeR 3 were also given type identification pictures.

On 22 May 1944 a *Mersu* (as the Finns called it) pair of 2/HLeLv 24 reconnoitred the Isthmus. Something like this had never before been seen. The report told:

"At 1740-1830 hours, altitude 1000-2000 m, planes MT-246 and 245. Route: Suulajärvi – Vuolejärvi – Peri – Levashovo – Kaukolanjärvi – Haapakangas – Valkeasaari – Siestarjoki – Revonnnenä – Suulajärvi. Roads leading to Lake Ladoga much more truck traffic than usual. On Hovinmäki-Lehtoinkylä-Vuole road about 100 loaded trucks mostly to north. On Kujala – Kaukola – Peri – Varsolovo road about 50 loaded trucks to both directions. Road Korpiselkä – Kissula – Sihvola crowded with trucks in both directions, estimated between 200 and 300 with 20–50 m distances. On Korpiselkä-Kissula-Sihvola road 75–100 loaded trucks mostly to north. On Haapakangas – Alasaari – Vaskisavotta road about 20 loaded trucks. Between Vaskisavotta and Kyläjatko 5 trucks. On Retukylä-Novoselki-Levashovo road a dozen trucks mostly to west. No worthwhile traffic on rail from Lahti to Vaskisavotta. Enemy air activity: at Kasimovo 12 fighters on both sides of the runway at the west end of the east-west runway. 12/2 in the shelters north of the airfield. In the south end of the north-south runway 2 fighters. No aircraft observed at Levashovo, Shuvalovo and Gorskaya. No enemy anti-aircraft fire observed. Weather: fully cloudy, 2000 m. Horizontal visibility poor, vertical good. Patrol leader 1Lt Olavi Puro."

At the end of May 1944, masses of Russian troops, tanks, artillery pieces and other materiel was in position on the Finnish front on the Karelian Isthmus, north-west of Leningrad. Fighter reconnaissance had no trouble seeing this, but getting the HQ to believe the reports was harder. In the new situation, the commander of LeR 3 redefined squadron tasks – HLeLv 24 was to block enemy air action in the south of the Isthmus and the corresponding coastal area of the Gulf of Finland. Reconnaissance and escort of ships and bombers, and fighter bombing was only to be performed by separate orders. HLeLv 26 was to block enemy air action on the central and northern part of the Karelian Isthmus. Interceptions, reconnaissance, and bomber escort in the south was only to be performed by

1/HLeLv 24 pilots on the way to Nurmoila on 12 May 1944. From the left: flight leader 1Lt Lauri Nissinen, 1Lt Vilppu Lakio, 2Lt Heimo Lampi, WO Viktor Pyötsiä, SSgt Paavo Koskela and Sgt Arvo Koskelainen. At right is the squadron CO, Maj Jorma Karhunen. (SA-kuva)

separate orders. HLeLv 34 will intercept enemy action directed at Helsinki and Kotka. Reconnaissance of sea areas, escort of ships and bombers, and fighter bombing was only to be performed by separate orders.

On 6 June 1944 Capt Hans Wind and his swarm of 3/HLeLv 24 protected a Pe-2 plane photographing the front lines. At Siestarjoki, about twenty Soviet fighters tried to stop such activity, but the MT escort fought well and shot down two fighters. MSgt Jouko Huotari and his patrol of the 3rd Flight shot down one Airacobra of a detachment of five over Lempaala. 1Lt Väinö Suhonen and his patrol of the 1st Flight shot down two of six La-5s around the same area.

Next day the swarm of 1Lt Urho Sarjamo of 2/HLeLv 24 was scrambled to intercept eight planes that appeared over Suulajärvi. In the fight that ensued, half of the Russians were shot down. The 1st Flight lost one MT in a fight over the front lines.

Great Attack

After the success achieved against Germany in the spring of 1944, the Soviet Union began in the Karelian Isthmus the fourth of ten strategic efforts; this was to be the only one which did not reach its goals. The offensive started on 9 June 1944, and on the next day the first Finnish line of defence was breached, forcing the Finns to retreat.

The massive attack on the Karelian Front was supported by the Soviet 13th Air Army with more than 1,100 planes, and the left flank was protected by 220 planes from the Baltic Fleet Air Force. These warplanes, totalling over 1,300, were concentrated on a 20 kilometre strip of land at the eastern Gulf of Finland.

Against this massive air armada aimed at the Karelian Isthmus, *Lentorykmentti* 3 could pitch the 14 Messerschmitts of HLeLv 24, the 16 Messerschmitts of HLeLv 34, and the 18 Brewsters of HLeLv 26. LeR 3 was tasked with reconnaissance of the entire Karelian Army area, the blocking of Russian air attacks over the whole of the Karelian Isthmus, and the escort of Finnish bombers over the Isthmus and the nearby seas.

The order of battle of *Hävittäjälentolaivue* 24 was then:
- Squadron commander Major Jorma Jarhunen, Suulajärvi
- 1st Flight, 1st Lieutenant Lauri Nissinen, Suulajärvi (5 MT)
- 2nd Flight, Captain Jouko Myllymäki, Suulajärvi (5 MT)
- 3rd Flight, Captain Hans Wind, Suulajärvi (4 MT)

On 9 June 1944, the first day of the assault, the Russians sent up 1,150 sorties. In the inclement weather, the patrol led by 1Lt Urho Sarjamo of 2/HLeLv 24 ran into twelve Airacobras; one was shot down, and then the patrol broke off. 1Lt Atte Nyman led five MTs of 2nd Flight to intercept over the front lines, where large Soviet formations were seen. The patrol shot down one bomber and two fighters.

Next day, after a preliminary barrage by hundreds of field artillery pieces and supported by strong ground attack plane formations and heavy tanks, the troops of the 21st Army had pushed in a wedge some 10 km deep in the direction of Mainila, and then went on to expand the breach. The 13th Air Army flew with 800 planes, and the intercepting HLeLv 24 shot down 11 Russian planes. 1Lt Urho Sarjamo reported of his second fight:

"At 0955-1100 hours. It is hard to name the initial location of the battle due to poor visibility. East of Kivennapa our swarm ran into twelve Pe-2s at 600 m, and we attacked. The chase and firing went on all the way to Lempaalanjärvi. The Pe-2s went gradually down to the deck, and the one I shot at went into the forest right after Lempaalanjärvi, felling trees as it crashed. During the return, after our swarm dispersed, I was jumped by a lone Airacobra, and I fought with it at Riihiö. After some curving around, between the deck and 500 m, I got a clear shot at the plane and gave it a burst with a distance of 200 to 40 metres. I saw its skin tear off and just then it belched black smoke, and then it went in the forest. I had a 12.7 mm hit in my radio, the bullet stopped in the fuel tank armour plate. My plane was MT-227."

1/HLeLv 24 Mersu MT-225, assigned to 1Lt Lauri Nissinen, after overhaul at Immola on 31 May 1944. The pilots around the plane belong to HLeLv 26 based at Immola. From the left: 1Lt Carl-Erik Bruun, SSgt Kauko Tuomikoski, 1Lt Nils Trontti and 1Lt Sakari Kokkonen.

On 13 June 1944 low clouds prevailed and HLeLv 24 flew just one interception. Capt Wind and his six *Mersus* of the 3rd Flight met in the Kämärä-Kaukjärvi area thirty bombers and twenty fighters as escort. The Finns dispatched one Airacobra and seven Pe-2s with no losses, of which four went to Wind. In his stoic way he reported:

"At 0900-1010 hours. On interception I met some 30 Pe-2s, of which I shot down four and got one to smoke. All planes burned. One plane also had its left wing break off at the engine. As I ran out of ammunition I had to break off. Battle altitude 2,000 m. My plane was MT-201."

On 14 June 1944 a pair of MTs of 2/HLeLv 24 had to fight their way out twice while on reconnaissance.

The Russians lost six planes. 1Lt Olavi Puro led both flights, and described the contact with an overwhelming enemy force on the latter flight as follows:

"At 1100-1210 hours. As I was on a recce flight, I neared the target area and found a continuous bombing underway. I dived to the target and was attacked with 1Lt Saarinen by strong enemy formations, which contained about one hundred bombers and a similar number of fighters. I managed to shoot down two La-5 fighters and one Il-2 ground attack plane which all fell right on the front line between Vammeljoki and Mustamäki. As the bomber formations retreated eventually it became calm enough for me to complete the reconnaissance too. As we left, new enemy formations arrived in three waves of maybe 70 planes each. Our own AA fired at the front line incessantly. In addition to the aforementioned planes, I fired at another Pe-2 and a couple of fighters. In such a crowded sky it is apparently hard for the Russians to tell two MTs from their own fighters, since the machine gunners of the bombers never fired. My plane was MT-246."

17 June 1944 was a black day for HLeLv 24. The commander of the 1st Flight, 1Lt Lauri Nissinen, took off with ten *Mersus* at 06.20 to attack dozens of bombers and ground attack planes which were escorted by numerous fighters and were harassing Finnish troops in the Kaukjärvi-Perkjärvi region, starting their attacks at 2,000 m. A Russian La-5 shot at the Finnish top pa-

trol, severing the wing of 1Lt Urho Sarjamo's MT as he led the top cover. The doomed plane fell on Nissinen's MT and both pilots were killed. Despite this, Finnish fighters shot down four *Shturmoviks* and four fighters. Capt Myllymäki and his 2nd Flight shot down four Russian fighters at Leipäsuo. 1Lt Joel Savonen was deputized to run the 1st Flight.

On the 19 June 1944 new Bf 109 G-6 deliveries from Germany began to ease the plane situation. HLeLv 24 and HLeLv 34 both came close to having the nominal strength of 25 serviceable planes.

In the evening of 20 June 1944, the Russians captured Viipuri, regrouped, and directed the attack towards the narrow terrain of Tali-Ihantala; this route had the only available tank route to the north-west. This was where the Soviet attack was finally stopped for good.

20 June 1944 was the peak of air combats. Before noon, the Messerschmitts of HLeLv 24 and 34 sent down 35 enemy planes in three large engagements. During the five more fights over the rest of the day, MT pilots downed 16 more enemy planes. The total catch of the day was thus 51 victories, of which 31 went to HLeLv 24 and 20 went to HLeLv 34, all without loss to the Finns. 1Lt Olavi Puro of 2/HLeLv 24 claimed three aircraft:

"At 0830-0920 hours. On an intercept mission we came across south-east of Viipuri a twenty aircraft strong Il-2 formation and a lot of fighters. In the ensuing combat I got hits in one La-5 from above and behind, when it went in a spin in the middle of Lake Suurijärvi south-east of Johannes. I kept hammering the formation and one Il-2 went down in flames into the forest between Römpötti airfield and Näränjärvi, lighting a forest fire. Finally I put one LaGG-3 into smoke crashing in a shallow dive in the forest at Lempaala. The Russians had taken older types into service like the LaGG-3 and Yak-1. My plane was MT-201."

22 June 1944 was the busiest day of the month for HLeLv 24, with 61 sorties. Capt Wind's 3rd Flight escorted bombers after midnight to the Viipuri area, and two La-5s were shot down. At noon in an engagement over Viipuri the Finns shot down

Late model Bf 109G-6 coded MT-456 of 1/HLeLv 24, running up at Lappeenranta on 29 June 1944. This fighter was assigned to 1Lt Otso Leskinen. Eighty minutes endurance without the 300 litre drop tank was sufficient for the missions in summer 1944. (SA-kuva)

two enemy fighters. Early in the evening, 1Lt Ahti Laitinen and his *Mersu* swarm of the 3rd Flight was scrambled to intercept the enemy over Tali, where the Finn fought with many bombers and fighters. One Pe-2 and three La-5s were shot down, but one MT was lost as well. On the last flight of the day, a patrol from the 3rd Flight shot down one more Pe-2.

On 23 June 1944 Capt Myllymäki took his flight of 2/HLeLv 24 to intercept the enemy over the Viipuri-Tali area, where they met 28 bombers and 13 fighters. The flight shot down six Il-2s, three La-5s, one Airacobra and a Mustang (actually a Yak-9).

Capt Wind was sent with three *Mersus* of the 3rd Flight to cover the ground forces at Viipurinlahti, where two dozen bombers and about ten fighters were attacking Finnish infantry. The Russians lost two DB-3Fs and three La-5s. The combat report of SSgt Emil Vesa reported:

"At 0730-0820 hours, altitude 50-1,000 m. I was assigned the task to fly with four planes a combat air patrol in Viipuri-Tali-Karisalmi area. We met south-east of Viipuri 8 R-10 and about 20 Il-2 ground attack planes escorted by 6 Mustangs and 5 La-5s. In the ensuing battle I fired from 200 metres straight behind at one R-10, which thinly smoking crossed over Viipuri. After this I began to fight with the Il-2s heading towards Tienhaara. I attacked from above and fired at several planes managing to put

one on fire (port side of the engine) and it went down in the beach willows at Ronkaa. The coolant liquids came out from another and it made a forced landing in the water at Havinlahti. My plane was MT-438."

The swarm of 1Lt Puro's 2nd Flight went on to Tali, where 50 bombers escorted by 15 fighters were found. The swarm attacked and shot down one *Shturmovik* and one DB-3F along with four La-5s.

While on morning intercept on 26 June 1944 to Viipuri, 1Lt Aulis Lumme and his swarm of 2/HLeLv 24 met eight Il-2s escorted by two La-5 fighters. One pair tied down the fighters and the other shot down five of the ground attack planes. Around noon 1Lt Saarinen of the 2nd Flight led six *Mersus* on a Blenheim bomber escort to Tali, where several fighters were found to be patrolling. Twenty Il-2s also appeared, of which four were shot down.

The commander of *Lentorykmentti* 3, LtCol Gustaf Erik Magnusson, became on 26 June 1944 a Mannerheim Cross holder. He had 5½ air victories, but the fighters he led had 800. Since the autumn of 1943, Magnusson had worked to develop a fighter control system operating by radio reconnaissance and visual observations, lacking radar. The regimental command centre ran the system in real time. This enabled limited interception forces to yield excellent results.

On 28 June 1944 HLeLv 24 set the highest daily victory score with 33 enemy planes shot down, in five engagements. Before noon, Capt Wind of 3/HLeLv 24 led eleven *Mersus* to intercept the enemy over Tali. There were 20 Il-2s and 20 Pe-2s there along with an escort of dozens of fighters. In the battle that ensued, the Finns sent down eight *Shturmoviks*, two Pe-2s and one La-5.

Half an hour later Capt Wind took his patrol to reconnoitre the Bay of Viipuri, where they were jumped by some 30 fighters; the Finns shot down four but Wind was wounded and reported the event as follows:

"*At 1030-1130 hours. As we were beginning our reconnaissance mission, some 20 Yak-9 s came at us at Juustila. We had no choice but to fight. I shot down two, and just as I got a third one to burn, a Yak hit me from the side and behind. I was badly wounded in the left arm and had to try to get home. With the last drop of strength left in me I managed to land. My plane was MT-439.*"

MSgt Katajainen who was the wing man on the same mission reported:

"*At 1020-1105 hours. I was flying wing on Capt Wind on a recce mission. We were attacked by some 20 fighters. I shot at one AC which flamed up, and Capt Wind saw it burn. I broke off and gained altitude off to one side, in a chase I shot at another AC and it started to smoke, but I did not see it fall to the ground. At the same spot south of Tali I saw an observation balloon, lunged through a flak barrage, and shot the balloon into flames. As I retreated to our side I saw seven Il-2s on my left. I attacked them*

and at Tali managed to make one smoke, and after it crashed on the ground it burned. I shot at yet another Il-2 but ran out of ammo. Battle altitude 10-3,000 m. My plane was MT-436."

1Lt Suhonen and his 1st Flight sent down four *Shturmoviks* and two Yaks over Tali. The 2nd Flight fought two separate engagements during the day, with dozens of Russian planes, and destroyed 12 planes. 1Lt Jorma Saarinen shot down three ground attack planes in the first engagement and reported the event:

"*At 1400-1500 hours. After getting rid of Russian Yak-9 escort fighters at 2000 to 3000 m, I took a dive at Il-2 formations which were crossing the front lines at Karisalmi and turning left, exiting over Tali. I attacked five times against groups of five Il-2s without being bothered by the "Mustangs" bouncing about here and there. I saw three Il-2s burn and crash. I ran out of ammo so I had to exit at the best part of the battle. In addition I shot at one "Mustang", but had no time to check my results. Only some Il-2s fired back. My plane was MT-452.*"

Commander of 3/HLeLv 34, Capt Hans Wind, and WO Ilmari Juutilainen of 1/HLeLv 34 were awarded the Mannerheim Cross for the second time, the only double knights created during the Continuation War. Both had 75 aerial victories at the time.

Next day 1Lt Mikko Pasila of 1/ HLeLv 24 led six *Mersus* on interception to the Viipuri-Tali area. There was already a battle raging between eleven HLeLv 34 Messerschmitts and some 200 Russian planes. Pasila and his group shot down four Russians, but the Finns lost one MT. 1Lt Lumme of the 2nd Flight took

Three aces of 3/HLeLv 24 confer in front of Mersu MT-460 *at Lappeenranta on 6 July 1944. From the left: SSgt Emil Vesa (30½ victories), flight leader 1Lt Kyösti Karhila (32½ victories) and SSgt Leo Ahokas (12 victories). (SA-kuva)*

Bf 109G-6 MT-456 of 1/HLeLv 24 flown by 1Lt Otso Leskinen in July 1944, based at Lappeenranta. (Karolina Hołda)

his eight *Mersus* out to help the ground troops in the Ihanta-la – Juustila area, where numerous Russians were patrolling. They shot down 3 Il-2s. 1Lt Teromaa and his 2nd Flight escorted bombers to Tali and shot down one Yak of a patrol of three.

The intensity of the battle of Tali-Ihantala reached its high point as June turned into July. The Russian advance was finally stopped in the IV Army Corps sector at the Viipurinlahti – Ihantalanjärvi – Vuoksi line. The Russians kept attacking the Finnish defence positions for another week but could not break the lines.

On 1 July 1944 1Lt Kyösti Karhila and his six *Mersus* of 3/HLeLv 24 covered shipping around Teikarsaari, where they met 10 Il-2s and 13 fighters. The Finns shot down one *Shturmovik* and three fighters in the dogfight that ensued. MSgt Nils Katajainen and his swarm of the 3rd Flight shot down an Airacobra at Juustila.

The base of HLeLv 24 at Lappeenranta and that of HLeLv 34 at Taipalsaari had escaped Russian air attacks so far, as had the German Detachment Kuhlmey's base at Immola. But on 2 July 1944 at 20.00 an attack by 16 Pe-2s and 36 Il-2s was made at Lappeenranta. Regimental radio surveillance had indicated that something was brewing, and most of the Messerschmitts of HLeLv 24 had replenished their ammo and fuel, and eleven had taken off after a Russian force, albeit a decoy one. Escort fighters tied up the *Mersus* in the air, but the Finnish pilots nevertheless shot down one Pe-2 and four Il-2s. Some planes remained on the ground. Two MTs were lost in the attack and four more damaged. Two Pe-2 photo reconnaissance planes of PLeLv 48 were destroyed.

On 3 July 1944 the Messerschmitts of LeR 3 were assigned to spend most of their time on bomber escort. In many cases flying escort for *Lentorykmentti* 4 bombers prevented fighters from attacking fighters. The Messerschmitts had a perfect record for escort, as not one bomber was lost to Russian fighters above the Karelian Isthmus.

In the morning, 1Lt Joel Savonen of 1/HLeLv 24 led nine MTs to attack forty Il-2s and forty Il-4s covered by 30 fighters at Tali-Ihantala. The Russians held their own and shot down one MT in exchange for just one Russian bomber and two fighters. Before noon, a swarm of 1st Flight MTs shot down a *Shturmovik* and a fighter while on escort but lost one of the MTs. Also before noon, 1Lt Karhila of the 3rd Flight entered the same area. They met "only" 15 *Shturmoviks* and 10 La-5s. The Finns prevailed and the Russians lost one fighter and three ground attack planes. 1Lt Väinö Suhonen took down a Pe-2 with his MT flight at Viipuri.

As the Finnish troops had retreated to the Vuoksi – Taipale line on June 20, Russian forces began on 4 July 1944 an attempt to cross this waterway. The Russians managed to cross Vuosalmi and reach its south-western beaches at the narrow spot of Äyräpää and five days later they started to cross Vuosalmi with a wide front. The battle raged for nine days without a breakthrough. The Russians were stopped now, permanently.

The six plane formation of 3/HLeLv 24 led by 1Lt Kyösti Karhila was on an escort mission to Viipuri when the Finns shot down one Yak-9 out of a mixed formation of some twenty planes. 1Lt Väinö Suhonen of the 1st Flight flew bomber escort with nine MTs over the Bay of Viipuri when they met many fighters escorting six Il-2s at Koivisto. One Il-2 and a La-5 were dispatched.

As eight *Mersus* of 3/HLeLv 24, led by 1Lt Karhila, went on interception on 5 July 1944 to Koivisto they saw about twenty enemy planes. In the battle that followed, three Il-2s and three Yak-9s were shot down. One MT that was damaged in the fight was destroyed in a crash-landing. The 2nd Flight sent eight MTs led by 1Lt Lumme to escort bombers to Viipurinlahti, where they met many enemy planes. Two LaGG-3s were shot down.

On 9 July 1944 HLeLv 24 had its busiest day of the summer with 68 sorties. 2/HLeLv 24's Capt Veikko Ala-Panula led 7 *Mersus* to cover infantry at Äyräpää. The detachment fought with eight Yak-9s and emerged victorious, shooting down three. 1Lt Suhonen of the 3rd Flight was also sent there to cover infantry with eight *Mersus*. Of eleven Russian fighters they met the Finns sent down one La-5 and one Yak-9.

On 10 July 1944 1Lt Karhila of 3/HLeLv 24 was at Äyräpää flying top cover for ground troops with eight *Mersus* in the afternoon, when fifteen Russian fighters appeared. The Finns responded to the attack and shot down five La-5s and one Yak-9. Early in the evening 2/HLeLv 24 sent Capt Ala-Panula and twelve MTs to escort bombers to Äyräpää, where numerous Russian fighters were snapping at Finnish bombers. 1Lt Olavi Puro submitted this report:

"At 1835-1935 hours. On bomber escort in bad visibility weather I saw some 20 Russian fighters arrive from Suulajärvi to our bombing target of Äyräpää. In the battle we managed to keep the Reds off the bombers despite superior numbers. I managed to shoot one La-5 down east of Äyräpäänjärvi, and two Yak-9s north-east of Vuosalmi. The latter Yak-9 blew up in front of me so close that I got oil and debris on my plane. In addition, when I pulled up I was able to hit one La-5 from below between the wings and the engine, with smoke and fire

Bf 109G-6 MT-431 of 2/HLeLv 24, flown by SSgt Pekka Simola in September 1944, based at Utti. (Karolina Hołda)

resulting, but as I stalled myself, I had no chance to see its fate. After that I was hit and had to strive to get back to base. My plane was MT-479."

On 11 July 1944 eight MTs of 3/HLeLv 24 led by 1Lt Karhila went on intercept to Äyräpää and met dozens of Russian planes there. Three fighters were dispatched, but one MT went missing in the battle.

On 15 July 1944 the war entered a stationary phase all over the Karelian Isthmus as Russian troops dug in for defence. The last great aerial battles of the Karelian Isthmus took place at this time. The LeR 3 Messerschmitts downed a dozen Russian planes in five engagements. 1Lt Jorma Saarinen of 3/HLeLv 24 protected infantry with eight MTs at Äyräpää. A few Il-2s es-

corted by some 30 Yak-9s fired at Finnish troops before the MTs could intervene, and when they did, they shot down one Il-2 and four Yaks.

During the great offensive, in a span of forty days Messerschmitts of HLeLv 24 and HLeLv 34 were credited with 425 Russian planes shot down and 78 damaged, during 355 missions and 2,168 sorties. Red fighters shot down ten MTs and three went missing. Russian flak claimed three shot down and ground attack planes dispatched two more. Eight Finnish pilots were killed and three were taken prisoner.

On 19 July 1944 1Lt Erik Teromaa led seven *Mersus* of 2/HLeLv 24 on interception to Vuosalmi where they met about twenty fighters. Four of these were shot down.

Pilots of 3/HLeLv 24 at Lappeenranta on 10 July 1944. From the left: 2Lt Per-Erik Ohls, 1Lt Jorma Saarinen, flight leader 1Lt Kyösti Karhila, Sgt Kosti Koskinen, SSgt Leo Ahokas, SSgt Emil Vesa, 2Lt Evald Estama, MSgt Jouko Huotari and Sgt Risto Helava. (SA-kuva)

Pilots of 1/HLeLv 24 at Lappeenranta on 16 July 1944. From the left: flight leader Capt Aate Lassila, 1Lt Mikko Pasila, 1Lt Joel Savonen, 2Lt Otso Leskinen, 1Lt Vilppu Lakio, SSgt Aimo Vahvelainen, Sgt Sten Toivanen, SSgt Paavo Koskela and SSgt Arvo Koskelainen. (SA-kuva)

Pilots of 2/HLeLv 24 at Lappeenranta on 16 July 1944. From the left: flight leader 1Lt Erik Teromaa, 1Lt Eero Riihikallio, 1Lt Aulis Lumme, 1Lt Lasse Kilpinen, 1Lt Veikko Koski, 1Lt Olavi Puro, 1Lt Atte Nyman, 2Lt Toimi Juvonen, SSgt Tapio Järvi, Sgt Eero Halonen and Sgt Kosti Kaloinen. (SA-kuva)

Pilots of 3/HLeLv 24 at Lappeenranta on 16 July 1944. From the left: flight leader 1Lt Kyösti Karhila, 2Lt Evald Estama, 1Lt Väinö Suhonen, MSgt Jouko Huotari, 2Lt Per-Erik Ohls, Sgt Kosti Keskinummi, SSgt Emil Vesa, Sgt Eero Pakarinen and Sgt Kosti Koskinen. (SA-kuva)

Next day 1Lt Teromaa took six *Mersus* of 2/HLeLv 24 to reconnoitre the area between Vuosalmi and Tali where they met fifteen fighters. Three were shot down in exchange for one MT lost.

On 22 July 1944 1Lt Puro and his pair from 3/HLeLv 24 reconnoitred the Viipuri-Seiskari area and found 20 *Shturmoviks*, 9 Pe-2s and 20 La-5s. The pair attacked and shot down two Il-2s and one La-5.

On 25 July 1944 *Mersu* swarms of HLeLv 24 and 34 took off together to intercept a large enemy detachment attacking Hamina. The Russians lost nine planes. This was the final engagement of HLeLv 24 in the Continuation War. 1Lt Väinö Suhonen led four *Mersus* of the 3rd Flight, and he reported the fight:

"At 1155-1305 hours. The swarm I was leading first met five Il-2s, escorted by four fighters, north of Lavansaari heading north. After this we saw four more similar detachments about one kilometre apart. Thus the enemy numbered about 30 Il-2s and some 20 escort fighters. I battled the Il-2 formations all the way from Lavansaari to Hamina and back to Lavansaari, but had to break off my attack many times because of enemy fighters. I bagged my first Il on the way in and the latter two as they were heading home. Enemy fighters did not cause much stress, although I saw them above me, until at Lavansaari after they turned to go home. My plane was MT-461."

A pair of *Mersus* of 2/HLeLv 24 were the last of the squadron to fly a combat mission when they reconnoitred the Viipuri region on the morning of 3 September 1944.

On 4 September 1944 the commander of the Air Force ordered the air regiments to tell squadrons to cease fighting at 7 am, ending the Continuation War. A ceasefire commenced and two weeks later it was confirmed by The Moscow Armistice.

Brief Analysis

HLeLv 24 flew almost 12,000 sorties during the Continuation War, claiming 743 enemy aircraft destroyed. The squadron lost 40 fighters, 27 in combat, 4 in air raids and 9 in accidents. 21 pilots were killed and 5 become prisoners.

1944 aces of HLeLv 24:

Rank	Name	Flight	Victories	Plane	Total victories
Capt	Wind, Hans**	3	36	MT-439	74½
1Lt	Puro, Olavi	2	27½	MT-449	33
SSgt	Vesa, Emil	3	21	MT-460	30½
1Lt	Saarinen, Jorma †	2	18	MT-452	23
SSgt	Järvi, Tapio	2	16	MT-450	25½
Sgt	Halonen, Eero	2	16	MT-202	17
1Lt	Suhonen, Väinö	1	15	MT-238	19½
MSgt	Katajainen, Nils*	3	15	MT-462	34½
1Lt	Riihikallio, Eero	2	10	MT-213	16½
2Lt	Lampi, Heimo	1	9	MT-464	14
1Lt	Laitinen, Ahti	3	8	MT-441	10
MSgt	Huotari, Jouko	3	8	MT-440	17½
WO	Pyötsiä, Viktor	1	6½	MT-235	21½
1Lt	Nyman, Atte	2	6	MT-465	6
1Lt	Teromaa, Erik	2	6	MT-470	19
1Lt	Lumme, Aulis	2	5	MT-206	16½
SSgt	Ahokas, Leo	3	5	MT-437	12
Sgt	Keskinummi, Kosti	3	5	MT-443	5½
Sgt	Koskelainen, Arvo	1	5	MT-455	5

* Mannerheim Cross, ** Mannerheim Cross twice, † *killed in action*

This table contains victories in the squadron only in 1944. They are rounded to nearest half. The right hand column shows the pilots all victories.

Late model Bf 109G-6 coded MT-504 of 1/HLeLv 24, parked at Utti in September 1944. The tactical number, yellow 1, is located on the fuselage between the cockpit and the machine-gun bulge.

Mersu MT-483 assigned to 1/HLeLv 24 leader Capt Aate Lassila at Utti in early September 1944. This plane was a Bf 109G-8 reconnaissance version, but it was flown without the cameras as an interceptor.

Three pilots of 1/HLeLv 24 on Mersu MT-477 at Utti in September 1944. From the left: 2Lt Heimo Lampi, 1Lt Mikko Pasila and 1Lt Otso Leskinen. This plane was assigned to Pasila and it was one of fourteen Bf 109G-6/R6 cannonboats.

The roots of *Lentolaivue 26* go back to Suur – Merijoki near Viipuri. *Erillinen Maalentolaivue* (Detached Land-based Squadron) was formed here on 16 December 1929, being only the second unit operating from land and starting the shift away from seaplane flying, according to the European practise.

In the major reorganization of the air arm on 30 June 1933 the unit names were homogenized and formed into six numbered air stations (*lentoasema*). *Lentoasema 5* was established at Suur-Merijoki and the previous single squadron was enlarged to two, a fighter and a reconnaissance squadron.

In the next organization change on 10 October 1934 the squadrons within the air stations were numbered. Thus the reconnaissance squadron of *Lentoasema 5* became *Lentolaivue 12* while the fighter squadron became *Lentolaivue 26*, commanded by Major Einar Nuotio.

After six months *Lentolaivue 26* became the most modern fighter outfit in Finland, when it received, on 20 March 1935, the first four Bristol Bulldogs. By the end of the next month all 17 acquired from Britain had arrived.

From July 1939 *Lentolaivue 26* began receiving Fokker D.XXIs. In the mobilization starting on 10 October 1939 the squadron had one flight with ten Bulldogs and two flights with five Fokkers each. On 23 October 1939 the ten Fokkers were delivered to *Lentolaivue 24* with their pilots, on a three month assignment to *Lentolaivue 24*.

Winter War

The Soviet offensive on Finland commenced on 30 November 1939. In poor weather the interceptors could not find the attacking Russian bombers.

At this point *Lentolaivue 26* was commanded by Captain Erkki Heinilä and it had ten Bulldogs at its disposal, naming this outfit simply as *Osasto Heinilä* (detachment). *Lentolaivue 26* belonged also to *Lentorykmentti 2*, which was tasked to prevent enemy flying over the Karelian Isthmus. *Lentolaivue 26* was based at Heinjoki and its operational area was Viipuri – Antrea – Valkjärvi – Kanneljärvi – Gulf of Finland.

On 1 December 1939 the first fighter contact was made at 11.45 hours, when six Polikarpov I-16 fighters of 7 IAP (fighter aviation regiment) jumped a Bulldog pair of LLv 26. While one Bulldog became separated the other, BU-64 piloted by SSgt Toivo Uuttu, was left alone to fight the Russians. After scoring hits on one *Rata*, he himself was shot down and crashed at Muolaanjärvi, being injured in the process. Uuttu's victim also came down and became the first aerial victory ever over Finland.

In the afternoon the Fokkers of LLv 24 took off in pairs for 59 sorties led by the squadron CO, Maj Gustaf Magnusson, sending down eleven bombers in the Viipuri – Lappeentanta area. Eight of these bombers came from 412 SBAP (fast bomber aviation regiment) and three from 24 SBAP. One Tupolev SB

Pilots of 1/LLv 26 on front of a Bulldog, at Suur-Merijoki on 13 August 1939. From the left: 2Lt Aarne Alitalo, 2Lt Tapani Harmaja, flight leader Capt Erkki Heinilä, Sgt Valio Porvari, 1Lt Paavo Berg and MSgt Jussi Tolkki.

Lentolaivue 26 was equipped with Bristol Bulldogs, lined up here at Suur-Merijoki for an aviation parade on 3 August 1935. From the left: BU-68, 69 and 70. Individual planes were distinguished by different colour propeller spinners. (Finnish Air Force)

LLv 26's intermediary equipment was the Gloster Gladiator. Here is one coded GL-258 of 2/LLv 26 at Ruokolahti in February 1940. It was shot down over the Karelian Isthmus on 25 February 1940, wounding the pilot Cpl Matti Sukanen.

bomber was entered in the records as shot down by 1Lt Jussi Räty on assignment from *Lentolaivue 26*.

On 6 December 1939 Major Raoul Harju-Jeanty took over command of the squadron. He immediately began preparations at Utti to prepare the unit, first to receive FIAT G.50 fighters bought from Italy and then train the personnel accordingly. But Germany prevented the FIATs being transported through their country and LLv 26 had to wait six weeks for their intermediary equipment, the Gladiators.

On 10 December 1939 *Lentorykmentti 2* gave orders to LLv 26 protect the IV Army Corps, which was fighting north of Lake Ladoga. For this purpose LLv 26 flew its six Bulldogs to Käkisalmi a week later.

The approaching winter, with snow storms, prevented flying until 19 December 1939. Then *Lentolaivue 24* flew to the Karelian Isthmus, flying 58 sorties and being engaged in combat on 22 occasions. The Russian lost seven SB bombers, of which six came from 44 SBAP, and five DB-3 bombers from other regiments. The squadron adjutant took notes of every combat. 1Lt Urho Nieminen, on assignment from LLv 26 reported:

"I took off at 1025 hours with four Fokkers in the Viipuri direction. At Äyräpää over the railway from Leningrad I observed seven SB bombers in a formation at 2,000 m altitude. We were 500 m higher. We turned to the right from behind in a shallow bank.

I took the plane at extreme left as my target and shot a burst to the fuselage while 2Lt Malmivuo fired over me bursts to the left engine, setting it in fire.

I left this plane and went to chase two planes, which separated from the formation and flew south along the railway. I fired the first burst from pretty far and after that several bursts to both engines. Both engines caught fire. I chased the plane up to Muolaanjärvi following its burning and descent. I saw the burning plane crash at 1050 hours between the stations of Äyräpää and Perkjärvi, just on the east side of the railway.

The planes were fast and only that they made a full evasive circle and our altitude advantage made catching them possible.

After the first bursts the landing gear came down from both planes. The fire in the engines did not increase even if I hit with several bursts. I saw the engines of two other planes on fire. Flying time 50 minutes. My plane was FR-111."

On 23 December 1939 1Lt Pentti Tevä's Bulldog pair of *Osasto Heinilä* scrambled from Käkisalmi. On the coast of Lake Ladoga it met a 13-plane squadron of 24 SBAP. In the ensuing encounter Tevä shot down one bomber while WO Lennart Mildh damaged another.

During the day *Lentolaivue 24* had 21 clashes and in fighter duels both 7 IAP and 64 IAP lost two I-16 fighters. From LLv 26 assigned pilots, Sgt Pentti Tilli claimed the former two and 1Lt Urho Nieminen one of the latter. *Lentorykmentti 2's* combat summary told thus of Tilli's claims:

"Between 1050 and 1120 hours Sergeant Tilli flew with 1Lt Luukkanen's swarm on an intercept mission south-east of Viipuri. At Heinjoki the swarm destroyed one bomber out of a three-plane patrol. After this Sergeant Tilli was engaged in a combat with several fighters, firing at many. One flipped on its back, began to draw smoke and crashed between Kämärä and Heinjoki.

When returning home he saw two I-16 fighters, believing these at first to be friendly (I-16s on skis) and was engaged in a duel, where he shot one down. The I-16 fell on its side and crashed in the Noskuanselkä area. Sergeant Tilli fired at the other at 50-100 m distance, but due a gun jamming and running out of fuel he made a successful forced landing at Kärstilänjärvi. His plane was FR-103."

On 27 December 1939 the Bulldog pair of 1Lt Paavo Berg of *Osasto Heinilä* was engaged in a combat, which the Russian archives later revealed that 68 IAP lost one I-16 fighter, as described by the *Lentorykmentti 2* summary:

"At 12 am Sergeant Porvari entered combat with 1st Lieutenant Berg over Käkisalmi with twenty I-16 fighters, when Sergeant Porvari got an opportunity to fire at one from 50 m distance. The I-16 dodged with a manoeuvre similar to a combat Immelmann, which was cut as the plane flipped into a dive. After a couple of spins the plane flew straight again. Sergeant Por-

The actual fighter for LLv 26 was the FIAT G.50. Here is FA-32 still in Sweden, after assembly and undergoing gun harmonization. This plane was flown to Finland on 12 March 1940, just missing the Winter War.

vari could not follow it longer due to the attack of other I-16s. His plane was BU-68."

By now it was clear that the Bulldogs were obsolete for front-line duties. On 28 December 1939 *Osasto Heinilä* flew the five remaining Bulldogs to Parola and three days later to a frozen lake at Littoinen, close to Turku. Their duty was to defend Turku and Pori harbours.

By the end of 1939 LLv 24 had destroyed 54 enemy aircraft for the loss of one Fokker. Of these victories twelve were recorded to pilots on assignment from *Lentolaivue 26. Osasto Heinilä* had got three victories in addition to two damaged planes, also for the loss of one of its own.

During the first two weeks of the new year *Osasto Heinilä* flew a number of fruitless missions. On 18 January 1940 the first four out of thirty (13 in January and 17 in February) Gloster Gladiator fighters reached *Lentolaivue 26.*

19 January 1940 saw the first ace of *Lentolaivue 26.* 1Lt Urho Nieminen, assigned temporarily to LLv 24, shot down over the Karelian Isthmus one SB bomber of 24 SBAP, as his fifth kill. On the following day another member of LLv 26, SSgt Pentti Tilli, claimed an SB bomber of 21 DBAP, his fifth victory. Sadly, shortly thereafter Tilli was engaged in a fighter duel with I-16s and died after being shot down in FR-107.

The State Aircraft Factory at Tampere was a frequent target for the Soviet bombers. Local fighter defence was organised by test pilots, occasionally assisted by front-line pilots picking up repaired aircraft. Such as in the afternoon of 29 January 1940, as reported by 2Lt Olli Puhakka, a member of LLv 26, showing exceptional gunnery skills:

"After an intercept with 1Lt Visapää we were about to land at Tampere airfield. He was about 200 metres ahead of me and almost above the runway, when we observed a large twin-engined aircraft flying a transverse course to us.

1Lt Visapää fired instantly from the side, but was left behind me in the climbing turn while I remained 500-700 metres behind the enemy plane. In the chase I noticed that the distance rather increased that the opposite. I started to fire short bursts. The first one went below. Judged by the cannon shell explosions I seemed to hit with the second burst around the port engine and the third burst hit the starboard side.

Probably either engine was damaged, because I instantly caught the plane, but now the guns did not work any more. After a few dives towards the plane, I left it between Lempäälä and Viiala, assuming that 1Lt Visapää on my starboard side would attack it, but he had lost it in a cloud almost simultaneously.

I assume that the shooting was seen by 1Lt Itävuori and 1Lt Visapää. The force landed plane was probably taken into custody by T-LentoR 2, so information from there would tell whether the landing was caused by cannon shells or mg bullets, the latter could also have been fired by 1Lt Visapää.

I flew FR-76 equipped with a 20 mm cannon in each wing, but only one was working. I used all 18 rounds in the drum. The fuselage mgs had fired about 60-70 rounds, when the synchronising gear broke down."

This aircraft was an Ilyushin DB-3M of 53 DBAP and the crew was captured after a forced landing at Urjala. The cannon hits gave credit for the downing to Puhakka.

On 31 January 1940 six of the ten pilots assigned to LLv 24 from LLv 26 returned to their unit, now equipped with Gladiators. Next day *Osasto Heinilä* was disbanded. The four remaining Bulldogs were sent to training units and the pilots converted to Gladiators.

The order of battle of *Lentolaivue 26* was then:
Squadron commander Major Raoul Harju-Jeanty, Utti
- 1st Flight, 1st Lieutenant Pekka Siiriäinen, Utti (4 GL)
- 2nd Flight, 1st Lieutenant Ensio Kivinen, Utti (7 GL)

The task was to protect the important road and rail junction at Kouvola and nearby industry.

After two weeks of familiarization flights, LLv 26 opened the Gladiator score on 2 February 1940 when 1Lt Paavo Berg of the 1st Flight fought against six I-15bis fighters of 38 IAP, shooting down one en route to their base. His combat report stated:

"At 1040 hours I saw at 2,000 m altitude, myself at 3,000 m, six aircraft arriving from the south, believing that these were SB bombers. I commenced the attack from behind. During the dive I observed that these were of a strange type and probably fighters. I approached from behind and started shooting, when both 3-plane patrols scattered and when looking back I saw three similar planes diving towards me. I dodged them and entered in a curve battle with them.

At the beginning it was easy, when there were several planes, but in the end when there were only three left, the dodging and shooting began to feel difficult, as there was always one behind my tail. I got then one pouring smoke and it departed towards Tammisaari. After one evasion I observed I was alone and the remaining two enemies were flying southwards. I did not go for a chase, but instead flew circles as I suspected a surprise from behind.

According to the local air surveillance boss there were about 20 enemy planes. One had landed on the ice between Bromarv and Petu and was in a pretty good shape. The planes had similar speed and agility as the I-16. My plane was GL-263."

Later SSgt Oiva Tuominen of the 2nd Flight chased two SBs and six I-16s to the Gulf of Finland, sending one fighter down at Kotka and another near the island of Suursaari, describing it thus:

"At 1520-1615 hours. Between Inkeroinen and Elimäki I caught two SBs and six I-16s, but could not attack the bombers as the fighters targeted me. Then I got to fire at the first I-16 from straight ahead, when the plane went into a spin and disappeared from my sight.

When shooting at the second fighter the others went out of my sight. I saw one flying towards Kotka. I caught it outside Kotka and fired from straight behind, when it landed on the ice. When the pilot tried to escape I shot him on the ice.

Then one I-16 got behind me, fired and turned to the sea. While it curved I caught it at the north tip of Suursaari, fired and it crashed on the ice.

The speed of the I-16 did not appear to be so great because with full throttle I caught them quite easily, starting from 1,000 m distance. I-16 is not so agile as the Gladiator. The first I-16 was on wheels, gear in. My plane GL-258 was intact."

On 3 February 1940 the *Lentorykmentti* 2 commander stated that the main task is to intercept the bombers. Front-line missions and transport escorts were covered by separate orders. LLv 26 to transfer within two days one flight (*Osasto Kivinen*) to Mensunkangas and begin protecting the IV Army Corps north of Lake Ladoga. Newly arriving Gladiators will stay at Utti.

On 11 February 1940 three Gladiators of LLv 26's *Osasto Kivinen* fought with fifteen I-16s of 49 IAP north of Lake Ladoga and claimed one destroyed.

The first FIAT G.50 arrived with LLv 26 at Utti. Thirteen more were to follow during the course of the month. They were formed into *Osasto Nieminen*, headed by 1Lt Urho Nieminen.

Next day *Osasto Kivinen*'s Gladiator pair of LLv 26 intercepted north of Lake Ladoga eight SB bombers of 18 SBAP and sent two down.

On 13 February 1940 six Gladiators of LLv 26's *Osasto Kivinen* were in combat with I-153s north of Lake Ladoga when nine SBs of 39 SBAP arrived on the spot, simultaneously with WO Lautamäki's pair which, undisturbed, attacked the bombers shooting four down in quick succession. SSgt Oiva Tuominen reported:

"At 1400-1430 hours. When I was patrolling with WO Lautamäki in Jänisjärvi station area, I noticed nine SBs arriving east of Suojärvi and heading west. I signalled WO Lautamäki and turned towards the enemy planes. The enemy formation banked to the east and east of Soanjoki I caught them and shot the port wing aircraft into fire. It crashed in the woods. So did the next one after a minute. And the third crashed into a small pond.

Then nine more bombers arrived from the Loimola direction joining the others. At first I thought they were fighters as the distance grew a little, but I caught them over Kivijärvi and shot at the starboard wing aircraft. When I fired a second burst it crashed in flames on the north bank of Kivijärvi lake.

Gladiator GL-270 belonged to 2/LLv 26 until 12 February 1940, when it was transferred to 1/LLv 12, assigned to the flight leader Capt Auvo Maunula (future Mannerheim Cross holder). The photo was taken at Karhusjärvi in early March 1940.

At the same time an I-15 took off from the ice, I shot it down immediately at the edge of the forest, where it crashed on fire. My fighter had two bullet holes in the wings fired from the ground. My plane was GL-255."

SSgt Tuominen's share was three and a half SB bombers plus a solitary I-15 and thus he became the first Gladiator ace. Other Gladiators then attacked the bombers and claimed another two destroyed.

On 18 February 1940 LLv 26's *Osasto Siiriäinen* destroyed two 54 SBAP bombers out of forty near Kouvola. They were credited to 1Lt Paavo Berg, reporting this way:

"At 1115-1145 hours. I met south of Kouvola an SB formation consisting of 40 bombers. I attacked the plane at the extreme right wing, which started to pour black and white smoke.

I received then fire from the side and dodged (probably I-153). In this manoeuvre I got in the middle of a 9-plane SB formation, of which I managed to shoot the left engine of one bomber into smoke. I shot also at others at close range, but without visual effect. The defensive fire was so dense that I evaded downwards, putting me behind the SB formation again.

There was one plane already smoking and left behind. I managed to cut the curve of the SB formation and caught it easily. The machine-guns had stopped working and I stayed to wait for the next plane. When it came close I saw that the left engine had stopped and the crew was alive because they were shooting at me. Since the guns stayed jammed I returned to base.

According to the air surveillance report one plane hit smoking the forest south of Kouvola and one on fire north-east of Kouvola. My plane GL-264 had twelve bullet holes."

On 19 February 1940 three Gladiators of LLv 26 engaged several fighters during a chase of a 32-plane strong bomber formation. Two I-153s of OAE (detached aviation squadron) were claimed shot down, one by 1Lt Paavo Berg. Next day saw the second Gladiator ace. 1Lt Berg of LLv 26 took off to attack 30 SB bombers of 6 DBAP with Kouvola as the target. He sent one down but was shot down by the defensive fire and bailed out, suffering burns.

25 February 1940 was a costly day on both sides on the Karelian Isthmus. LLv 26 sent three Gladiators to drive off nine R-5 artillery fire control planes, escorted by six I-153s of 13 OIAE. After downing four R-5s in the ensuing combat with 68 IAP *Chaikas*, two Gladiators were shot down and a third damaged in a forced landing.

FIATs

Urgently needed on the front, LLv 26 used only ten days in FIAT training before the baptism of fire. On 26 February 1940 1Lt Olli Puhakka took his pair to combat bombers and fighters south of Kouvola. He shot down one I-16 of 49 IAP and his wingman 2Lt Kauko Linnamaa sent down one DB-3 bomber of 1 AP, KBF (Baltic fleet). Puhakka's first FIAT combat report stated:

"At 1440-1520 hours. I chased DB-3 bombers having bombed Kouvola. North of Kouvola I observed below me three I-16s attacking a lone Gladiator. When I dived on the scene I saw only two I-16s, which I attacked, pulling up after every dive. I managed to score hits in both, especially one received long bursts from behind and above behind. After one pull up I saw it exit in a glide. Also the other disappeared from my sight as I had to pay attention the enemy planes above me.

After that I chased two I-15s, firing at one several bursts straight from behind. Finally it appeared that its engine was not running properly. It flew straight in a shallow dive towards the sea while the other tried to protect it. I left it at the coast between

These five pilots of 3/LLv 26 on 25 June 1941 downed ten Tupolev SB bombers of 72 SBAP. Seen in front of FA-11 at Joroinen, from the left: Sgt Ilmari Pöysti, SSgt Onni Paronen, MSgt Valio Porvari, flight leader 1Lt Urho Nieminen and 2Lt Sakari Kokkonen. (SA-kuva)

FIAT G.50 FA-15 of 2/LLv 26, flown by Sgt Sulo Suikkanen in June 1941, based at Joroinen. (Karolina Hołda)

FIAT FA-15 of 2/LLv 26 at Rantasalmi on 28 June 1941. The assigned pilot was Sgt Sulo Suikkanen, who is sitting here on the cockpit sill. He almost became an ace, with 4½ aerial victories.

Kotka and Pyhtää, when 15 fighters approached from Suursaari direction.

In combat against the I-16 fighters participated another FA plane, flown by 1Lt Nieminen. The GL plane was probably piloted by Cpl Paunu.

My plane FA-4 had a few bullet holes, the oxygen bottle punctured among other things. Especially a Fiat with enclosed canopy is easy to take by surprise, since the visibility to the sides and rear is extremely poor."

On 28 February 1940 *Lentorykmentti* 2 gave the following instructions to LLv 26: 1) attack the highest flying escort fighters from above, 2) doing this in a group, 3) gaining altitude by pull up and 4) aiming to create maximum losses to the fighters. This method was believed to give better access to attack the bombers.

February 29 1940 was the saddest day for the fighter arm, when Soviet fighters carried out a series of air raids on *Lentolaivue* 24 and 26 bases. Detachment *Luukkanen* of LLv 24 had three weeks earlier moved to Ruokolahti and two Gladiator flights of *Lentolaivue* 26 were there put under its command.

68 IAP fighters had already shot down one Gladiator in the morning. At noon a bomber formation was announced to be approaching Ruokolahti, but it turned out to be the 68 IAP

with six *Chaikas* and eighteen *Ratas*, which took by complete surprise the Gladiators taking off. Three were instantly destroyed and in a low-level combat a further two Gladiators plus a Fokker D.XXI went down. Only one I-16 was shot down and another hit the trees while manoeuvring at low altitude. Though the Finnish loss of six fighters was heavy, the Soviets tripled them as the 68 IAP claimed ten Bulldogs and eight Fokkers shot down.

Kouvola road and railway junction was under attack by 132 aircraft during the course of the same day. FIATs of LLv 26 were given the task to prevent these bombardments. In the morning a FIAT pair had destroyed two DB-3 bombers from a 9-plane group. Later they engaged a 9-plane SB formation of 57 AP, KBF heading towards the town and shot down one and damaged another.

On 1 March 1940 LLv 26 delivered the remaining Gladiators to LeR 1 squadrons LLv 12 and 14 and became fully equipped with the FIAT G.50 fighters.

From 4 March 1940 the Russian managed to cross the ice at Viipurinlahti and form a couple of bridgeheads in the Finnish rear. Though the Fokkers of *Lentolaivue* 24 flew most of the strafing missions, LLv 26 participated a few times on these attacks.

On 9 March 1940 at noon six Moranes of LLv 28 led by 1Lt Veikko Karu and a Fiat pair of LLv 26 led by 1Lt Olli Puhakka attacked a truck convoy between Koivisto and Reposaari, shooting up a few vehicles. In the afternoon Capt Eino Jutila's six Moranes and 2Lt Kauko Linnamaa's FIAT pair strafed the same convoy. Linnamaa observed two I-153s harassing a Morane, firing at one straight from behind and then in from the side in a bank, when it crashed at the shore line at Revonsaari in a flash. Right after he fired at the other from straight ahead, but could not observe the effect.

On 11 March 1940 Lentolaivue 26 carried out ten strafing sorties to Viipurinlahti. In the defence of Kouvola the FIATs recorded LLv 26's last air victory in the Winter War, as told by 1Lt Olli Puhakka:

"At 1340-1420 hours. Just north of Kouvola I attacked an enemy formation, which consisted of five DB-3 bombers. I followed them south to Elimäki attacking the plane at the right wing. After a long burst the right engine appeared to catch fire, but went soon out. The plane began to leave a dark trail of smoke, remaining in its formation.

I moved to go after the plane at the left wing because it had separated a little from the formation. I first shot in the fuselage, which got several hits judging by the flashes of the incendiary bullets. After that I fired at the left engine, which began to pour white smoke.

I ran out of ammunition and banked away. Then flying higher and behind me two I-153s went in a dive to attack me, but I broke away by diving. At the beginning MSgt Tuominen in my patrol was attacking the same bombers. Ammunition spent 600 pieces. My plane was FA-21."

The Winter War ended on 13 March 1940 at 11.00 hours, having lasted 105 days.

Brief Analysis

Lentolaivue 26 fighters had flown 1,170 sorties (excluding Fokker sorties by LLv 26 pilots with LLv 24), claimed 73 aircraft shot down (BU 4, FR 27, GL 34 and FA 8) and lost 12 aircraft and 7 pilots in combat.

Winter War top scorers of *Lentolaivue 26*:

Rank	Name	Flight	Victories	Plane
SSgt	Tuominen, Oiva	2	9	GL-255
1Lt	Puhakka, Olli	3	7	FR-117
1Lt	Nieminen, Urho	3	6	FR-111
1Lt	Berg, Paavo	1	5	GL-279
SSgt	Tilli, Pentti	2	5	FR-103
Sgt	Aaltonen, Lasse	3	4	FA-4
Cpl	Joensuu, Ilmari	2	4	GL-256

Between the wars, as in all other squadrons, training continued. The new base was set at Joroinen, where the Fiats were flown on 29 and 30 May 1940. As the G.50 was a very difficult plane from a servicing point of view (engine used castor oil for lubrication, among other things), the technical personnel had quite a challenge, making the machines work satisfactorily within a year.

Continuation War

The German offensive on the Soviet Union, Operation Barbarossa, was revealed to the Finnish high command four weeks before its launch. Keeping this mind a full-scale mobilization commenced five days earlier, on 17 June 1941. *Lentolaivue 26* was also made combat ready.

The order of battle of *Lentolaivue 26* was then:
Squadron commander, Major Raoul Harju-Jeanty, Joroinen
- 1st Flight, 1st Lieutenant Mikko Linkola, Joroinen (7 FA)
- 2nd Flight, Captain Ensio Kivinen, Joroinen (9 FA)
- 3rd Flight, 1st Lieutenant Urho Nieminen, Joroinen (10 FA)

The task of *Lentorykmentti 2* was defined as the protection of the mobilization of the field army and traffic junctions. The area for *Lentolaivue 26* was specified as Huutokoski-Simpele-Savonranta-Huutokoski, in case the army would remain in defence. If it would go on the offensive the area would be Jääski – Käkisalmi – Sortavala – Soanlahti – Korpiselkä – Joensuu – Sulkava – Puumala – Jääski.

On 25 June 1941 the first observations of large bomber formations entering Finnish air space were made at Turku at 6 am. During the day the Soviets flew over Finland with 263 bombers and 224 fighters. The Continuation War had started.

LLv 26 was flying combat air patrol all morning. A swarm of 1Lt Lauri Hämäläinen's 2nd Flight has just landed at Joroinen, when fifteen bombers of 72 SBAP attacked the airfield from 1,000 metres at 1145 hours. With almost empty fuel tanks two Fiats took off again in the bomb rain and rapidly caught the SB bombers, shooting down three.

1Lt Urho Nieminen led his 3rd Flight past the bombers without noticing them, when his six FIATs received a radio message to return. Having the advantage of higher altitude the Finns struck at 1155 hours and in twenty minutes downed ten SB-bombers. Nieminen described his encounter:

"I engaged with my flight a 15-20 aircraft strong SB formation above Tuusniemi. With the speed gained from the altitude we got straight into the business. I shot at two aircraft on the right flank (both engines on fire). I improved two smoking cases. At Kerisalo Island I got again into a favourable position behind one SB. Its fuel tanks exploded in the air. The aircraft crashed. I continued to fire at the remaining aircraft. I ran out of ammunition, 4-5 aircraft were still flying in the formation, also the smoking one I had fired at. The planes made independent evasive manoeuvres within the formation and the composition of the formation changed all the time. My plane was FA-11."

On 29 June 1941 the air force commander gave an order according to which LeR 2 was to fly top cover for the 100,000 man strong Karelian Army, which was formed to take back the area lost in the Winter War. Army co-op squadrons LLv 12 and 16 were subordinated for the use of the Karelian Army and bombers of LeR 4 were to be used where needed. *Lentorykmentti 3* would remain in the defence of southern Finland. The main offensive was planned to occur north of Lake Ladoga. The LeR 2 commander designated the new areas of operations, which were for LLv 26 Joensuu – Tohmajärvi – Pälkjärvi – Soanlahti – Korpiselkä.

On 4 July 1941 MSgt Oiva Tuominen of 1/LLv 26 scrambled to encounter an SB bomber formation of 72 SBAP attacking Joensuu. At 1,000 m altitude he attacked as follows:

"At 1100-1120 hours. I observed the aircraft shortly after the release of the bombs and flak explosion clouds. I came in just under the clouds. I flew against them and hit the left wing aircraft straight into the engine and thereafter from 50 metres behind, when it caught fire. I fired at the second a bit from the port side, when it began to dive heavily smoking. Then I shot at three aircraft, two of

which began to smoke heavily and at the same time I got an explosive shell in my own fuel tank. Thereafter I shot the gunner who had fired at me, when the aircraft dived smoking into the clouds. Only the other gun worked, since the ammunition belt snapped off after the first burst from the other gun. My plane was FA-3."

To Karelia

At this point the Soviet land forces had the 23rd Army on the Karelian Isthmus and the 7th Army north of Lake Ladoga, with front responsibility continuing up north to Uhtua. The air forces of the 23rd Army consisted of 5 SAD (7 and 153 IAP plus 65 and 235 ShAP. 65 ShAP was transferred in August 1941 to 7th Army air forces) and respectively the 7th Army air forces 55 SAD (72 SBAP, 155, 179, 197 and 415 IAP plus from August 1941 on 65 ShAP).

On 10 July 1941 the offensive of the Karelian Army commenced from the Kitee-Ilomatsi area towards the north-western coast of Lake Ladoga. The CO of LeR 2 specified the operational areas and tasks for its squadrons. The sector of LLv 26 was Pälkjärvi – Kakunvaara – Värtsilä – Saarivaara – Vuotjärvi – Petäjäjärvi and it was to escort planes of LLv 12 and LLv 16 on their missions.

On 13 July 1941 1Lt Olli Puhakka's pair of 3/LLv 26 shot down near Jänisjärvi a solitary SB of 72 SBAP in the early afternoon. Puhakka described his later evening clash, actually with MiG-3s of 5 AP, KBF, this way:

"At 1835-1905 hours. I took off to chase bombers passing the airfield. I lost visual contact with them and saw 5 or 6 small monoplane fighters circling between Värtsilä and Pälkjärvi at 1,000 metres altitude. First I thought they were friendly aircraft

Four pilots of 2/LLv 26 in front of FA-35 at Lunkula in August 1941. From the left: Cpl Klaus Alakoski, Sgt Ilmari Joensuu, 1Lt Lauri Hämäläinen and SSgt Sulo Suikkanen. FA-35 was the assigned plane of SSgt Lasse Aaltonen.

LLv 26 top brass in front of FA-33 at Lunkula in early August 1941. From the left: 3rd Flight leader 1Lt Urho Nieminen, squadron CO Maj Raoul Harju-Jeanty and 1st Flight leader 1Lt Mikko Linkola.

having already shot down the bombers. I went right next to them to see and noticed the red stars on one and tried to get into a firing position. The others began then shooting at me. I tried to get above them, but they followed without difficulty. After that I tried to break off by diving, but they caught me right at the surface. I tried to turn against them, but there was always one who got behind my tail. I managed to shoot at one from ahead with a small deflection. The aircraft went east in a shallow glide at 100 metres slightly smoking near Havuvaara,

I made steep banks and controlled false manoeuvres. When I made one bank so tight that I ended flying on my back at only a few tens of metres altitude and managed to level right at the tree tops, I saw that the one that followed me could not level from the false manoeuvre and instead crashed uncontrolled into the forest.

I thought at first that the type was I-16. It appeared to have a radial engine and a similar wing to I-16, but the nose looked slimmer from the front. The aircraft was faster and climbed better than the Fiat. The enemy planes dropped their wing tanks before the combat. During the combat my plane had probably one main wheel partly out. My plane had three bullet entry holes. Ammunition expenditure about 20 rounds. My plane was FA-1."

While the swarm of 1/LLv 26 was flying combat air patrol on 14 July 1941 over Suojärvi, MSgt Oiva Tuominen observed three SB bombers of 72 SBAP, sending two down:

"At 0930-1030. When patrolling in the Tolvajärvi-Loimolanjärvi area three SBs came from the Suojärvi direction. I began firing, one started promptly smoking and all turned back to Suojärvi. I chased them, when the smoking aircraft caught fire, two Russians bailed out and the plane fell into a lake.

I shot also at a second one and it dived into the same lake. Then three I-16s arrived, but they evaded into the clouds after a couple of turns. The bomber on the left wing laid out some 100 metres of cable, which swung very much behind it. Combat height 1,800 metres. My plane was FA-26."

Four pilots of 2/LLv 26 in front of FA-35 at Lunkula on 4 September 1941. From left Cpl Kauko Tuomikoski, 1Lt Lauri Hämäläinen, Sgt Tage Bergman and Sgt Ilmari Joensuu. (SA-kuva)

Four FIATs of LLv 26 have landed to Lunkula on 4 September 1941. The planes are from left FA-3, 35, 27 and 6 and they were flown by the four pilots above, on a mission to River Svir direction. (SA-kuva)

On 24 July 1941 the VI Army Corps of the Karelian Army reached the intermediary target at the Tuulosjoki line on the east coast of Lake Ladoga and stopped.

On 30 July 1941 WO Oiva Tuominen's pair of 1/LLv 26 engaged on a search mission over Olonets for two SB bombers of 72 SBAP, sending both down. Tuominen related:

"At 0835-0920. I flew toward Vitele and observed two SB bombers between Vitele and Tuulos, flying away 1,000 metres higher towards the sun. I caught the planes about 10 km east of Tuulos. There was no gunner in the wing aircraft and I shot a burst straight over the rudder, when it pulled straight up and went into a spin.

I shot two bursts at the gunner of the other and after that from above a burst to the back of the pilot and the same followed as with the first aircraft. The armour piercing bullets appeared to go right through the pilot's armour plate. Combat altitude 4,200 m. My plane was FA-26."

On 1 August 1941 1Lt Mikko Linkola's pair of 1/LLv 26 reconnoitred in the Säämäjärvi area, where one I-16 of 7 IAP was shot down. Later 1Lt Urho Nieminen's swarm of 3/LLv 26 scrambled for an intercept to Tuulos, where three *Chaikas* of 197 IAP were patrolling. One of these was destroyed. Two days later 2Lt Lauri Sihvo's pair of 1/LLv 26 encountered at the mouth of River Svir three Beriev MBR-2 flying boats of detachment Hrolenko, sending all three down.

On 5 August 1941 1Lt Olli Puhakka of 3/LLv 26 led six FIATs on a scramble mission to Tuulos, where three I-15*bis* and three I-153 aircraft of 65 ShAP (assault aviation regiment) were harassing the Finnish troops at the front-line. In a short combat all I-15*bis* aircraft were destroyed.

On 13 August 1941 LLv 26 flew escort for an artillery fire control plane over Aunuksenjoki. At 2 pm, when the swarm of 1Lt Lauri Hämäläinen was changing the shift with the swarm of 1Lt Olli Puhakka, nine *Chaikas* (of 195 and 197 IAP) appeared to harass the fire control aircraft. In five minutes the FIATs shot down all enemy aircraft. Puhakka damaged one and shot down another, but the air force HQ credited them as two destroyed for him:

"At Hoski I saw two I-153s at about 200 m altitude, I attacked and shot the latter quite close and saw the bullets hit the fuselage. Then I saw two more above me and pulled up. When I looked back to the scene I saw already two Fiats having attacked and saw one I-153 parting company and sliding to the south-east. It came down at Alavoinen catching fire.

After that I tried to get behind the tail of another. The other two were already fired at by other Fiats. The whole battle occurred in a small area. When I finally got behind one and shot very near, I saw after pulling up and going down again that no Chaikas remained. My plane was FA-1."

On 15 August 1941 Capt Urho Nieminen flew with a Brewster – as a reinforcement to 3/LLv 26 – out to Lake Ladoga on an alert mission. In the area two *Chaikas* escorted a lone SB bomber. Nieminen fired at both fighters, which broke off and after that he shot the bomber down. In the evening WO Lauri Lautamäki's pair of the 1st Flight was scrambled for an intercept to Olonets, where eight I-153 fighters circled. One was destroyed.

On 18 August 1941 the victory score of Warrant Officer Oiva Tuominen of 1/LLv 26 had gone up to 20 and he became the first Knight of the Mannerheim Cross of the air arm.

On 3 September 1941 the Karelian Army began the advance from Tuulosjoki towards the River Svir. The advance was rapid as the spearhead arrived at the stream four days later.

On 3 September 1941 2Lt Lauri Sihvo's and MSgt Onni Paronen's pairs of LLv 26 reconnoitred the Säntämä – Aunus area, encountering two I-16s and three I-153s. In the ensuing combat the Russians lost all their aircraft. Paronen's combat report stated:

"At 1150-1240 hours. I shot first at one I-16 from straight behind and thereafter in a bank. The plane caught fire and smoked heavily. I pulled away, when WO Tuominen shot at the plane when it crashed to the ground. After this I attacked an I-153 from above and behind. The aircraft did not dodge at all, but crashed to the ground in a shallow bank and caught fire. After this I fired at the last I-153, which was set aflame by 2Lt Bruun. I saw altogether four aircraft being shot down. My plane was FA-6."

FIAT FA-25 of LLv 26 being re-fuelled at Helsinki Malmi in February 1942. The pilot in front of the plane is Sgt Klaus Alakoski, who caused the dent on the cowling on 8 January 1942, banking in the landing and collapsing the port gear.

This was for LLv 26 the last engagement of the year. On 16 September 1941 LLv 26 was separated from *Lentorykmentti 2* and transferred to Immola for a rest and subordinated to *Lentorykmentti 3*. The strength of the squadron had worn down to six Fiats in the battles north of Lake Ladoga. The poor serviceability was caused by the difficulty in keeping the engines running properly.

The operational area was defined as the eastern Karelian Isthmus, where the land forces had arrived on 2 September 1941 to the old border and a stalemate war had started. The main task was to protect the armed force on the Karelian Isthmus and the crossings of the waterways at Vuoksi river. The side task was to protect the industrial area at Vuoksi and traffic junctions on the Karelian Isthmus. The air space over the Karelian Isthmus remained very quiet for the rest of the year.

On 21 September 1941 LLv 26 moved to Sakkola on the eastern Karelian Isthmus. Here one flight carried out flying duties in one month tours while the rest of the squadron did garrison service at Immola.

On 14 December 1941 one flight was transferred to Helsinki Malmi, taking charge of the protection of the capital. The number of flights was reduced to two, which had one month tours in the defence of Helsinki. The rest of the squadron remained in garrison service at Immola.

Brief Analysis

In 1941 LLv 26 had gained 55 aerial victories losing two fighters, of which only one was in combat.

1941 top scorers of *Lentolaivue 26*:

Rank	Name	Flight	Victories	Plane	Total victories
WO	Tuominen, Oiva	1	13	FA-26	13 + 8
1Lt	Puhakka, Olli	3	7	FA-1	7 + 7
Capt	Nieminen, Urho	3	6	FA-11	6 + 6
WO	Porvari, Valio	3	4½	FA-20	4½ + 3
SSgt	Suikkanen, Sulo	2	3½	FA-15	3½ + 1

The right hand column shows the 1941 victories, those gained in the Winter War are after the + sign

1942

During the first four months the air space was very quiet over southern Finland. No contact was made with enemy aircraft. The two flights of LLv 26 took one month tours in the defence of the capital. During the winter the State Aircraft Factory was working hard to bring as many FIATs as possible into working order.

On 23 April 1942 *Lentolaivue 26* was reorganized at Immola. The 3rd Flight was re-established. Fully repaired and overhauled FIATs were received from the aircraft factory and field air depot at Immola. The main task of the squadron was set as reconnaissance and interception on the Karelian Isthmus and the auxiliary task was the protection of Helsinki.

On 3 May 1942 the sector of *Lentorykmentti 3* was specified as the area between the coastline of the Gulf of Finland and the Valamo – Uusi – Laastokka line and the tasks were: 1) aerial reconnaissance, 2) interception of enemy aircraft and 3) protection of larger transports, transfers and bombers. LeLv 26 continued to fly on the eastern Karelian Isthmus and in protection of Helsinki and LeLv 30 on the western isthmus.

On 23 May 1942 LeLv 26 made the first enemy contact of the year. WO Oiva Tuominen's pair of 1/LeLv 26 scrambled towards Loviisa, where one Hurricane was met and shot down.

In June and July 1942 only a few encounters occurred. As an example, on 9 July 1942, when 1Lt Olli Puhakka's swarm of 3/LeLv 26 was sent for interception to Miikkulainen, where four I-15bis fighters were encountered. Two were sent down. Puhakka wrote in his combat report:

"*At 0825-0910 hours, altitude 1,000-50 m. While leading a swarm of four aircraft, based on radioed information, I circled via Miikkulainen to the north. At Tappari I saw four I-15bis coming towards me. I was able to climb higher than them and attacked the last one. Even if the hits were seen on the fuselage, no other specific effect was observed. After the pull-up I attacked the rearmost wing plane. The machine caught fire on the right side behind the engine and the pilot bailed out. I did not see the chute open.*

I instantly attacked the first plane, which went to the deck toward the Miikkulainen river. My guns were out of order. During the pull-up I did not see in the air anyone other than the Fiats and this enemy plane. I repeated the attack, but the guns were still jammed. I saw 2Lt Trontti attack the same enemy aircraft. I pulled up, circled the spot and headed for base, three other Fiats followed me. The speed of the enemy aircraft was about the same as Fiats. They did not take advantage of their good climb rate nor did they turn to the opposite direction in order to get into a firing position. They only dodged with swift turns, losing altitude. My plane was FA-25."

Lake Ladoga

In August 1942 parts of a German-Italian naval detachment had arrived at Lake Ladoga. Its task was to hamper the Leningrad supply traffic in the southern part of the big lake. The detachment operated on Lake Ladoga for three months and the Fiats of LeLv 26 were busy flying top cover.

The order of Battle of *Lentolaivue 26* was then:
Squadron commander, Major Eino Carlsson. Kilpasilta
- 1st Flight, 1st Lieutenant Mikko Linkola, Kilpasilta (7 FA)
- 2nd Flight, 1st Lieutenant Lauri Hämäläinen, Malmi (5 FA + 3 HC)
- 3rd Flight, 1st Lieutenant Olli Puhakka, Kilpasilta (7 FA)

On 24 August 1942 1Lt Olli Puhakka's seven FIATs of 3/LeLv 26 were sent to intercept in the Konevitsa area, where a combat ensued with six Il-2s escorted by four I-16s. The Russian losses were the four fighters. WO Oiva Tuominen's and 2Lt Reino Stenberg's pairs were scrambled to Miikkulainen, where they engaged twelve I-16s in an air battle, the Soviets losing two fighters.

On 26 August 1942 1Lt Olli Puhakka's seven FIATs from the 1st and 3rd Flights were scrambled to Konevitsa, where a formation of seven I-16s and I-153s, six LaGG-3s and four Il-2s was engaged. In a fifteen minute combat the Finns claimed three Soviet aircraft shot down. WO Oiva Tuominen reported:

"*At 1105-1200, altitude 1,800-100 m. On the return the radio at Saunasaari informed us that four planes were coming straight towards us. I observed the planes when the Russian flak shot at them. I climbed at the side. At 1,800 metres I was at the same level as the last plane. I attacked it when all four I-153s came at me. I shot at one during the curving and the altitude fell to 200 m. Three planes gained a little altitude when I was able to shoot at the fourth and it started to pour a little smoke and after a while exploded. The three others attacked me at the same time,*"

FIAT FA-19 of LLv 26 banks over Helsinki Malmi airfield in March 1942. The aircraft wears 1ˢᵗ Flight blue colours. Typical of FIATs, the paint chipping was excessive.

when after rolling a while with them I got to fire one in a bank, when it crashed in a 45° angle to the ground. Two fled to the south. I returned since I was low on fuel. My plane FA-18."

When MSgt Onni Paronen of 3/LeLv 26 was ferrying a serviced FIAT from Immola to Kilpasilata on 30 August 1942, he engaged a thirty plane strong detachment, sending one I-16 down. Paronen reported:

"At 1140-1150 hours, altitude 200-1,500 m. When I was returning from Immola to the base I flew south of Käkisalmi towards Konevitsa. Then I observed the flak of Sortanlahti firing and saw at the same time the enemy aircraft flying to south-east in several separate formations. I soon reached a large formation flying at about 500 m altitude, consisting of 5-10 Il-2s, which were escorted by at least 10 I-16s. I attacked these twice but on both occasions the mgs did not work. I pulled up and made several cocking attempts and after which I got the left gun in a working order. I attacked from the rear and above one plane flying at the side. I managed to shoot at this by surprise from 50 m distance. The target did not dodge and began to go in a shallow dive, turned on its starboard wing and crashed in the forest. Obviously the pilot was hit. After this I turned to the return flight. Above me were all the time 4-6 Hurricanes, but they did not make any attacks. Only one Hurricane fired from a distance of about 500 m and broke off. The I-16s tried all the time to fire in opposite directions and shoot their rockets. My plane was FA-33."

On 21 September 1942 nine FIATs of LeLv 26 took off for an interception mission to Morje. Seven enemy fighters were

FIAT FA-17 of 2/LeLv 26 at Helsinki Malmi in August 1942. It was assigned to 2Lt Lauri Kalkkinen. This flight stayed in the defence of the capital until March 1943. (Finnish Air Force)

1/LeLv 26 at Kilpasilta in late July 1942. On the knee from left: E. Laiho, T. Riivari, E. Jauhiainen, E. Liukkonen and U. Tarvainen. Standing from the left: K. Melander, L. Pasi, R. Karpala, T. Lintunen, x, K. Penttinen, O. Tuominen, M. Linkola, O. Länsivaara, L. Lautamäki, A. Lassila, E. Mynttinen, K. Tuomikoski, x, V. Leskinen, E. Lyykorpi, A. Suolanen, x, x, N. Launiainen, P. Koivunen, E. Frimodig and E. Lehtonen. The pilots are shown in italics. Behind is FA-26.

Five pilots of 2/LeLv 26 sitting on the wing of FIAT FA-17 at Helsinki Malmi in August 1942. From the left: 2Lt Lauri Kalkkinen, 1Lt Lauri Hämäläinen, 2Lt Carl-Erik Bruun, MSgt Lasse Aaltonen and 2Lt Orvo Helenius.

patrolling in the area. Two were downed, both claimed by WO Oiva Tuominen:

"*At 1850-1940 hours, altitude 2,200 m. I took off for interception, when I saw north of the airfield seven enemy fighters. I caught them over Miikkulainen above the clouds. I got to shoot at one I-153 obliquely from behind and from slightly above. I fired two bursts from 50 metres, when it caught fire. At the same time all the Russians pulled into the clouds. I dived after the burning plane below the clouds and saw it fall into the ground.*

I continued my dive in order to get out of the flak fire. When I levelled the dive, one I-15 was a short distance in front of me, I gave a burst from about 20 metres, when it pulled up at once and fell from there in a dive. When it hit the ground it caught fire. When I was watching the fall of the I-15 the others managed to get out of my sight. My plane was FA-26."

On 22 October 1942 1Lt Aate Lassila's six FIATs of 1/LeLv 26 observed two *Chaikas* on a search mission to Lake Ladoga, shooting one down. On the next mission the same formation engaged over Lake Ladoga DB-3 bombers with a fighter escort. A swarm tied the Russian fighters in combat and two enemy fighters were sent down. The bombers were now without escort and one was shot down. WO Oiva Tuominen downed one *Chaika* on the first mission and a *Rata* on the second, reporting the latter:

"*At 1350-1500 hours we engaged on a search mission enemy aircraft ammounting to two bombers and eight I-16 fighters. We attacked with 1Lt Lassila from straight ahead. After turning around some time I saw 1Lt Lassila shoot at one I-16, which went vertically down. I followed it and saw it dive into Lake Ladoga. The I began to chase another, which while dodging came*

Four pilots of 3/LeLv 26 at Kilpasilta on 14 August 1942. From the left: SSgt Ilmari Pöysti, 2Lt Nils Trontti, MSgt Onni Paronen and flight leader 1Lt Olli Puhakka. All but Pöysti were or became aces. Behind is FIAT FA-29. (SA-kuva)

FIAT FA-15 of 3/LeLv 26 about to take off from Kilpasilta on 1 October 1942. This plane was assigned to SSgt Tage Bergman. He later trained to become a bomber pilot. (SA-kuva)

right in front of me. I fired a burst from the side at 30-50 m, when it crashed in Lake Ladoga. I continued to chase three I-16s. I caught them, but then more enemy fighters came from Morje. While dodging them I tried to climb all the time to get in the clouds, because my guns did not work since the pressure bottle was empty. I was able to break off in the clouds and return to the base. They fired air torpedoes (rockets). My plane was FA-26."

After the departure of the Italo-German naval unit the air space over the eastern Karelian Isthmus and Lake Ladoga grew silent and no clashes occurred. The arriving winter had also its impact on flying.

1942 top scorers of *Lentolaivue* 26:

Rank	Name	Flight	Victories	Plane	Total victories
WO	Tuominen, Oiva	1	10	FA-26	23 + 9
Luutn	Puhakka, Olli	3	5	FA-25	12 + 7
Vänr	Trontti, Nils	3	4	FA-13	4
Ltm	Lautamäki, Lauri	1	2	FA-28	4 + 2½
Vääp	Paronen, Onni	3	2	FA-33	4½ + 2

This table shows the victories obtained in 1942

The right hand column shows total victories, with the Winter War score after + sign

Pilots of 3/LeLv 26 at Kilpasilta on 3 September 1942. From the left: SSgt Tage Bergman, MSgt Onni Paronen, 2Lt Nils Trontti, SSgt Ilmari Pöysti, flight leader 1lt Olli Puhakka, 1Lt Lauri Sihvo, WO Valio Porvari and 2Lt Kaarlo Stenberg. (SA-kuva)

1943

The winter months were very quiet on LeLv 26's sector on the eastern Karelian Isthmus.

On 18 March 1943 internal rearrangements were carried out in LeR 3. LeLv 26 was relieved from the protection of the capital and the 2nd Flight was transferred to Kilpasilta, where the whole unit was now based. The squadron's equipment, FIAT G.50 fighters, were by now obsolete.

On 2 May 1943 *Lentolaivue* 26 carried out one of the last clashes with FIATs. 2Lt Nils Trontti's pair of 3/LeLv 26 took off to reconnoitre Lake Ladoga, where they engaged four I-15*bis* planes, later joined by a pair of I-153 fighters. The Russians lost one each of both types. These proved to become the last air victories scored by Fiats, as described by Trontti:

"At 0950 hours we met, as the ground station informed us, four I-15s, which were flying south. They flew at about 200 metres and we attacked from above and behind. In the ensuing battle the enemy remained mostly passive and tried to head towards Morje harbour. After having fought some 15 min I observed two I-153s coming from the south. These took part in the battle over Kaavinankylä.

I saw one I-153 left behind by the others and I attacked it. The plane did not take any evasive actions and I was able to close down to 10-15 m firing all the time. I saw how the bullets hit the plane, but was not able to follow it further since the other I-153 had also dropped back and fired at me straight from the side. I dodged and saw then below me in the forest quite an explosion, so I assumed that the earlier plane was destroyed. Even if I had seen six aircraft during the battle, when departing for home I saw only four, which circled under the cover of Morje flak. The enemy dropped small bombs into the sea during the combat. One hit my plane, FA-29."

After this not many aircraft were seen over LeLv 26's area of operations. Quite luckily, as the FIAT was no match for more modern enemy fighters. On 14 February 1944 the unit was named *Hävittäjälentolaivue* 26 (HLeLv 26).

An attempt to improve the plane situation in the squadron took place on 26 February 1944, when a State Aircraft Factory manufactured *Myrsky* fighter was received for trials. It was a pre-series plane coded MY-4. About twenty pilots managed to get acquainted with the Myrsky, until 17 March 1944 when its wing tips broke off in a dive test and the plane crashed, killing the pilot. This led to the refusal of the type by the fighter squadrons.

FIAT G.50 FA-21 of 1/LeLv 26, flown by 2Lt Reino Sartjärvi in March 1943, based at Kilpasilta. (Karolina Hołda)

Hawker Hurricane I HC-452 of 2/LeLv 26 flown by Sgt Aarno Siro in March 1943, based at Helsinki Malmi. (Karolina Hołda)

2/LeLv 26 also had a few Hawker Hurricanes. Here is HC-452 at Helsinki Malmi, shortly before it was flown to Kilpasilta on 18 March 1943. This plane was assigned to Sgt Aarno Siro.

FIAT FA-28 of 1/LeLv 26 at the end of the runway at Kilpasilta in summer 1943. The plane was assigned to 2Lt Olli Riekki. Though the camouflage pattern was standard to all types, some FIATs were an exception with individual variations.

FIAT FA-19 of 2/LeLv 26 after thorough overhaul and re-painting at the field air depot at Immola in mid-October 1943. This flight was distinguished by the two bars on the rudder. The regular pilot was SSgt Paavo Saarni.

HLeLv 26 on 26 February 1944 received one Myrsky pre-series machine coded MY-4 for trials. However, it broke up in a dive mid-air on 17 March 1944, killing 1Lt Jaakko Marttila, causing the fighter units to reject the type.

New Old Planes

On 8 May 1944 HLeLv 26 received its first nine Brewsters and during the month a further nine. The FIATs were transferred to HLeLv 30 and T-LeLv 35 for advanced trainer use. The gradual deliveries had begun in February and were done, except for one, by 1 June 1944. The last one was given away on 27 June 1944.

The squadron transfer from Kilpasilta to Heinjoki was completed by 9 May 1944. On 25 May 1944 the unit was re-organized into two flights, each with nine Brewsters.

On 27 May 1944 the squadron began combat missions with the Brewster. A four plane swarm was sent for an intercept in the Lempaala direction, on the Karelian Isthmus. There it met twelve La-5s and Il-2s, though neither side scored any victories.

Aerial reconnaissance had observed huge troop and artillery concentrations on the Karelian Isthmus, close to the Gulf of Finland. Though nothing special had occurred on the Lake Ladoga side of the isthmus, the LeR 3 commander issued on 28 May 1944 an order to HLeLv 26: 1) reconnaissance on my orders, 2) repelling enemy aircraft over the central and northern Karelian Isthmus and 3) escorting bombers on my orders.

The order of battle of *Hävittäjälentolaivue* 26 was then: Squadron commander Major Lauri Larjo, Heinjoki
- 1st Flight, Captain Juhani Ruuskanen, Heinjoki (9 BW)
- 2nd Flight, Captain Aate Lassila, Heinjoki (9 BW)

The Great Attack

The Russian began a major offensive on 9 June 1944 on the Karelian Isthmus. It was supported by the 13th Air Army while the Baltic Fleet air force was in charge of the left flank over the Gulf of Finland. These units possessed a total of 1,552 combat aircraft, which were concentrated on a strip just 20 kilometres wide. This massive air armada was opposed by just 50 fighters of LeR 3.

HLeLv 26 found itself in intercept action right from the first day of the offensive, above the defensive areas of the III and IV Army Corps. A swarm of six Brewsters, led by the squadron commander Maj Lauri Larjo, shot down a solitary Pe-2 over Kivennapa. A few hours later, 1Lt Erik Teromaa of 1st Flight took his five Brewsters to attack two DB-3F formations, which were covered by large fighter formations. One bomber was shot down.

On 10 June 1944 a four plane swarm of 1/HLeLv 26 led by 1Lt Erik Teromaa shot down one Pe-2 out of six. A short while later, the swarm ran into a 15 plane mixed formation. One DB-3F plane was damaged. In the afternoon, the Russians shot down one Brewster at Raivola.

On 16 June 1944 HLeLv 26 was transferred further to the rear and flew to Käkisalmi on the eastern Karelian Isthmus. Next day LeR 3 gave orders to operate in the area Sairala-Käkisalmi-Elisenvaara-Ilmee-Sairala. The front-line was not to be crossed.

On 17 June 1944 HLeLv 26 sent 12 Brewsters led by 1Lt Erik Teromaa to fight 10 Airacobras and 35 Pe-2s as they attacked Hiitola and Käkisalmi with bombs. One Pe-2 and one P-39 were shot down.

On 18 June 1944 Pe-2 dive bombers with a fighter escort tried once more to bomb Hiitola, but 1Lt Erik Teromaa and his 10 Brewsters of HLeLv 26 shot down two bombers and one fighter. Teromaa reported the clash this way:

"At 0745-0905 hours, altitude 3,000-50 m. I acted as the formation leader during the battle. South of Hiitola we met about 30 bombers and 10 fighters. I managed to shoot from behind at one bomber, but had to pull away when the escort fighters attacked me. After this I could not engage the bombers because of the slowness of the Brewster.

Later over Kiviniemi I observed five La-5s flying southwards about 500 m below me. In the ensuing curve battle I got a good deflection burst into one La-5, after which it dived smoking into the ground.

Due to the agility of the Brewster I could turn easily with the La-5 on the deck. My plane was BW-364."

Then the Eastern Karelian Isthmus was quiet for a week. Only on 26 June 1944 did a Brewster patrol of HLeLv 26 meet an enemy plane, and shot down the solitary U-2 over Lake Ladoga.

On 7 July 1944 1/HLeLv 26 moved to operate from Mensuvaara and it was subordinated to *Lentorykmentti* 1. The task of the flight was to cover railway lines in the rear of the VI Army Corps, especially the rail yards of Sortavala, Jänisjärvi and Läskelä, and to intercept Russian air attacks against the settlements of Sortavala and Lahdenpohja. On specific orders the flight was also to escort the Blenheims of TLeLv 12 on daylight reconnaissance and photographing missions. 2/HLeLv 26 remained at Käkisalmi to operate in the area of the III Army Corps.

The VI Army Corps retreated near Lake Ladoga and reached the Pitkäranta-Loimola line (also known as the U line) on 10 July 1944, and settled down for defence. The U line withstood all Russian attacks and the fighting subsided a week later.

On 14 July 1944 the commander of HLeLv 26, Maj Lauri Larjo, led six Brewsters of the 1st Flight on an interception to Nietjärvi, where they had to duel with ten Russian fighters. They reported two damaged, but modern research has shown two Yak-9s of 14 GIAP were lost.

On 26 July 1944 HLeLv 26 moved to Värtsilä as air activity was directed to the Loimola-Tolvajärvi area. Two days later the squadron was operationally subordinated to *Lentorykmentti* 2. Intense fighting had whittled the plane complement down to five Brewsters.

On 29 July 1944 the Brewster pair of HLeLv 26's Maj Lauri Larjo entered combat above Tolvajärvi with 17 Yak-7s, many Airacobras and Kittyhawks and eight Il-2 planes. One Airacobra was dispatched but Maj Larjo was shot down too. His replacement as HLeLv 26 commander was Maj Erkki Metsola.

After a few clashes the air space north of Lake Ladoga calmed down and the war became stationary. HLeLv 26 flew its last mission on 31 August 1944 when a pair of Brewsters reconnoitred the area of Kuolismaa.

On 4 September 1944 the commander of the Air Force ordered the regiments to tell squadrons to cease fighting at 7.00. A ceasefire commenced and two weeks later it was confirmed by The Moscow Armistice.

Brief Analysis

HLeLv 26 flew over 4,500 sorties during the Continuation War, claiming 108 enemy aircraft destroyed. The squadron had lost 19 fighters, 7 in combat, 4 in air raids and 9 in accidents. Six pilots were killed.

Lapland War

The agreement of a cease fire between Finland and the Soviet Union included a requirement that German troops in northern Finland were to be either stripped of arms, or expelled from the country by 15 September 1944. An impossible task, which led to a short war with the Germans.

The Cyclone engine of Brewster BW-384 of 1/HLeLv 26 has just started at Immola on 15 June 1944. After three years of continuous fighting the Brewster was still a good firing platform and could hold its own, relying on its manoeuvrability. (SA-kuva)

Brewster 239 BW-375 of 1/HLeLv 26, flown by 2Lt Lauri Mattson in June 1944, based at Immola. (Karolina Hołda)

Brewster 239 BW-368 of 2/HLeLv 26, flown by 1Lt Reino Sartjärvi in June 1944 based at Immola. (Karolina Hołda)

Brewster BW-355 of 2/HLeLv 26 at Mensuvaara in mid-July 1944, being assigned to the flight leader Capt Aate Lassila. Just visible below the tarpaulin is the inscription NOKA, which had survived for three years.

Lentorykmentti 2 and 4 plus TLeLv 14 were placed in charge of the air operations against the Germans. LeR 2 consisted of two fighter squadrons, HLeLv 26 and 28. The former was commanded by Major Erkki Metsola and the ten plane Brewster flight was led by 1Lt Erik Teromaa.

The III Army Corps was formed in Lapland. Real action began when the Finns made a surprise landing in the rear of the Germans at Tornio on 1 October 1944. Next day *Lentorykmentti* 2 was tasked with interception, reconnaissance along main roads, and escort of LeR 4 bombers.

On 3 October 1944 1Lt Erik Teromaa and his eight Brewsters of HLeLv 26 directed its flight to Oulu, from which one swarm navigated north along the coast. At the Bay of Bothnia there was a III Army Corps shipping convoy en route to land at Tornio, where advance parties had already fought intense battles. The swarm met a German Ju 87 D detachment apparently flying towards the convoy. The *Stuka* formation opened fire and in the battle that ensued, two Ju 87s were shot down and three fled the scheme smoking. The Germans on the other hand reported having had only six planes in the air and none lost. Teromaa wrote:

"*At 1315-1545 hours, altitude 500-50 m. As I was on reconnaissance with 8 BWs along the Kemi-Rovaniemi road we met 12 Ju 87s southbound. At first we wondered whether to fight but as the Stukas opened fire, we decided to fight. I fired at the far left*

1/HLeLv 26 leader 1Lt Erik Teromaa in front of Brewster BW-361 at Käkisalmi on 26 June 1944. Teromaa became an ace in two squadrons, accumulating a total of 19 air victories.

wingman, its engine began to smoke, and it broke formation to descend and ultimately crash-land in a swamp between Ristijärvi and Louejärvi.

Some planes fired at me and I was hit in the fuel tank and an explosive bullet hit the cockpit, wounding me slightly with shrapnel in my hand and thigh. After this I fired at another Ju 87 but had no results.

The Stukas dropped their bombs as we attacked them. The Stukas held formation very well all through the battle, machine gunners gave concentrated fire and it was very accurate. SSgt Oiva Hietala saw the Stuka crash in the swamp. My plane was BW-361."

The swarm led by 1Lt Carl-Erik Bruun was flying from Oulu out to the Bay of Bothnia, when it met a solitary Ju 88

diving on a ship, attacked it and left one engine smoking. Later a crashed Ju 88 was found east of Kemi and it is probable this was the plane the Brewsters had damaged.

After this no clashes occurred with German aircraft, as the Soviets started a major offensive in the north on 6 October 1944. Within days all German flying units were transferred away from Finland to oppose this threat.

The Brewsters and Myrskys, which arrived later in Lapland, flew only reconnaissance missions until 23 January 1945, when all were flown south to peace-time bases.

In Lapland the Brewsters of HLeLv 26 claimed two or three aircraft shot down, but lost in this unnecessary war six aircraft and four pilots, most to the extremely accurate flak.

VL Myrsky MY-16 of 2/TLeLv 12 flown by the flight leader Capt Oiva Tylli in October 1944 based at Kemi. (Karolina Hołda)

In the Lapland War HLeLv 26 was reinforced by Myrsky flight 2/TLeLv 12. Here MY-26 is seen at Kauhava on 24 October 1944, just before the transfer to the north by the flight leader Capt Oiva Tylli.

Lentolaivue 28

Winter War

The 1937 five-year development plan for the air force included the equipping of a third long-distance (bomber) squadron. When the Soviet Union attacked Finland on 30 November 1939, starting the Winter War, the third bomber squadron was changed on 8 December 1939 into a fighter squadron. This saw the establishment of *Lentolaivue 28* with Maj Niilo Jusu in command. The base was set at Säkylä in south-western Finland and the unit began building up awaiting the arrival of the Morane Saulnier MS.406 fighters, 50 of which France had promised to Finland.

On 4 February 1940 the first two Moranes arrived with LLv 28 and the rest of a total of only 30 by the end of the month. The squadron was tasked with the defence of Turku and other ports in south-western Finland. During the following two weeks several intercept mission were flown with the familiarization flights, though without contact with the enemy.

On 17 February 1940 the Russian bomber targets were the ports of Turku, Rauma and Reposaari outside Pori. Before noon 1Lt Tuomo Hyrkki of 2/LLv 28 was on a search mission and engaged a 9-plane DB-3 formation heading to the sea south of Pori. The Moranes had at this point three 7.5 mm machine-guns. Hyrkki's centre gun was not working and he emptied his two wing guns, puncturing a bomber's fuel tanks, but forced to leave his victim to continue its flight.

In the afternoon the unit's victory score was opened. 1Lt Tuomo Hyrkki's pair engaged a lone DB-3 bomber of 53 DBAP (long-range bomber aviation regiment) claiming it destroyed. Hyrkki related it thus:

"*At 1410-1515 hours. Flying at 3500 m altitude I observed above Rauma a solitary bomber flying southwards 1000 m lower than me. I dived after it and recognized it as a DB-3. First I shot the gunner from below and behind and continued to the left engine. When I did not get it into fire, I pulled steeply up from below and gave a burst to the cockpit, when from the right side of the pilot flashed a sharp flame. Soon after this the whole fuselage was on fire. The bomber exploded mid-air and fell in three large pieces on the ice south-west of Korppoo. The pilot bailed out using his parachute.*

Each time when I got closer than 50 m, the bomber pilot pulled up his plane by shutting the throttle and then banking to either side. I got the impression that the pilot intended to cause a crash after noting that escaping was not possible. My plane was MS-301."

On 20 February 1940 LLv 28 shot down two SB bombers out of six approaching to bomb the port of Rauma. Later one Morane chased nine DB-3 bombers of 53 DBAP towards Estonia. 1Lt Veikko Karu caught the formation just at the Estonian coastline and sent two down:

"*At 1415-1530 hours. I was on an intercept mission with two Moranes in Kokemäki area flying at 3000 m. When I observed a 9-plane enemy formation it was 25-30 km from me and 3-4000 m higher. I caught the enemy very slowly, even when flying at full throttle (and using a few times the additional boost). When approaching the Estonian coast and about to engage the enemy formation, I observed an enemy fighter formations climb-*

Morane Saulnier MS.406 coded MS305 of 3/LLv 28 taking off from the ice at Pyhäniemi in Hollola on 13 March 1940, the last day of the Winter War. The Moranes flew 288 missions in this conflict, claiming 14 Soviet planes shot down. (SA-kuva)

2/LLv 28 leader 1Lt Reino Turkki (centre) and his mechanics in front of his assigned plane MS307 at Säkylä in late February 1940. Turkki became a general and Finnish Air Force commander 20 years later .

ing towards me, first six planes 1-2000 me below me and under this a similar detachment.

I attacked first the bomber at the extreme right wing and then moved on to the next. I attacked from behind and below, staying constantly under the cover of the tail. The enemy planes caught fire easily after shooting at close range (maybe 20-30 m) three short bursts, making my attack to last just a moment. After the attack on these two bombers I saw the enemy fighters threateningly close behind my tail.

I don't remember if the enemy fighters fired at me, but anyway they were very close, so I thought that is was best to quit and took a dive via a half-roll. The combat occurred at 7500 m. During the vertical dive I took first an 180 degree aileron turn and before levelling a second. I began the pull up at 2000 m. My plane seemed to hold well that heavy dive. I did not observe any kind of vibrations in the wings or tail. I saw no enemy fighter anywhere. Just to be sure I flew with full throttle close to the Finnish coast.

I came to the conclusion that my attack must have come as a total surprise, because I did not observe any counter fire from the bombers nor did I observe any evasive actions for defence. Though I did not pay much attention to this matter as I would not have had time to fire at the gunners first. Very likely the condensation stream left by the bombers had covered my approach.

I had to stay above 5000 m a fair half hour, of which time 15-20 minutes above 7000 m without oxygen equipment. My condition was not pleasant, but not so bad that I would have aborted the attack on the bombers. In my opinion experience is quite significant in flying at higher altitudes. During the previous week I had frequently been up to 6000 m without oxygen gear during intercept missions, so this case did not feel more unpleasant. My plane was MS-321."

On 26 February 1940 the fighters of 10 ABr (aviation brigade) of the Soviet Baltic fleet air forces claimed having destroyed four Spitfires in the Turku region. The enemy had clearly noticed a new type of fighter, but had misidentified the

Moranes. The Russian claims were also fictitious and the clash ended with the opposite result.

On 2 March 1940 LLv 28 sent down two bombers in south-western Finland and chased one fighter to Estonia, shooting it down. The Moranes of the 2nd Flight engaged in two encounters. The pair of 2Lt Pauli Massinen destroyed one DB-3 bomber of 1 AP, KBF out of a large formation west of Turku. Massinen wrote in the combat report:

"At 1325 hours. The enemy formation had arrived by surprise. I was south of the city waiting for the formation to return. I flew 1000 m higher that the enemy escadrille returning southwards. I approached from the side above and dived from rear behind. By additional speed I caught the formation easily. I attacked the last plane, which was a little separated from the others. I fired a short burst to the fuselage from slightly behind and below, when the gunner ceased firing. Thereafter I fired at the right engine making a few short dives. It started to leak fuel and finally caught fire.

Now an enemy fighter was diving towards me and I considered it best to break off the duel by a dive, especially when only the centre gun had ammunition left. According to air surveillance the bomber had exploded mid-air and the remnants fallen to the archipelago. The bomber was a DB-3.

As my personal opinion I dare to state that the Morane is a good fighter when it comes to speed and manoeuvrability, but the armament is weak. The optical sight is malfunctioning too often, the machine-guns lose their aim and the ammunition count is too small. The bullets did not penetrate the enemy armour. My plane was MS-318."

In the afternoon Cpl Urho Lehtovaara of 2/LLv 28 claimed one SB bomber out of a nine plane formation destroyed. He reported:

"I scrambled at 1430 hours towards Loimaa, to where a bomber formation was approaching from the south. When arriving above Loimaa I observed a formation of nine bombers, which flew in a tight echelon northwards, changing the course

immediately to south. I approached the formation from the rear climbing behind thin clouds above it.

I targeted my first attack to the plane on the right wing closing in a 45 degree dive, the enemy gunner opened fire as soon as I went in the dive. I approached my target to 150 m straight behind and gave a short burst, when the gunner disappeared and the gun remained pointing up. During this time the left wing – 5 planes – of the formation had slowed down straight to my side and opened fire, when I had to pull up and break off the fight.

I repeated my attack instantly to the last plane on the left wing which was flying about 100 m behind the others. I approached in a steep dive from the right behind, when the gunner and the left engine came to the same aiming line. I fired a short burst, when I observed that the left engine produced a stream of smoke, banked to the left down and disappeared from my sight.

After first pulling up I dived after the falling bomber, to make sure that it will become shot down. But due to big difference in altitude I neither caught the bomber nor did I see it fall down to the ground. During the dive I had lost so much altitude that I would not have caught the other bombers before the Estonian coast, so I gave up the chase and returned to base. My plane was MS-326."

Lehtovaara, the future Morane top scorer, got his victory confirmed by both the air surveillance and later Russian records, which showed that 35 LBAP (light bomber aviation regiment) lost one SB bomber.

Viipurinlahti

On 4 March 1940 the Soviet troops had managed to cross the Gulf of Viipuri over the ice and formed a bridgehead in the Finnish rear at Vilaniemi and Häränpääniemi. Troops and columns flowed across the ice from Pulliniemi and Tuppura. All regiments were thrown in to repel this serious threat.

By 7 March 1940 the situation had become critical and two flights of Moranes of LLv 28 were transferred closer to the front, on the Karelian Isthmus. They joined immediately the strafing attacks over Viipurinlahti.

On 9 March 1940 LeR 1 kept closed watch of the enemy ground movements while LeR 2 and LeR 4 kept on the strafing

Morane MS329 still in 3/LLv 28 markings, as seen at Turku in summer 1940. In the Winter War it was flown by 1Lt Juhani Ruuskanen.

LLv 28 also possessed a flight of Hurricanes for three months. Here is HC453 at Turku before the handover to LLv 30, which took place on 29 August 1940.

and bombing of the troops in the invasion area and attacked the supply columns. LLv 28 claimed two fighters shot down over Viipurinlahti, 2Lt Jouko Myllymäki relating thus:

"At 1110-1300 hours. When participating on a strafing mission to Viipurinlahti and returning to the base I observed three I-16 fighters behind me. I instantly took a dive via a half-roll and after an aileron turn took the previous course flying at the deck. Now they were inside the shooting distance and started firing. I took evasive action by banking. These three fighters stayed all the time pretty well tightly together. During the curving I did not pay enough attention to the terrain and when flying down at the deck I became disoriented. After a while I recognized having arrived at the west coast of the Karelian Isthmus.

Now I got two three plane I-16 patrols more after me. These three patrols flew in a line, wing trios 50 m higher. When I began curving to left and right, the patrol on the curve side pushed down and started firing. I started looking for a higher hill, around which I could turn and escape to the north, as I was already over the Karelian Isthmus.

When I arrived to a frozen lake, which had an enemy base, one I-16 took off towards me. I shot the remaining 20-30 rounds into it, aiming at the pilot. The plane pulled up and while banking in the shelter of the lake shore I saw it crash on the ice.

I got now the course to the north and pulled up into the clouds. They were so thin that the enemy planes followed me. When I went down to the deck I managed to increase the distance, when seven enemy planes gave up the chase, both wing patrols and one plane from the centre patrol. I broke quite easily off from these two enemies by my slightly greater speed as I did not need to perform any greater evasive turns. I came back to Viipurinlahti, where these two enemies gave up the chase because the distance had grown enough that there was no point in continuing. I returned to my base at Utti. My plane was MS-330."

On 9 March 1940 eight Hurricanes arrived at LLv 28's base at Säkylä. They were originally intended for LLv 22, but remained with LLv 28 until the end of the hostilities.

On 11 March 1940 the last combats were fought over southern Finland, where enemy formations as large as 200 were ob-

Morane MS316 of 2/LLv 28 on the dry beach sand at Pyhäselkä near Joensuu in July 1941. This plane was assigned to Sgt Eero Lajunen.

Morane MS-608 of 1/LLv 28 about to take off from Naarajärvi on 28 June 1941. The assigned pilot was SSgt Matti Leinonen. (SA-kuva)

Four pilots of 2/LLv 28 at Joensuu on 17 July 1941. From the left: Sgt Urho Lehtovaara (41½ victories), flight leader 1Lt Reino Turkki (4 victories), 1Lt Pauli Massinen (5 victories) and Cpl Urho Jääskeläinen (2 victories). (SA-kuva)

served. LLv 28 shot down three DB-3s of 7 DBAP, as the last victories in the Winter War. 2Lt Martti Inehmo filled in the combat report:

"At 1445 hours I took off for a combat air patrol led by 1Lt Linkola. The start took place at Hollola and we flew to Lahti-Kouvola railway, which the Russians had been bombing the whole day. We climbed to 3500 m searching from both the north and south side of the rails. When approaching Koria station from the west we observed a large enemy formation (70-80 planes) bombing just this station from an altitude of 2500 m.

The lead plane had obviously not observed the two nine plane formations following the main part and when attacking became a target of the fierce machine-gun fire sent from the nose turrets of these planes. This firing stopped when I banked into a dive towards them and attacked the left wing plane. Now the fire from the fuselage turrets was aimed at me.

I dived too soon towards the enemy planes leaving very little time to shoot. I had to break off behind the DB-3 bomber, down left to the sunny side. During this attack the next two enemy planes dropped back to the side of my target and fired at me all the time.

I stayed with them at about three kilometres distance, until they were again in an echelon. Now I attacked again the same enemy plane on the left wing. I managed to fire at close range long bursts to the left engine, when it at first started to pour smoke and then also flames.

Also during this attack the next two planes dropped back on my wing, firing, One burst hit my engine cutting oil lines and I was forced to break off to the side and down. I made a forced landing on the ice of Ruuhijärvi, when my plane MS-304 was damaged more after ending on its nose."

Having lasted 105 days the Winter War ended on 13 March 1940 at 1100 hours, in the peace negotiated in Moscow.

Brief Analysis

Lentolaivue flew combat missions for only seven weeks. During this time it performed 260 intercept, 20 strafing and 8 reconnaissance sorties. On 22 airworthy days the Moranes were engaged in combat on 28 occasions, where 14 enemy aircraft were destroyed. One Morane was shot down by flak and no pilots were lost.

On 4 April 1940 *Lentolaivue* moved to Turku with twelve Moranes. The Hurricanes of LLv 22 were handed over to LLv 28 on 27 May 1940, only to be delivered three months later to *Lentolaivue* 30.

Continuation War

In December 1940 the German leader Adolf Hitler decided to invade the Soviet Union. Operation Barbarossa was intended to commence right after the spring thaw. But a delay was caused by the conquest of Yugoslavia and the Balkans. The beginning of Barbarossa was postponed to 22 June 1941. By this date the Germans had persuaded on to their side all countries with a border with or close to the Soviet Union, i.e. Hungary, Romania, Bulgaria and Finland. In this situation Finland commenced a full-scale mobilization on 17 June 1941.

The Soviet air forces had at this point 2,530 combat aircraft in Leningrad and Baltic military districts, positioned between the Arctic Sea and occupied Poland. Soviet intelligence had discovered large numbers of German aircraft on Finnish airfields and suspected major aerial attacks against Leningrad. To counter this threat the Russians on 25 June 1941 began a six day bomber offensive against Finland. The Continuation War had started.

The order of battle of *Lentolaivue* 28 was then:
Squadron commander Captain Sven-Erik Sirén, Naarajärvi
- 1st Flight, Captain Timo Tanskanen, Naarajärvi (7 MS)
- 2nd Flight, 1st Lieutenant Reino Turkki, Naarajärvi (10 MS)
- 3rd Flight, 1st Lieutenant Erkki Lupari, Naarajärvi (10 MS)

On 25 June 1941 the first observations of large bomber formations entering the airspace of southern Finland were made, at Turku at 6 o'clock in the morning. The Soviet targets in south-eastern Finland were Joensuu and Joroinen. Though LLv 28 was based away from these, the patrol of the 1st Flight met over Rantasalmi a lone, apparently disoriented, SB bomber and shot it down. This was described by Sgt Antti Tani:

"*At 1250-1300 hours. I observed the enemy plane over Rantasalmi 15 kilometres west of the railway flying in direction 135 degrees. I instantly turned after it. When I got within 75-50 m behind it slightly to the right, I opened fire aiming at the fuselage turret and left engine on the same line. Soon after the left engine burst into flames, which went equally soon out, only pouring thick smoke from the engine. I shot three more bursts but could not see any additional results. My plane was MS-311.*"

The air force HQ entered this claim in the records as destroyed. Modern research confirmed that 10 SBAP had lost that particular plane.

On 29 June 1941 the air force commander gave an order, according to which LeR 2 was to fly top cover for a 100,000 strong Karelian Army, which was formed to take back the area lost in the Winter War. The main offensive was planned to occur north of Lake Ladoga. LeR 2's commander gave the new areas of operations, which was for LLv 28 Savonranta – Sortavala – Suistamo – Korpiselkä.

On 8 July 1941 enemy intelligence observed the Finnish troop concentrations on the south-east border and began air raids against these.

On 9 July 1941 the swarm of 1Lt Pauli Massinen of 2/LLv 28 was engaged in a combat at Räisälä with five MiG-3 fighters, shooting two down. On the return flight five SB bombers were observed and two were destroyed in spite of the interference of the escort fighters. Sgt Urho Lehtovaara fought like this:

"*At 1440-1500 hours. After observing five enemy fighters I signalled my lead aircraft and dived instantly towards the enemy fighters. I entered immediately into a turning battle and after 5 minutes I fired a burst hitting the enemy fighter, which instantly dived to the ground catching fire. After breaking off I flew towards Elisenvaara, one enemy fighter following me.*"

Morane Saulnier MS.406 MS-601 of 3/LLv 28 flown by 2Lt Reino Ilmonen in July 1941, based at Joensuu. (Karolina Hołda)

Morane MS-601 of 3/LLv 28 at Joensuu, before being shot down by flak on 10 August 1941, killing the assigned pilot, 2Lt Reino Ilmonen. When a flight had more than ten planes, tactical number zero was used.

After arriving to Lumivaara I observed five enemy SB bombers, which flew in a tight echelon straight west. I attacked the wing aircraft on the right flank and shot at the starboard engine, when it immediately caught fire and the plane crashed on the ground. The enemy fighter which was following me fired at me all the time and pulled over me, banking to the right.

After noticing being left alone for a while, I moved again behind the wingman of the right flank of the formation and gave a short burst into its fuselage and next to starboard engine, which also caught fire and the plane dived in flames into the forest. The remaining three bombers changed course to north-east. The enemy fighter followed me up to Elisenvaara, where it turned to the south. I could not participate in combat since I had run out of mg ammunition. My plane was MS-327."

The bombers had belonged to 202 SBAP and the MiG-3 fighters to 7 IAP. One of the latter was claimed by 2Lt Martti Inehmo:

"At 1440-1510 hours. When flying at 4,000 m altitude I observed five enemy planes below and left to me, I went into a dive towards them. Now the enemy planes turned towards me and started firing, but the deflection was too small. Then started a curve battle and at one point I managed to shoot at one enemy in a bank from slightly above, the cannon shells ripping off the rudder and right elevator, the plane going into an uncontrolled spin. Then another enemy fighter was firing from behind, so I could not follow the fate of my victim.

The enemy was faster than my plane, maybe also climbed better, but the agility of my plane was at least as good. The enemy fighters had obviously heavy machine-guns if not cannons. My plane was MS-602."

To Karelia

After the opening of the hostilities the Soviet land forces had 23rd Army on the Karelian Isthmus and the 7th Army north

Four pilots of 1/LLv 28 at Läskelä on 6 August 1941. From the left: SSgt Antti Tani, 2Lt Aarre Linnamaa, 1Lt Aarne Alitalo and MSgt Jorma Norola. Tani and Linnamaa became later aces. (SA-kuva)

1/LLv 28 at Joroinen on 10 August 1941. Pilots are sitting in the front row, from the left: 2Lt Paavo Reinikainen, SSgt Matti Leinonen, 2Lt Aarre Linnamaa, 1Lt Tuomo Hyrkki, flight leader Capt Timo Tanskanen, 1Lt Aarne Alitalo, MSgt Uuno Karhumäki and MSgt Jorma Norola. (SA-kuva)

of Lake Ladoga with front responsibility continuing up to Uhtua. The air forces of the 23rd Army consisted of 5 SAD (7 and 153 IAP plus 65 and 235 ShAP. 65 ShAP was transferred in August 1941 to 7th Army air forces) and respectively the 7th Army air forces had 55 SAD (72 SBAP, 155, 179, 197 and 415 IAP plus from August 1941 on 65 ShAP).

On 10 July 1941 the offensive of the Karelian Army commenced from the Kitee-Ilomatsi area towards the north-western coast of Lake Ladoga. LeR 2 specified the operational areas and tasks, for LLv 28 it was Saarivaara – Korpijärvi – Kolosenjärvi – Mannervaara – Tohmajärvi – Pälkjärvi – Kakunvaara – Kaurila – Matkaselkä. Air superiority was to be held in this area in turn with LLv 24.

On 16 July 1941 the VI Army Corps of the Karelian Army arrived at the north tip of Lake Ladoga and continued along the coast to south-east. Next day 1Lt Aarne Nissinen's pair of the 3rd Flight surprised two MiG-3 fighters in the Elisenvaara area , one escaped but the other was shot down. 1Lt Reino Turkki's swarm of the 2nd Flight patrolled in the Jänisjärvi area and discovered three fighters escorting two DB-3 bombers. One pair tied up the fighters and the other shot down both bombers. 1Lt Pauli Massinen claimed one bomber this way:

"At 1005-1030 hours. We observed at 2,000 m altitude over the north tip of Jänisjärvi three enemy fighters which were escorting two bombers. The first pair led by 1Lt Turkki engaged the escort fighters and Cpl Jääskeläinen and I attacked the bombers. They began to descent to the deck. I first shot the gunner of the rearmost plane and thereafter the engine. The landing gear slipped out. Then we fired in turns at the same plane. On my second round I managed to put from 30 m distance the left engine in flames. The burning plane crashed instantly from 20 metres to the ground.

Then I shot all my remaining rounds at the other bomber (killing the gunner), which Cpl Jääskeläinen eventually downed. My plane was MS-314."

On 23 July 1941 the VII Army Corps of the Karelian Army arrived at Säämäjärvi land Marshal Mannerheim stopped its advance. Next day the VI Army Corps of the Karelian Army reached the intermediary target at the Tuulosjoki line and stopped.

On 23 July 1941 2/LLv 28 of 1Lt Reino Turkki encountered on a reconnaissance mission three SB bombers belonging to 72 SBAP. Two were sent down and two more damaged. Sgt Urho Lehtovaara claimed both bombers:

"At 1315-1330 hours. On returning from Manga we met three bombers. I attacked the right flank wing plane. It caught fire after the first burst and crashed to the ground.

A fellow pilot fired at the centre plane, hanging the landing gear, but got no results. When he exited I moved instantly behind this bomber. I fired a burst with the cannon shells exploding behind the right engine and fuselage. The plane went into a shallow dive and bellied on a field.

Judging by the rate of fire the nose gunner appeared to have a heavy machine-gun or a cannon. My plane was MS-314."

On 12 August 1941 the swarm of 1Lt Aarne Alitalo from 1/LLv 28 was engaged in a combat with six I-15bis biplanes over Vieljärvi. One, piloted by Capt M.P. Krasnoluchiy of 65 ShAP, collided with MS-301 flown by MSgt Jorma Norola. Krasnoluchiy made a forced landing and claimed a *taran*, ramming victory, but Norola flew back to base with a large dent in the starboard wing. One more I-15bis was shot down and the others damaged. The victor was 2Lt Aarre Linnamaa:

"At 0815-0945 hours. By the swarm leader's signal I dived down and behind, where at least four I-15s flew with two higher up. The uppermost turned towards me and I shot from ahead. The plane went in a shallow dive, the pilot bailed out and the plane fell in the forest and caught fire.

I shot at another I-15, which took a tight curve in front of me, a long burst and I saw the cannon shell explosions in the plane. I then pulled up. I got to fire at one more from behind, but I had exhausted the cannon rounds. Then there were three enemies left and they were breaking off.

The four plus two I-15s stayed well together. In a curve they always missed when shooting, not having enough deflection. They were not any more manoeuvrable than the Morane. My plane was MS-611."

On 17 August 1941 2/LLv 28 swarm of 1Lt Reino Turkki was patrolling in the Lahdenpohja area, and engaged two I-16s,

Morane Saulnier MS.406 MS311 of 1/LLv 28, flown by SSgt Antti Tani in September 1941, based at Lunkula. (Karolina Hołda)

Morane Saulnier MS.406 MS-310 of 2/LLv 28, flown by 1Lt Jaakko Puolakkainen in October 1941, based at Viitana. (Karolina Hołda)

of which one was shot down. The swarm headed to Lake Lado-ga, where two MBR-2 flying boats were seen taking-off in the shelter of the warships. The Moranes remained circling further off. The boats took off, after which the swarm attacked and downed both in flames. 1Lt Pauli Massinen claimed the other:

"At 1900-1905 hours. We flew north-east of Lahdenpohja over Lake Ladoga at 3,000 m altitude when we observed two flying boats on the deck heading north-east. We dived after them. 1Lt Turkki shot first at the rearmost plane and myself at the oth-er. On the first burst the engine began to smoke and on the second burst the plane exploded into pieces.

The cannon did not work. The gunner had armour and he could easily fire directly to the rear, when the pilot slid the plane slightly sideways. My plane was MS-304."

On 21 August 1941 a swarm of 2/LLv 28 attacked six SB bombers at Maaselkä, escorted by two I-16 fighters. One bomber was destroyed. Later Capt Urho Nieminen of 3/LLv 26 led with one Brewster a three-plane Morane patrol of 1/LLv 28 to Suojärvi, where nine I-15*bis* aircraft were striking against Finnish positions. 1Lt Jaakko Puolakkainen's pair of 2/LLv 28 arrived simultaneously on the scene. The Finns attacked and the strength of the Russian detachment was decreased by three. Sgt Antti Tani claimed two fighters:

"At 1800-1840 hours. I was wingman for the lead plane when I observed nine I-15 planes about 2 km east. I gave a sign to the lead plane and banked against the enemy. In the cover of a cloud I approached two I-15s, which were dive bombing the road. During the pull up I got 10 m behind one, fired, when the tail burst off. I shot at the other from 75 m straight behind. It

went in a shallow dive and went side first to the forest. I dam-aged one more, which 1Lt Puolakkainen eventually shot down. My plane was MS-311."

On 2 September 1941 1Lt Tuomo Hyrkki's swarm of 1/LLv 28 flew top cover to the ground forces at Säämäjärvi. Six I-16 fighters appeared on the scene and three of them fell victim to the Finnish pilots. Hyrkki reported:

"At 1345-1520 hours. When the swarm attacked from ahead and above via a half roll against five I-16s, one fled straight to the east. The I-16 pulled up every once in a while and fired simulta-neously. I got the engine to smoke and the plane made a forced landing into a swamp north of Suojujoki. The combat occurred at 50-300 metres and the I-16 tried to flee at the surface. The pilot made a safe forced landing. My plane was MS-607."

On 3 September 1941 the Karelian Army continued its ad-vance from the Tuulosjoki line towards the River Svir between Lake Onega and Lake Ladoga, arriving at the stream four days later. The VI Army Corps continue east along the River Svir and the VII Army Corps headed north towards Petrozavodsk.

On 9 September 1941 a Morane swarm of 2/LLv 28 en-gaged nine *Chaikas* and nine I-16*bis* fighters on a combat air patrol to the River Svir, shooting six down. On the return the Moranes encountered an 8-plane mixed formation, destroying one I-153. In the clash SSgt Urho Lehtovaara shot down three I-16 fighters of 155 IAP, two of them in just a few seconds and reported thus:

"At 1000-1010 hours. After arriving to the specified area led by 2Lt Inehmo we encountered nine I-153 and nine I-16bis fight-ers. Two I-16bis made an attack against the latter pair of the

swarm firing from straight behind. I made at attack from straight ahead against these two shooting from ahead the left side aircraft, which got hits in its engine and made a very quick pull up.

I shot immediately at another I-16bis from straight ahead, it was hit by a long burst and the aircraft crashed directly in the ground.

Simultaneously I observed that my first victim went down in a shallow dive and disappeared with the engine strongly smoking.

Behind these planes there was still a three-plane patrol, with which I entered in a curve battle and managed to shoot at one in a bank, which went with high speed in the forest. One of the remaining ones pressed to the deck and broke off the fight and the other pulled in the clouds. My plane was MS-304."

2Lt Martti Inehmo also scored a triple, but in two encounters:

"At 1000-1045. First combat area: Mätysova – Pitmojärvi – Grishinkaya – Svir. When flying top cover to ground forces with four planes (2Lt Lehtonen, SSgt Lehtovaara and Sgt Jääskeläinen) we encountered nine I-16bis and nine I-153s which came from above through the clouds. Of these one I-16bis attacked from behind, but after a steep curve I got behind it just at the tree tops. When I fired it took a false manoeuvre and hit the forest.

I managed to shoot at another from behind at 100 m altitude, when the cannon shells hit the fuselage and left wing lip, probably aileron. This plane sent straight into the forest in the first combat area.

Second combat area: Ostrov – Karelskaya. When returning with 2Lt Lehtonen we met three bombers escorted by five I-15s. These flew partly inside the clouds at 1000 m altitude. We followed them below and straight behind in order to arrive within firing distance by surprise. When in firing position I shot at the leftmost I-15, when it separated from the others and began to smoke and sway. The others pulled up into the clouds after first dropping their bombs. I dived to the side and in spite of searching no enemy aircraft were found.

I-16bis does not match the Morane in turning, but it is superior in speed and climb at 100 m altitude. My plane was MS-613."

On the morning of 12 September 1941 2Lt Aarre Linnamaa's swarm of 1/LLv 28 engaged five DB-3 bombers near Pyhäjärvi, heading towards Prääshä. Three of them were shot down and one damaged. Linnamaa's combat report said:

"At 0710-0830 hours. Our three-plane patrol met 5 DBs over Pyhäjärvi on the way to bomb in the Prääshä direction, altitude 3,000 m. I shot one of the two rearmost aircraft. It went down burning and on its back.

We continued our attack against the other, which pressed down to the deck. We shot at it like we did with the previous aircraft. Then the other three DBs came to our side and I moved behind their tails. Then 2Lt Myllylä and SSgt Tani got their aircraft down, probably in the Lohijärvi area. I climbed a bit and waved to the boys, but they did not notice and turned away.

At the same time about ten I-153s were coming towards them at the surface. They did not notice me and I continued after those three DBs. I got one to smoke, when another one got to my side. I got it also to smoke and spent all my ammunition. I pulled to the side and observed that the plane I had at first fired was smoking and one landing gear was down. The other plane began to smoke heavily. It banked to the shore of Lake Onega and bellied down to a swamp in Soksu area. The others continued to eastsouth-east over Lake Onega.

On the return flight I met a flying boat at Bubnova. I got 4-5 holes in my aircraft. One went through an attachment point of the fuselage tube. My plane was MS-607."

On 15 September 1941 2Lt Paavo Myllylä's swarm of 1/LLv 28 was engaged in a combat with a bomber escorted by five MiG-3 fighters of 179 IAP, near Prääshä direction. The Moranes shot three of the fighters down, 2Lt Paavo Reinikainen claiming two.

On 26 September 1941 a 3/LLv 28 swarm of 1Lt Jouko Myllymäki encountered on a search mission a ten plane mixed formation, where five fighters escorted five bombers. In the ensuing battle one SB bomber of 72 SBAP was sent down and

Morane MS-310 of 2/LLv 28 at Viitana. The assigned pilot was 1Lt Jaakko Puolakkainen. The upper kill bar denotes Puolakkainen's shoot down of an I-15bis on 7 November 1941.

one MiG-3 of the escort was damaged. Myllymäki described the event:

"*At 1340-1530 hours. Over Matrossa I saw a solitary plane arriving from the south. I turned my swarm instantly to the west and went around a rain cloud from its north side and then to the south, when I observed five SBs and five I-18 planes. I gave the attack signal and turned towards Matrossa, where the enemy were heading too. I went through a cloud when one I-18 appeared in front of me. I fired from behind right at a 30 degree angle. I saw the cannon shells burst the cockpit and wing root from 50-60 m distance.*

Behind this I-18 flew one SB, which got a cannon burst to the right engine. Then one I-18 got behind me and started firing. I pulled instantly into a cloud and could not observe what happened to the planes that I had fired at. When I came out of the cloud I could not see a single enemy plane.

The escort planes flew lower and loose around the bombers. The whole detachment flew in a dense formation just below the clouds. My plane was MS-603."

Though Myllymäki did not see the fate of his victims, 1Lt Aarne Nissinen with the swarm saw and reported that one SB came behind the rain cloud about 10 kilometres north of Matrossa. It crashed burning into a forest patch on a swamp. The aircrew bailed out. Additionally he saw two smoke clouds, which the crashes could have caused.

Northwards

On 1 October 1941 the VII Army Corps of the Karelian Army occupied Petrozavodsk on the west coast of Lake Onega.

On 9 October 1941 a Morane pair of 2/LLv 28 over Suopohja jumped a climbing MiG detachment of six aircraft. When the combat started another Morane pair arrived, and all the Russians were shot down. 2Lt Martti Ihehmo downed two and damaged one:

"*At 1115-1125 hours. When we flew about 8 km north to Suopohja I observed in Suopohja direction an enemy fighter in a climb. I gained altitude and attacked from straight ahead, but I had to pull aside. More enemy aircraft were taking off all the time and at one point I observed at least five I-18s.*

In the ensuing curve battle I shot one straight from behind at 20-30 m distance, when I observed the cannon shells hit its fuse- lage, *which promptly puffed thick smoke and the aircraft jerked up going right in a cloud. I began curving to see it come out of the cloud, but I was instantly attacked by another fighter.*

I continued the curve battle with it and after a while another I-18 joined it. These tried to shut my entry to the west and I ended during evasive manoeuvres all the way to Soralahti, where I got to shoot one from straight ahead. It then pulled up and went into a dive crashing in the water near an island.

I continued the curve battle with the other at 600 metres at first, but it forced me below 100 metres. Finally I managed to use a deflection shooting from left below and behind, when it banked smoking and crashed in the forest. There it exploded and caught fire. My plane was MS-327."

On 19 October 1941 1Lt Aarne Nissinen's swarm of 3/LLv 28 flew a search to Poventsa, engaging one R-Z recce plane, which was sent down. Two *Chaikas* appeared on the scene and they both were shot down, too. Sgt Toivo Tomminen downed two:

"*At 1310-1320 hours. I observed south of Poventsa one R-Z heading towards Poventsa. I went in a dive after it and shot one burst, when both port wings broke off and the plane crashed in a lake. Then I observed two I-153s taking off from Poventsa airfield.*

I attacked the latter as soon as it had lifted off from the ground. I shot a burst from behind, when it crashed to a field on the nose. I obviously hit the pilot. While shooting at the I-153 pieces of it threw around, one piece of plating hit my starboard wing and stuck to it. No other damage. My plane was MS-315."

Then the poor weather of the coming winter cut down the flying on both sides.

On 4 December 1941 a swarm of 3/LLv 28 was engaged in combat with three I-18s over Maaselkä. Both parties lost one aircraft in a mid-air collision. SSgt Pekka Vassinen participated in this fight:

"*At 1300-1305 hours. When we approached Maaselkä from south-west at 1000 metres (I flew on the left wing and at the moment about 400 m aside) I observed two I-18s climbing from below and behind in about 45 degrees angle, the enemy having considerable speed advantage still at our altitude. Sgt Tomminen had also observed one I-18, which was shooting at SSgt Jussila from 100 metres behind. Sgt Tomminen shot at the I-18 straight from the side getting hits behind the engine to the cockpit.*

The I-18 suddenly pulled up, when Sgt Tomminen hit the wing of the I-18 snapping it off. Sgt Tomminen's plane flipped on its back, flying a while in this position and then the nose slowly sank into a vertical dive. A moment later I saw the aircraft burning on the ground.

I went to the surface, but did not observe any parachutes hovering The aircraft shot by Sgt Tomminen crashed in flames. The climb rate of the I-18 was obviously good since a while earlier Sgt Tomminen had been to the surface without observing anything. The Russians opened fire from a relatively close range. My plane was MS-620."

The other party in the mid-air collision was SnrLt N.F. Repnikov, flying a Hurricane of 152 IAP, also killed in this incident.

On 5 December LLv 28 fought the last combat of the year. It occurred above Maaselkä where a 1st Flight swarm engaged an eight plane mixed detachment. One Hurricane was shot down from the formation, as described by 2Lt Paavo Myllylä:

"At 1300-1435 hours. When escorting one Blenheim two I-18s attacked from behind, when I entered in combat with them. A while later the other I-18 broke off, when I managed to shoot at the other from the side in a bank. After the second burst the plane started smoking and went in a shallow dive into the forest about 15 kilometres east of Romantsy station.

The wings of the enemy planes had dark blue or black circles, inside which was a red spot*. My plane was MS-317."

The last stage of the advance of the Karelain Army was on 6 December 1941, the conquest of the town of Poventsa on the north shore of Lake Onega. The C-in-C stopped the offensive here and a stalemate of two and a half years commenced.

*Probably the original RAF roundels.

Brief Analysis

In 1941 LLv 28 had achieved 70 aerial victories losing eleven Moranes, of which nine were in combat. Three pilots were killed in action.

1941 top scorers of *Lentolaivue* 28:

Rank	Name	Flight	Victories	Plane	Total victories
MSgt	Lehtovaara, Urho	2	10	MS327	10 + 1
2Lt	Inehmo, Martti	2	7	MS318	7 + 1
Sgt	Tomminen, Toivo	3	6½	MS329	6½
1Lt	Massinen, Pauli	2	4	MS314	4 + 1
MSgt	Tani, Antti	1	4	MS311	4

The right hand column shows the 1941 victories, those gained in the Winter War are after the + sign

1942

At the beginning of 1942 *Lentolaivue* 28 possessed twelve serviceable Moranes while another eight were under overhaul. The operational sector of *Lentolaivue* 28 on the Maaselkä Isthmus and Lake Onega proved to be the quietest.

During the whole year of 1942 LLv 28 recorded only five confirmed air victories. There were two major reasons: 1) not many contacts were made and 2) the 20 mm Hispano-Suiza engine-mounted cannons were exhausted and the replacement 12.7 mm Colt or captured Berezin UB machine-guns were far less effective and also subject to malfunction.

The first encounter took place on 5 February 1942 when a Morane swarm of 1/LLv 28 led by 2Lt Aarre Linnamaa made a reconnaissance mission to Osta at the eastern end of the River Svir. En route two R-5 planes were met and both shot down, as told by Linnamaa:

"At 0945-0950 hours, altitude 10-20 m. I flew with a four-plane swarm along the river valley from Androvkaya to the southwest shooting at trucks. I saw one R-5 fly in the valley to southwest in the surface fog. I fired at it by surprise from 70-20 m distance from above and behind. The plane glided to the ice of the river, when two more aircraft shot at it, though it was already

Morane MS-304 of 3/LLv 28 in a servicing tent at Solomanni near Petrozavodsk on 20 April 1942. The kill bars on the fin were by SSgt Urho Lehtovaara, when the plane was earlier with the 2nd Flight. (SA-kuva)

Morane MS-606 of 2/LLv 28 leader 1Lt Reino Turkki at Viitana in early 1942. He flew this plane for over a year, claiming one victory out of his four.

Moranes MS-325 and MS328 of 1/LLv 28 about to take off from Viitana on 17 March 1942. The former was assigned to Sgt Unto Silvonen and the latter to Sgt Ensio Terho. (SA-kuva)

SSgt Oskari Jussila of 3/LeLv 28 in front of 1st Flight Morane MS328 at Solomanni in July 1942. Jussila was almost an ace, with four kills and one damaged.

Morane MS-325 of 1/LLv 28 taxiing at Viitana on 17 March 1942. The pilot on this occasion was the squadron CO Maj Sven-Erik Sirén. (SA-kuva)

Morane MS-622 of 2/LeLv 28 at Hirvas in October 1942, raised on jacks to test the landing gear function. The assigned pilot was Sgt Pentti Piispa. (Finnish Air Force)

The commander-in-chief Marshal Carl Gustaf Emil Mannerheim (left) greets LeLv 28 commander Maj Auvo Maunula at his Mikkeli HQ. Maunula was awarded the Mannerheim Cross on 8 September 1942. (SA-kuva)

unable to fly. The plane hit the ice hard (probably the pilot dead) and went on its back.

I continued to south-west and met another R-5, in the valley as with the earlier one. I fired first from aside and soon after that from behind at 20 m, when it went in a dive to the forest and flipped on its side, when Capt Blomqvist (LLv 28 supply officer) shot at it from a distance, but the aircraft had already disappeared in the forest.

The first aircraft was also fired at by MSgt Tani before and over me, but probably did not hit, because the plane continued as before. I did not observe any men leave the first plane. The second one broke into pieces when hitting the forest. I did not observe the gunners firing at me. My plane MS-621."

On 25 March 1942 a Morane pair of 1/LLv 28 scrambled to Petrozavodsk after two Pe-2 bombers and managed to shoot one down. MSgt Antti Tani reported his claim this way:

"After an engine test run I took off directly from the blast pen. At 08.25 when I had reached the altitude of 1500 m I received a radio message: two 2/2 coming from Soksu straight to the north. I was then in a steep climb right towards Petrozavodsk and at the same time climbed through the clouds. Then I took a 360° climbing turn, after which I observed the enemy plane about 2 km away and 300-500 m higher.

When the distance decreased to 250-300 m I fired but the deflection was too small. I pulled more deflection, fired and increased again the deflection. Now the burst seemed to hit. After this I turned behind the plane at 300-400 m and fired three bursts. After the last one the plane made a steep 180° turn and began a shallow descent. It appeared to me that the plane lost speed considerably.

I tried to fire again, but the Colt jammed. While fixing this the plane made a steep 90° turn towards me and I dodged with a combat Immelmann. During that manoeuvre the gun began to work again and at the moment of passing I observed that the enemy went into a spiral. Then it seemed that the pilot cut off the engines. I thought that this was a bluff so I followed the plane in a spiral 800-1000 m higher and waited for the plane to level off. When the plane reached 300-500 m altitude, I observed that it cannot any more pull up and hit the ground right after that. My plane was MS-619."

Morane Saulnier MS.406, MS-623 of 3/LeLv 28 flown by SSgt Erkki Alkio in August 1942, based at Hirvas. (Karolina Hołda)

Morane MS-623 of 3/LeLv 28 at Hirvas in August 1942. It was assigned to SSgt Erkki Alkio. This plane was one of eleven fitted with a fixed radiator, a standard feature on the MS. 410.

On 3 May 1942 the air regiments were re-organized on geographical basis. *Lentorykmentti* 2 continued to operate for the Maaselkä Group. In addition to LeLv 24 and 28 it was annexed to LeLv 16. In the new sector of LeR 2 the left border was set at Lieksanjärvi – Kuusiniemi – Vojatsu – Virma and the right one at Lohijärvi – Derevjannoje – Ääninen – Volodarskaja – Vytegra.

The regiment task was specified as 1) reconnaissance, artillery fire control, bombing and transport missions, 2) interception of enemy aircraft and 3) protection of transports, transfers and bombers. These missions were to be flown at the request of Maaselkä Group, Onega Coastal Brigade and Olonets Group. In practice LeLv 16 reconnoitred the closer areas at Maaselkä and LeLv 24 took care of the interception and long-range reconnaissance of the whole front. LeLv 28 handled Lake Onega and the areas south of it.

On 12 July 1942 SSgt Oskari Jussila's Morane pair of 3/LeLv 28 was sent for interception to the Bulayeva area, meeting two LaGG-3 fighters, one of which was shot down.

During mid-July 1942 a 700-man strong partisan brigade had slipped through the lines undetected east of Ontajärvi lake

and managed to advance by the end of the month to 40 km south-west of lake Segozero, into the rear of the Maaselkä Group. In the chase the brigade retreated along the same route and its remains crossed the waterways at Lake Jolmozero.

On 18 August 1942 a Morane swarm of 1/LeLv 28 was sent to Jomozero to strafe Russian troops crossing the waterways. They were protected by a four Hurricane detachment, from which the Moranes shot one down and damaged two more. On a later mission eight Moranes were patrolling at Jolmozero and were attacked by a similar number of Tomahawks. One Russian was damaged, but one Morane was lost as shot down.

On 24 August 1942 Major Auvo Manula took over the command of *Lentolaivue* 28. Two weeks later he was awarded the Mannerheim Cross for his earlier achievements as a commander of reconnaissance squadron *Lentolaivue* 12.

On 1 September 1942 three Pe-2s bombed Hirvas air base by surprise. When LeLv 28's Morane swarm on duty became airborne. 29 Tomahawks, Hurricanes, I-153s and I-15s appeared all of a sudden from the north and began to strafe the airfield with machine-guns. Two Moranes were alerted from

Morane Saulnier MS.406 MS328 of 1/LeLv 28, flown by Sgt Martti Vihinen in September 1942, based at Hirvas. (Karolina Hołda)

Äänislinna to assist, but the Moranes were slow and faced jamming problems with their heavy guns and only one Tomahawk was damaged. One Morane was shot down. After the raid a single Pe-2 photographed the target.

On 11 October 1942 the aircraft inventory of LeLv 28 improved when Moranes – bought from Vichy France and refurbished by the factory – began to arrive. The 2nd Flight, which was disbanded on 26 July 1942 due to aircraft shortages, was re-activated.

1943

At the start of 1943 LeLv 28 had 16 serviceable Moranes while another seven were in overhaul.

The order of battle of *Lentolaivue* 28 was then:
Squadron commander Major Auvo Maunula, Hirvas
- 1st Flight, Captain Pekka Siiriäinen, Solomanni (4 MS)
- 2nd Flight, Captain Reino Turkki, Hirvas (6 MS)
- 3rd Flight, Captain Tuomo Hyrkki. Hirvas (6 MS)

The operational area and the task remained the same. Likewise the air space remained quiet most of the time, though the modernization of the Red air forces began to show.

On 12 January 1943 a Morane swarm of 2/LeLv 28 was on an escort and search mission in the Nopsa area, encountering four Pe-2 bombers escorted by three Hurricanes. One of the escorts was shot down as told by MSgt Urho Lehtovaara:

"At 1235-1445 hours, altitude 7000-4000 m. While patrolling between Liistepohja and Karhumäki we observed north of Maaselkä at a very high altitude one Pe-2 aircraft, which glittered and gained altitude. We commenced a chase instantly and after about 10 minutes I got in a position to fire at it from about 600 m below and behind.

But simultaneously an HC fired at me from above and behind and I had to give up the chase. A curving battle developed then with three HCs and myself. I managed to shoot at one HC in a bank and scored a hit with the burst, when the HC made a sudden twist and continued into a spin down to the earth, where it exploded and burned.

Right after this I was able to fire at the second HC, which evaded smoking. The third HC broke off the battle in a dive and disappeared to north-east. My plane was MS-627."

On 24 February 1943 a Morane MS.406 patrol of 1/LeLv 28 encountered over Sautjärvi a lone Pe-2 of 119 RAE (reconnaissance aviation squadron), which was caught after a chase and sent down by MSgt Antti Tani:

"At 1255-1355 hours, altitude 3000-1000 m. While on an intercept mission I was radioed about a plane, which flew from Maasjärvi to south. I was flying in the same direction observing it after about 4 minutes. At Tokari the plane turned north. The I managed to fire from ahead on the right side, when the starboard engine began to smoke, first weakly and then stronger all the time. I followed the plane until its crash. In this period I shot a further three bursts in it. My plane was MS-619."

On 5 March 1943 a Morane pair of 2/LeLv 28 scrambled after a bomber north-west of Karhumäki. MSgt Urho Lehtovaara worked thus:

"At 1355-1500. I took off on an alarm mission and climbed to 5,000 m north-west of Karhumäki. I was told of one aircraft north from Savujärvi. I saw immediately below me against the clouds one twin-engined aircraft heading north. I dived after it, caught it soon and recognized it as a Pe-2. I shot from straight behind at the starboard engine, which instantly caught fire and the

plane crashed to the ground, where it remained burning. During the dive one man bailed out from the burning plane.

I dived after the burning plane to pick up the exact location of the crash site, but at about 1000 m altitude I was attacked by four I-16s. A fierce aerial battle arose with the four enemy planes and myself. I had received so much oil to the windscreen from the Pe-2 that it obstructed my sight so much that several of my bursts missed the target. I finally got behind and slightly below one enemy fighter and gave a long burst, when some of the ventral plating came loose and it fell to port and hit the ground.

I also tried to fire at the other enemy fighters, which still kept attacking me, but the machine-guns of my aircraft stopped working, so I was forced to break off the combat. My plane was MS-641."

On 16 March 1943 commandos destroyed the storage at Jeljärvi, west of the White Sea, free of Russian aerial interference. The five Moranes of 2/LeLv 28 escorting the commandos shot down two Tomahawks out of a formation of five.

On 8 May 1943 the LeLv 28 commander Maj Auvo Maunula's Morane pair bounced a detachment of four I-16s, on a search mission to Seesjärvi. In the ensuing combat two Russian were sent down immediately and a third a while later. The fourth took evasive action. Maunula reported thus:

"At 1917-1924 hours. When I was flying a search mission with a patrol I met at Suontele about 1500 m below four low-flying I-16s. I surprised them as they were shooting at the ground. I chose a target, but when I dived towards it another observed me and began to climb and bank towards me. I interrupted the attack and got after climbing and turning over and behind it. The plane quickly dodged by turning into a dive. I went after, got a good aim and fired a short burst from obliquely behind.

1/LeLv 28 pilot MSgt Antti Tani (left) and his mechanic Risto Hiltunen pose by Tani's Morane MS-619 at Hirvas, after Tani shot down a Pe-2 of 119 RAE on 24 February 1943, denoted by the upper kill bar.

Morane Saulnier MS.406 MS-624 of 2/LeLv 28 flown by Sgt Uolevi Jaakkola on March 1943, based at Hirvas. (Karolina Hołda)

I could not see the results since from above and the right side another plane was attacking me. By pulling up I was able to get above it on the side. Then the plane began to wind, making rolls, some of which in a 30 degree dive, approach the deck. I managed to fire a short burst, but observed then 300-400 m above me on the port side two aircraft. By climbing and banking I was able to get above the planes and slightly behind, when they commenced a dive from about 1000 m down to the deck and headed east. I followed one aircraft, which continued to fly on the deck easily dodging. I fired short bursts, but could not get a good aim. Finally the plane tried to break off by taking a climb and bank. During this manoeuvre I shot a long burst from 100-75 m distance hitting the front fuselage. Then the plane dived into a bay below and disappeared under the sludge. My plane was MS-615."

On 8 June 1943 Capt Tuomo Hyrkki leading a 1/LeLv 28 swarm escorted FK dive bombers of 3/LeLv 16 north of Maaselkä, engaging over Sumeri station two I-16 fighters. One was shot down while the other escaped due to the malfunctioning of the heavy machine-guns of the Moranes, but recent research proved that 197 IAP lost two fighters. Hyrkki's report stated:

"At 1150-1305. After the FK planes bombed a train 2 km north of Sumeri station, two I-16s appeared on the scene, apparently planning to attack the FK planes. I pressed my plane into a dive and bank when the enemy planes turned towards me. I shot at both from straight ahead, but without any obvious results. I took a fast turn, but ended too close in a position, where the enemy could not fire at me properly and I could not fire at the enemy plane either. After passing the plane I made a swift turn and ended up in the opposite direction again from the ene-

Morane MS-624 of 2/LeLv 28 at Hirvas after white winter camouflage application, which was done on 20 February 1943. Sgt Uolevi Jaakkola was the assigned pilot.

3/LeLv 28 pilots in front of a Morane (yellow 9) at Hirvas in summer 1943. From the left: SSgt Erkki Alkio, 2Lt Juha Räisänen, 1Lt Jaakko Kalliomäki, SSgt Oskari Jussila, Sgt Torsti Louhenlinna and Cpl Kaarlo Valpas.

my planes, which were flying in a row about 200 m apart from each other. I got a good aim and opened fire from about 200 m distance, when I observed the burst hit the engine and the enemy plane went in a dive and soon after that caught fire. The pilot bailed out.

I managed to shoot at the other one twice from relatively close range, but due to gun malfunctions I was forced to break off the combat. At one point black smoke came out of the enemy plane, but that stopped before I had to leave the battle. The enemy aircraft were using rocket projectiles, which after firing from the wings exploded 50-10 m behind the plane! My plane was MS-657."

The air force commander gave LeR 2 an order to save fuel on 19 June 1943. Other than scrambles, LeLv 28 was permitted to take off only by the order of LeR 2 commander. But nothing worthwhile happened in the air on the LeLv 28 sector. In a few encounters the Moranes entered twice in a fruitless combat. The only major operation of LeLv 28 in the last half of the year was to shoot up the fuel dump at Saharevskaya with a swarm of Moranes, after an attack by LeLv 44 Junkers Ju 88 dive bombers.

Last War Year

At the beginning of 1944 LeLv 28 had 24 serviceable Moranes while another nine were in overhaul. The air space around Lake Onega remained quiet. On 14 February 1944 the front-line squadrons were given a task prefix, *Lentolaivue 28* becoming *Hävittäjälentolaivue 28* (HLeLv 28).

The Moranes were becoming obsolete, as the first encounter on 6 March 1944 would show. While a Morane pair of 3/HLeLv 28 was reconnoitring the roads in Uikujärvi direction it was attacked by two Tomahawks over Petrovkiy Yam. In the ensuing combat both enemy fighters were damaged, but the Morane of 1Lt Kalliomäki received hits wounding the pilot. After running out of fuel Kalliomäki made a good forced landing at Pyhäniemi. 2Lt Estama's plane received hits in the

engine and collided with the other enemy fighter, damaging his tail. Estama made a forced landing at Karhumäki without further damage.

On 7 March 1944 the commander of HLeLv 28, Maj Auvo Maunula, took six Moranes on a reconnaissance mission to Lintujärvi and on to Kärkijärvi, where they met seven Tomahawks. Two of these were shot down, and three more were damaged. Maunula wrote in the combat report:

"*At 1105-1120 hours, altitude 30-2000 m. I was reconnoitring the airfields with six planes. When approaching Maaselkä airfield, I observed at low level two Tomahawks, which were attacked then by two MS planes. Immediately after this one Tomahawk took off followed by another two. I observed below on the starboard side one plane in a climbing bank, shot at it, but could not get a good deflection because I could not see the light of the sight clearly (the tinted glass was in down position). I observed again below on the starboard side another plane in a climbing bank firing two bursts at it, hits in the engine, which puffed smoke. The plane stopped banking and went into a shallow dive. I got behind the tail and fired several bursts from close range until it went into a vertical dive and then I passed it. I followed its path after the pull up, saw the pilot bail out and the plane crash in the forest catching fire.*

I saw below and ahead of me one aircraft pulling steeply up. I took a dive and went after it managing to open fire at a distance of about 75 metres, making hits on the fuselage. I could not observe the results since I had to go into a dive after losing my speed. The enemy went down on its wing. I observed one MS fire at one plane at low level. The plane poured a light fog type of smoke trail, when I observed that the MS was losing distance. The aircraft flew towards the airfield, I took a dive after it and got about 100 metres behind it, fired, but only the wing guns had ammunition left. I emptied the guns and when reaching the edge of the airfield I turned away. The aircraft continued to fly westwards.

During the final stages of the combat two additional planes took off, so based on what I saw there were seven enemy aircraft

SA-Kuva

Two Moranes of 1/LeLv 28 on a forward landing strip at Latva on 9 September 1943. Closer one is MS-619, assigned to Sgt Bengt Ringbom, and behind is MS-643, flown by 1Lt Kalle Uola. (SA-kuva)

Sgt Bengt Ringbom of 1/LeLv 28 taxis his Morane MS-619 at Latva in August 1943. The victory markings belong to the previous assigned pilot, MSgt Antti Tani.

in the air. No planes landed during or after the combat. When I radioed the order to break off and set about re-assembling the group (the aircraft were scattered in the final phase of the combat since the Tomahawks flew individually apart), I observed only two enemy aircraft in the air. My plane was MS-653."

The HLeLv 28 Morane planes were ageing badly, and in speed tests it was found that none of the planes could reach a speed of 350 km/h on the deck. The Russians had brand new Lavochkin La-5 and Yakovlev Yak-9 fighters which were much better in most respects. The Morane squadron fought back by flying always at least six plane formations and making use of their fighter's agility. This did not always work, as the encounter on 8 May 1944 showed. Of three Moranes of the 1st Flight sent to reconnoitre the Vytegra direction, Airacobras of 773 IAP shot down one, one Airacobra collided mid-air with 2Lt Kuusela's Morane, Kuusela becoming POW, and the third barely escaped into the clouds.

On 22 May 1944 a Morane swarm of 1/HLeLv 28 led by Capt Reino Turkki reconnoitred Vytegra in the south end of Lake Onega. Eight Airacobras jumped the Finns. One Red fighter was shot down, but one Morane was also lost, and another was badly damaged. The guns of the MS planes malfunctioned.

Three days later, when Capt Reino Turkki's Morane swarm of 1/HLeLv 28 was reconnoitring the Vytegra area again, eight Airacobras attacked the Finns. One red fighter was shot down, but also one Morane was lost and another received severe damage. The weapons of the Morane were not functioning properly.

On 6 June 1944 a swarm of Moranes from 3/HLeLv 28 led by 1Lt Jouko Timonen reconnoitred Segesha in the north and got into a fight with six Tomahawks. One was shot down and one more damaged.

Soviet Offensive

The offensive of the Soviet Army on the Karelian Isthmus had no initial effect on the sector of *Lentorykmentti 2*. Nevertheless the 2nd and 3rd Flights of HLeLv 28 were combined to form Detachment Sovelius on 10 June 1944. It was transferred to *Lentorykmentti 3*, set up base at Utti within a week and took charge of the rear on the Karelian Isthmus.

The rapid movement of the front line on the Karelian Isthmus caused the HQ to order troops at Maaselkä to move closer to the Finnish borders. *Lentorykmentti 2* units were ordered on 17 June 1944 to retreat and this was begun on a squadron basis.

On 25 June 1944 when one Morane of 1/HLeLv 28 was taking off, the pilot Sgt Lars Hattinen noticed a lone Boston bomber flying over low. He caught it, and sent it into the forest in flames. Hattinen related:

"I scrambled at 1250 hours after having received a message: One plane towards Solomanni from the west. Right after take-off I saw the enemy plane, which curved over Petrozavodsk from south to west. The right landing gear of my Morane did not retract so the distance stayed at 500 m. I got the gear in and caught the plane. I fired from straight behind from 100-150 m to both engines. I did not observe any effects so I flew closer and fired at the fuselage from 20 m, when the plane went into the forest uncontrolled.

The dorsal and ventral gunners were firing all the time. The enemy was flying at the tree tops. My plane was MS-308."

On 26 June 1944 a three Morane patrol of 1/HLeLv 28 flew an intercept mission to the Kumsa area. They met a solitary Il-2 ground attack plane, which managed to escape. At the end of a chase a lone Tomahawk (actually a Yak-9 of 197 IAP) was jumped and shot down, as told by Sgt Lars Hattinen:

"I took off at 19.30 hours for an intercept mission flying as the wingman. Over Kumsjärvi we encountered one Il-2 plane, which we attacked. First SSgt Laakso, then 2Lt Laitinen and I was the last, getting closest to the enemy. I shot at it from straight behind but the bullets did not have any effect on the Il's armour. The plane started to fly northwards. I followed it at a close range shooting every now and then. It flew right on the deck. While changing randomly its course I could not quite follow where I was in a strange terrain and the Il took me right over Aitalampi air base.

I observed the base in front of me and at the same moment a Tomahawk taking off. With a small change in course I was right behind the Tomahawk, fired into the fuselage from 30-40 m distance. The burst hit well and the plane caught fire, crashing at the edge of the base 200-300 m away.

By now the four already airborne Tomahawks took me by surprise, shooting from above and behind but missed. The clouds were at 200 m, where I got by a steep climbing turn. The enemy following me to my home base. My plane was MS-639."

On 1 July 1944 HLeLv 28 received its first Bf 109G-2 and ten more would follow during the course of the month, equipping the 2nd and 3rd Flights.

LeR 2 squadrons had retreated as planned through Hirvas and Kuutamolahti to Värtsilä, where they were finally joined by the HQ of LeR 2 on 7 July 1944. During the transfer, the squadrons had been able to perform all reconnaissance tasks and discover the movements of Russian troops in their area of responsibility. Two days later the commander of the air force defined the new area of operation for LeR 2: the left border was set at Vuonislahti-Paatene-Kärkijärvi and right border at Heinävesi-Soanlahti-Sotjärvi-Puujoki.

The VI Army Corps had withdrawn along the north coast of Lake Ladoga and arrived at the Pitkäranta-loimola line (U line) on 10 July 1944. The U line withstood all Soviet efforts of break through and the fierce battles were silenced in a week.

On 11 July 1944 1/HLeLv 28 received its first "Ghost Morane" at Värtsilä. This version had an 1,150 hp Klimov M-105 engine, which gave a much better speed and climb rate. But the factory managed to deliver only three of these before the end of hostilities.

On 16 July 1944 the Ghost Morane of 1/HLeLv 28 saw action for the first time with SSgt Lars Hattinen as he scrambled from Värtsilä:

"I took off at 18.00 hours to intercept planes heading west from Ägläjärvi. At Tolvajärvi I saw four fighters, two at 1000 m and two at some 3000 m, and on the deck I saw six Il-2s. I attacked the lower pair of fighters which flew in an agitated manner. They evaded right away and a curve battle formed up, which the upper pair also joined. The planes were very agile and equal to the MSv, it was hard to get an aim on them. After some curving around, the top pair went for the deck and the other pair attempted to break off east.

The battle went on for some 15 more minutes, until I got a clean shot at the plane I had targeted first. I gave it a burst but it kept evading. On the second burst it caught fire and fell in a swamp from 10 m. I tried to get the other Yak, but he had a speed advantage of maybe 20 kms per hour. It broke off. Then I got to the Ilyushin formation flying at one side of me, and fired from the side, but my cannon was out of action. I gave them a burst with my wing guns but knowing they have no effect on Ils, I gave up the chase. My plane was MSv-617."

On 23 July 1944 Detachment Sovelius moved to Värtsilä and now the entire HLeLv 28 was at one base. Its subordination to *Lentorykmentti 3* was terminated and it was re-subordinated to *Lentorykmentti 2*. Its task was interception above Karelia, north of Lake Ladoga, where large Russian detachments were seen all through July. On arrival the detachment had a strength of five Messerschmitt Bf 109G-2s.

On 26 July 1944 a 1st Flight Morane scrambled to Suistamo, where it met an artillery spotter U-2. This was shot down into the forest. A swarm of 3/HLeLv 28 Messerschmitts took off for interception in the Loimola-Uomaa area. The Finns attacked a flight of Il-2s escorted by five Airacobras at Pitkäranta, dispatching one ground attack plane and one fighter. These were the first Messerschmitt victories of the squadron. The first one was claimed by MSgt Erkki Alkio thus:

"At 1105-1200 hours. I met on an intercept mission two enemy Airacobras. I managed to take them by surprise from behind and below. After firing at one it caught fire and started to descend towards the ground. After a while the fire went out, but smoking all the way the plane fell in the forest east of Kaukojärvi.

Then I met between Loimola and Uuksujärvi an Il-2 pair. I managed to shoot at one both from above and below behind on the left side. I saw the cannon rounds hit but had to quit due to cannon jamming, When I got the cannon working I had lost the sight of the plane and could not tell more about its fate. When I last saw it, it was in a shallow dive to the ground. My plane was MT-208."

The arrival of the Messerschmitts meant that the Russian harassment of slower Moranes in this sector had come to an end.

On 29 July 1944 three *Mersus* of 3/HLeLv 28 led by 1Lt Antti Heimo fought a 30 plane mixed formation over Tolvajärvi. One Il-2 and one Airacobra were shot down. Later the same day, 1Lt Jouko Timonen, also of the 3rd Flight, went to intercept Russians with his swarm, and he met many enemy planes. The Russians lost three Airacobras shot down.

On 30 July 1944 Capt Ala-Panula with his four *Mersus* of 3/HLeLv 28 fought a twenty-plane Russian fighter formation at Tolvajärvi. In the ensuing battle the Finns shot down three La-5 fighters of 760 IAP.

SSgt Lars Hattinen joined the previous combat in his Ghost Morane, managing to shoot down two Airacobras. He proceeded with attacking 694 ShAP Il-2s when his plane was shot down, and he bailed out. The squadron adjutant 1Lt Mauno Maunula filled the combat report on behalf of Hattinen:

"At 0905-1020 hours. SSgt Hattinen reported being over the airfield at 4000 m altitude when he got at 0920 hours an order to fly to the south of Tolvajärvi. When arriving in the area he observed more enemies, about 30 Il-2s and 20 Airacobras. SSgt Hattinen attacked the Airacobras, shooting two down.

Then he went after the Il-2s, when the rear gunner of one hit his fuel tank, which exploded.

In a report from the infantry SSgt Hattinen's plane went in a dive, pulled up and went in a dive again, when a large chunk

departed from the plane. This was SSgt Hattinen who bailed out successfully. His plane was MSv-617."

On 1 August 1944 three *Mersus* of 2/HLeLv 28 escorted bombers to Loimola. A squadron of La-5s attacked and forced the 109s into battle, in which one La-5 was shot down.

At nightfall, nine Bostons bombed the Värtsilä airfield and right after that Il-2s strafed the field, with 10 fighters remaining as top cover. One Morane was destroyed on the ground and six Moranes, two Messerschmitts and one *Chaika* (of TLeLv 16) received minor damage.

On 6 August 1944 a *Mersu* swarm of 3/HLeLv 28 scrambled to Tolvajärvi where they met an LaGG-3 flight. Two enemy planes were shot down.

On 9 August 1944 Capt Ala-Panula of 3/HLeLv 28 led five *Mersus* to fight the Russians while flying top cover to Finnish infantry at Kangasjärvi. Five Il-2s and five Kittyhawks attempted to attack the Finnish infantry, but the MTs managed to intercept and shoot down two Il-2s, which turned out to be the final confirmed aerial victories of Finnish pilots in the Continuation War. Ala-Panula wrote of the fight:

"At 1515-1625 hours we met five KHs and five Il-2s east of Ilajanjärvi. The fighters climbed and I managed to dive straight on an Il-2. I fired three bursts and saw it dive straight in. As I pulled up and right I saw a strong fireball to the left of me. The Russians had white spots at the national insignia (red star in the middle). From afar it looks like a Finnish insignia. My plane was MT-232."

On 3 September 1944 the final mission of HLeLv 28 was flown by a swarm of *Mersus* to reconnoitre in the direction of Kuolismaa.

The commander of the Air Force ordered the air regiments to tell squadrons to cease fighting on 4 September 1944 at

1/HLeLv 28 pilot 2Lt Kalle Kyllönen is thrown in the air at Solomanni on 25 May 1944, after downing an Airacobra of 773 IAP. At right, the flight leader Capt Reino Turkki look on, next to his Morane MS-650.

Bf 109G-2 coded MT-232 was one of ten which HLeLv 28 received during the Continuation War. It was assigned to 3/HLeLv 28 leader Capt Veikko Ala-Panula and is seen here at Värtsilä in late July 1944.

7 am. A ceasefire commenced and two weeks later it was confirmed by the Moscow Armistice.

As the only squadron being part of *Lentorykmentti 2* all through the Continuation War, HLeLv 28 had flown just over 6,000 sorties and accrued in this war 123 aerial victories (MS 108 and MT 15). 24 Moranes had been lost in combat. Seventeen pilots had been killed, and three had become POW.

When the Lapland War commenced on 1 October 1944 the five Ghost Moranes on 1/HLeLv 28 inventory were sent north to participate in the conflict against the Germans. Their stay was not very long as on 18 October 1944 they flew the last mission, not having seen during the few reconnaissance missions a single German aircraft in the air. On 23 November 1944 the Ghost Moranes were flown south to the peacetime base at Rissala near Kuopio.

The other new type for HLeLv 28 was the Mörkö Morane (MS. 406 fitted with 1,100 hp Klimov M-105P engine). Here is MSv-624 parked at Kemi in November 1944, flying no more missions in the Lapland War.

When the static air stations were reformed into mobile air regiments on 1 January 1938, *Lentorykmentti* 1 was based at Suur-Merijoki with four squadrons: *Lentolaivue* 10, 12, 14 and 16, the first two existing and the other two re-numbered. *Lentolaivue* 14 was formerly LLv 34 and it arrived at Suur-Merijoki on 19 January 1938 with Capt Jaakko Moilanen in command. The aircraft were initially Blackburn Ripons, which during the next months were replaced by Fokker C.Vs and C.Xs. Like all *Lentorykmentti* 1 squadrons, *Lentolaivue* 14 was a reconnaissance outfit.

In the Winter War the task of LLv 14 was reconnaissance, photography and harassment bombardment on the eastern Karelian Isthmus. The arrival of new fighters allowed one flight to be formed into a fighter unit. This was to be 2/LLv 14 headed by 1Lt Tauno Ollikainen, which received the first of its eight Gladiators on 12 February 1940. The Gladiator also performed day reconnaissance, reinstated after two weeks of type training, while the Fokker C.X planes continued the nocturnal operations.

When the Soviet advance seemed to halt by March 1940 to the rearmost defence line of the Finns on the Karelian Isthmus, just south-east of Viipuri, the Red Army decided to attack to the rear of the defences by advancing over the frozen Viipurinlahti, commencing the crossing on 4 March 1940.

On 5 March 1940 LLv 14 was transferred from the eastern part of the Karelian Isthmus to repel the invasion at Viipurinlahti. In the afternoon the 2nd Flight received an order to shoot down an artillery fire control plane between Äyräpää and Pölläkkälä. 1Lt Tauno Ollikainen took off with six Gladiators at 1325 hours for a 55 minute mission, as told by the flight mission report:

"Planes and crews: GL-267 1Lt Ollikainen, GL-273 1Lt Pitkänen, GL-274 1Lt Kuula, GL-276 2Lt Malinen, GL-278 SSgt Perälä and GL-279 Cpl Roine. At 1400 hours the artillery fire control plane was not met.

Between Kihlasaari and Kuussaari two I-153s attacked SSgt Perälä's plane. In the ensuing curve battle SSgt Perälä shot

Gloster Gladiator coded GL-276 served with LLv 26 until 1 March 1940, when it was handed over to 2/LLv 14. Seen here at Utti around that time. On 7 March 1940 2Lt Kalervo Malinen shot down an I-153 of 148 IAP whilst flying this plane.

one of these down with a short burst. The altitude was then about 150 metres. 7.7 mm ammunition spent 200 rounds."

The *Chaika* was also later confirmed as lost, belonging to 68 IAP.

On 7 March 1940 2/LLv 14 flew escort for Finnish troop transfers west of Viipuri and six Gladiators were involved in a combat with three I-153s of 148 IAP, two of which were shot down at the cost of two damaged Gladiators. The mission report stated as follows:

"Planes and crews: GL-267 1Lt Ollikainen, GL-273 1Lt Pitkänen, GL-274 1Lt Kuula, GL-276 2Lt Malinen, GL-278 SSgt Perälä and GL-279 Cpl Roine. Mission between 1435 and 1610 hours.

When patrolling over Tienhaara two pairs of I-16s were observed, no contact. Over Tervajoki village met three I-153s, which pulled over, turned and attacked the second pair. The first pair turned around and started the combat, which was carried out between 1515 and 1525 hours, from 300 metres to the deck. The wind shifted the participants west of Tervajoki.

Two I-153s were shot down and they both crashed in the forest. 1Lt Ollikainen claimed one witnessed by 1Lt Kuula. The other was claimed by 2Lt Malinen and witnessed by 1Lt Pitkänen. Cpl Roine made a successful emergency landing to Löytöjärvi after being hit in the fuel tank. SSgt Perälä´s wing touched the wing of the I-153 breaking the starboard tip of the upper wing."

On 8 March 1940 the peace negotiations began in Moscow, ending the Winter War on 13 March 1940 at 11 am.

Brief Analysis

Lentolaivue 14 had flown in the Winter War 36 sorties with the Gladiator, claimed three enemy aircraft shot down and lost one plane and its pilot.

After the Winter War *Lentolaivue* 14 reverted back to a pure reconnaissance squadron. Even if the Gladiators remained in its inventory they were now obsolete as front-line fighters.

In the four first months of the Continuation War, beginning on 26 June 1941, LLv 14 continued in the army co-op role working on the Karelian Isthmus and then from 26 October 1941 five hundred kilometres north at Tiiksjärvi in Rukajärvi direction reconnoitring for the 14th Division.

On 1 November 1941 LLv 14 received a fighter element, when former the 1/LLv 10 was renamed as *Lentue Kään* and was subordinated to it from operational point of view. The flight was led by Capt Pekka Kään and it was formally 3/LLv 30 equipped with Wasp-powered Fokker D.XXIs.

On 9 January 1942 six Hurricanes of 152 IAP attacked Tiiksjärvi air base. Flight Kään managed to scramble with three Fokkers and destroyed one of the attackers. The victory was claimed by Cpl Hemmo Leino, a future ace.

On 19 January 1942 1Lt Veijo Taina's pair from *Lentue Kään* performed a reconnaissance over Ontajärvi. They found one R-5 on the ice which was perforated and another which became airborne and was sent down. From Taina's combat report:

"At 0810-0910 hours, altitude 200 m. We attacked with two FR-planes, the other was FR-146 of Cpl Leino, 300 m higher and

2Lt Kullervo Virtanen of Lentue Kään heads towards Fokker D.XXI coded FR-140 at Tiiksjärvi on 4 November 1941. Lentue Kään was the fighter element subordinated to LLv 14 from 1 November 1941 for nine months. (SA-kuva)

Pilots of Lentue Käär *at Tiiksjärvi on 31 May 1942, about to take of for the daily evening mission. From the left the flight leader Capt Pekka Käär, Sgt Eino Ruutsalo, 2Lt Heikki Keso, Sgt Eero Määttä, Sgt Hemmo Leino and 2Lt Leo Ukkonen. (SA-kuva)*

2/LeLv 14 deputy leader 1Lt Veli Hakola flew his 100th mission on 2 July 1942 piloting Fokker FR-116. He is seen here at Tiiksjärvi with a wreath, congratulated by the squadron CO Maj Ragnar Magnusson. (SA-kuva)

slightly behind on the right. I pushed down a bit and opened fire from 100 m. The plane banked immediately to the left and the nose went down. Cpl Leino made a similar attack after me. The engine stopped and the prop blades stood still. Before hitting the ice we had time to repeat our attacks. The observer fired back. When the plane hit the ice the pilot and observer jumped off. My plane was FR-156."

On 29 April 1942 when 1Lt Martti Kalima's pair of *Lentue Käär* was reconnoitring in the Tunkua direction, a nine-plane fighter detachment of 152 IAP surprised the Finns. In a fierce air battle Sgt Aaro Nuorala shot one of the Hurricanes down. Both Fokkers succeeded in breaking off, but both aircraft were riddled by about fifty bullet holes. The aircraft returned with the tires punctured by bullets and landed without more damage.

Most of the fighter tasks were carried out by a 10-plane Brewster detachment of LLv 24. This *Osasto Ahola* had arrived at Tiiksjärvi in January 1942. The detachment could attain air superiority when needed and when the air space became quiet they departed in November 1942, rejoining their squadron.

On 22 July 1942 after midnight three of *Osasto Käär's* FR-planes of LeLv 14 were on an ambush mission of Russian R-5 supply aircraft, meeting two of them. One was shot down and the other was able to flee under the cover of the twilight. 2Lt Mikko Salomaa wrote in the combat report:

"At 0040-0140, altitude 50 m. The search was done over lake Voijärvi. I saw the plane when it flew over Pekonvaara hill, heading west at tree top level of 5-10 m. I turned behind its tail and fired a burst, during which the observer was killed and the plane

Fokker D.XXI, FR-140 of 2/LeLv 14 flown by 1Lt Aaro Virkkunen in August 1942, based at Tiiksjärvi. (Karolina Hoda)

Fokker FR-140 of 2/LeLv 14 about to take off from Tiiksjärvi in August 1942. By 1 August 1942 it served with Lentue Käär and was frequently piloted by 1Lt Aaro Virkkunen.

caught fire. I pulled up and circled above the plane. When I saw the burning plane trying to dodge the bursts of my comrades, I shot a new burst, after which it flipped into the forest. My plane was FR-156."

Morane Fighters

On 1 August 1942 *Lentolaivue 14* was reorganized. *Lentue Käär* was merged with it and its Wasp-Fokkers were handed over to the 2nd Flight. The pilots were transferred to the 1st Flight, which was equipped with Morane 406s and became a fighter outfit.

On 1 Septembet 1942 LeLv 14 flew its first mission with the Moranes, when 1Lt Martti Kalima's pair of the 1st Flight reconnoitred the east side of lake Ondozero. Russians were not met in the air, but one observed camp area was strafed.

On 5 November 1942 the Moranes scored for the first time. 1Lt Martti Kalima's pair of 1/LeLv 14 flew a reconnaissance mission to Segozero, where they bounced a single LaGG-3. Another LaGG-3 appeared on the scene and both were sent down. Since five more Russian fighters were observed to approach, the Finns decided to break off and returned to Tiiksjärvi. The squadron's first Morane encounter is told here by Kalima:

"At 1155-1340. I patrolled with Sgt Leino at 2500 m altitude west of Voijärvi heading eastwards. From north came one LaGG-3 flying almost at Leino's wing, who was on my right side and about 200 metres below. Leino tried to get behind the Lagg, but it dodged towards me in a climbing turn.

The pilot obviously failed to see me so I got easily behind it. It pulled into a shallow dive to the east and tried to escape. I was about 100 m away from it, took a glimpse in the rear mirror ob-

serving the nose of an enemy aircraft behind me and at the same time a burst passed over me.

I planned to dodge, but then the plane in the mirror flipped to the starboard and puffed black smoke. I took aim again on the one ahead of me and fired a short burst. Pieces tore off behind the cockpit and the plane pulled slightly up banking to the starboard side. I shot a burst from 60 m, when the plane flashed into fire and went down in a spin. The enemy fighters had smoke tracers. The planes appeared to be clumsy. No damage in my MS-326."

On 23 December 1942 an FR-pair of 2/LeLv 14 shot down one R-5, as told by 1Lt Veli Hakola:

"At 1250-1350, altitude 100 m. We met an R-5 over the shore of Nuokkijärvi, about 8 km east of Pääkköniemi. First I curved behind and above the plane and after seeing the enemy markings,

I shot a burst at it, when the plane began to pour white smoke (obviously fuel from the punctured upper tank) and tried to head towards Nuokkijärvi losing altitude. There it flew at low level around a higher split and during this turn 2Lt Lilja fired at it. Now I observed that the plane tried to land next to the split. I fired another burst into it before it managed to reach the ground. Then it flashed all over into flames and when the skis touched the snow it flipped over and was all in flames. After this I observed an airman pull his wounded comrade into the forest for cover. My plane was FR-135."

On 1 January 1943 *Lentolaivue* 14 had seven serviceable Moranes with two in overhaul. 1/LeLv 14 was back under the long-time leadership of Capt Matti Tainio.

On 14 march 1943 a Finnish commando detachment began the destruction of the storage of Russian partisans in Jeljärvi vil-

Morane Saulnier MS.406 coded MS-603 of 1/LeLv 14 running up at Tiiksjärvi in early spring 1943. The regular pilot was 1Lt Esa Anttonen, who painted the spinner of the Ratier propeller red.

1Lt Mikko Salomaa approaches the war photographer with Morane MS-646 of 1/LeLv 14 at Tiiksjärvi on 9 June 1943, after returning from a 20 minute test flight. This plane has a Hamilton Standard propeller. (SA-kuva)

Lentolaivue 14 *commander Maj Ragnar Magnusson in the cockpit of Morane MS-640 belonging to the 1st Flight, seen at Tiiksjärvi in summer 1943. He led the unit over two years, from February 1942 to April 1944.*

Morane MS-642 of 1/LeLv 14 flies over the wilderness near Tiiksjärvi on 21 August 1943. The tactical tail marking is a white 4. This fighter was usually flown by Sgt Yrjö Hakulinen. (SA-kuva)

lage. LeLv 14 flew top cover for the next couple of days for the transport of the detachment on the Rukajärvi-Ontajärvi road.

Two days later the commandos destroyed the storage at Jeljärvi, free from Russian aerial interference. LeLv 14 took off for 35 sorties along the day. In the afternoon Capt Matti Tainio's Morane flight managed to surprise a ten-plane I-15bis detachment, sending five instantly down and during the combat a further two. Sgt Hemmo Leino claimed two aerial victories and his combat report stated:

"At 1425-1435 hours, altitude 50 m. I observed below me three I-15s heading east. I attacked the lead plane and shot at it until it fell in the forest. After this the wingmen banked away and now I began shooting at the aircraft flying on the starboard side. It caught fire, but the fire went out, after which it turned on its back and disappeared from my sight since I had to pull up to avoid the collision. My plane was MS-319."

On 23 March 1943 1/LeLv 14 Morane pair of 1Lt Martti Kalima shot down one I-16 fighter east of lake Ontajärvi. Kalima reported the action:

"At 8.40-8.45 hours, altitude 200 m. I met with Sgt Nuorala two eastbound I-16bis planes east of lake Ontajärvi. We managed to take them by surprise from behind and when closing in (distance 50-70 m) I fired at the port wing plane, which flashed into fire and dodged, but the fire went out. Early in the curve battle we got with Sgt Nuorala behind the I-16bis, when we both fired at short range bursts, causing the said aircraft to catch fire and crash down. My plane was MS-326."

On 25 May 1943 Capt Matti Tainio's ten Moranes of 1/LeLv 14 flew a reconnaissance mission on the route Tunkua-Sorokka-Kotshkoma-Jeljärvi. Three fighters took off from Jeljärvi airfield and one ditched in the nearby lake while the others avoided contact.

Several months passed without a contact with the enemy in the air. On 25 October Capt Martti Kalima was posted to lead 2/LeLv 14, which began transition to Moranes.

On 1 January 1944 *Lentolaivue* 14 had 14 Moranes on duty and four in overhaul. On 14 February 1944 in the re-naming of the front-line squadrons, *Lentolaivue* 14 became officially

Moranes of 1/LeLv 14 about to take off from Tiiksjärvi on 26 May 1943. From the front the planes are MS-640, MS-315 and MS-603. Most of the unit's Moranes had individually coloured spinners. (SA-kuva)

Sgt Sauli Valkeiskangas of 1/LeLv 14 takes off in MS-640 at Tiiksjärvi on 26 May 1943 for an early evening reconnaissance mission. This plane had the most common Chauviere propeller. (SA-kuva)

a reconnaissance unit, *Tiedustelulentolaivue* 14 (TLeLv 14), operating in the Rukajärvi sector.

On 13 April 1944 the Morane swarm of 2/TLeLv 14's Capt Martti Kalima went to Rukajärvi for an interception. They met two LaGG-3s, of which one was shot down and one escaped thanks to its speed.

On 26 May 1944 a flight of 2/TLeLv 14 led by Capt Kalima was on reconnaissance to Jeljärvi, when it met four LaGG-3s. Though they tried to break off, one of them did not manage to do that but fell prey to Finnish guns.

On 2 June 1944 Morane pairs of 1/ and 2/TLeLv 14 reconnoitred Kuutsjärvi where they met ten LaGG-3s. One Morane was hit but managed to return to Tiiksjärvi. The Russians lost three planes. Capt Martti Kalima shot down two and wrote:

"*At 1750-1900 hours, altitude 1500-50m. I was top cover leader on recce mission, when at Jeljärvi four LaGG-3s attacked us, two going for lower patrol and two for my patrol. In the ensuing curve battle I fought two Russians, because some fighters that joined the fray forced my wingman away from me. After the initial pass, both Russians circled at about 1000 m altitude.*

Pilots of 1/LeLv 14 sitting on the wing and fuselage of MS-326 at Tiiksjärvi on 10 August 1943. From left SSgt Pekka Simola, Sgt Sauli Valkeiskangas, 2Lt Kaarlo Temmes, 1Lt Martti Kalima, 1Lt Mikko Salomaa, 2Lt Johan Durchman and 1Lt Esa Anttonen. (SA-kuva)

Morane MS-605 of 1/LeLv 14 seen at Tiiksjärvi shortly before it was sent to the aircraft factory for full overhaul, which took place on 21 May 1943. SSgt Pekka Simola frequently flew this plane.

I surprised one and got at its neck from above, and I zoomed and boomed it. After my fourth such pass the plane in front of me spun in and went in the forest.

After this I gained altitude because I saw three more enemy planes coming in from south-west about 500 higher than me. My wingman Capt Anttonen then attacked the solitary enemy, whose wingman I had shot down just before Capt Anttonen joined me. I was in a shallow rising curve when out of the sun yet two more planes attacked me. I was slow to evade and I got hit in the wing and my landing gear dropped down.

I tried to break off into a cloud, but did not make it and had to go to the deck. We fought on at the treetop level and I wound up some 12 to 15 km south-west of Ontajärvi. Finally I had no choice but to try head on, and I did hit the enemy directly at the face and below. The Russian exploded at some 30 m just as I went under it. Its remains fell in a swamp.

At this point there was one Russian at about 1500 m altitude, which had followed the fight but not gotten into it. I made it to our side without that enemy plane attacking me even once. Two 20 mm and two 12.7 mm hits in my plane, which was MS-622."

On 6 August 1944 2/TLeLv 14 had to fight nineteen Airacobras above Ontrosenvaara. One of these was destroyed, as was one Morane. The final aerial victory of the squadron is described here by 1Lt Matti Niinimäki:

"At 1845-1930 hours, altitude 200 m. As I flew away from the cloud base I saw two Airacobras on the deck, and after an 800 m dive managed to fire at one of them at about 100 m, hit it, and made it dive in the forest at a 45 degree angle. As I pulled up to avoid another AC pair, I saw a plane crash in the woods. My plane was MS-629."

On 3 September 1944 the final casualty of the Air Force happened when a Morane of 2/TLeLv 14 was hit in the engine by flak at low level and caught fire. The pilot bailed out and was rescued. The last mission of the Air Force in the Continuation War was flown by a Lysander of 3/TLeLv 14, which reconnoitred camp fires at Jyskyjärvi and dropped bombs on them. LY-116 landed at 21.10.

The commander of the Air Force ordered the air regiments to tell squadrons to ccase fighting on 4 September 1944 at 7 am. A ceasefire commenced and two weeks later it was confirmed by The Moscow Armistice.

Though being for the major part of the Continuation War a reconnaissance squadron TLeLv 14 had achieved 23 air victories. It had lost nine fighters and four pilots in combat.

When the Lapland War commenced on 1 October 1944 the eight Moranes on the 2/TLeLv 14 inventory were sent north to participate in the conflict against the Germans. The stay was not very long as on 18 October 1944 they flew the last mission, not having seen during the few reconnaissance missions a single German aircraft in the air. On 27 November 1944 the Moranes were flown south to the peace time base at Rissala near Kuopio and the unit disbanded a week later.

The last pilots of TLeLv 14 at Tiiksjärvi in mid-September 1944. From left 2nd Flight leader Capt Toivo Vuorinen, 1Lt Matti Niinimäki, squadron CO Maj Kyösti Kurimo, 1Lt Heikki Keso, 1Lt Åke Hohenthal and 2Lt Jorma Malmberg.

Lentolaivue 32

When the Winter War broke out on 30 November 1939, urgent aircraft purchases quickly resulted in establishment of new units. Forty-four Brewster Model 239 fighters were bought from USA. For these a fourth squadron, *Lentolaivue* 22 with Capt Erkki Heinilä in command, under *Lentorykmentti* 2, was founded on 16 January 1940. The transfer base was set at Säkylä while the actual base was prepared at Hollola.

The first four Brewsters arrived on 1 March 1940 and the last of 17 was delivered on 21 March 1940. LLv 22 also received, between 7 and 10 March 1940, ten Hawker Hurricanes as interim equipment.

The Winter War ended on 13 March 1940 in the Moscow peace treaty. Neither the Brewsters nor the Hurricanes saw any combat.

On 29 March 1940 the name of *Lentolaivue* 22 was changed to *Lentolaivue* 32, Capt Heinilä continuing in command. It was also placed under the new fighter regiment, *Lentorykmentti* 3, commanded by LtCol Einar Nuotio.

The original idea was to place the Brewsters with LLv 24, but with the Winter War in a critical situation, training on a new type was not possible. But five weeks after the conflict the change was made. On 17 April 1940 LLv 24 flew their Fokkers for a full overhaul at the factory and received the Brewsters from LLv 32 within two days.

On 8 May 1940 the first Fokkers were flown to the new base of LLv 32 at Siikakangas. On 27 May 1940 the Hurricanes were delivered to LLv 28. In mid-1940 LLv 32 had 17 serviceable Fokkers D.XXIs and nine being overhauled. The task of LLv 32 at this point was to arrange Fokker courses for both regular and reserve personnel, training over 100 pilots by the time of the next conflict.

Continuation War

When Germany informed the Finnish military leaders of their offensive on the Soviet Union, a full-scale mobilization was initiated in Finland on 17 June 1941. The German attack began on 22 June 1941 and three days later the Red air forces bombed Finland, starting the Continuation War.

The Finnish fighter force was then positioned in two regiments, *Lentorykmentti* 2 commanded by Col Richard Lorentz and *Lentorykmentti* 3 led by LtCol Einar Nuotio. The latter had three squadrons: *Lentolaivue* 30 with Fokker D.XXIs and Hurricanes, *Lentolaivue* 32 with Fokker D.XXIs and *Lentolaivue* 34 with advanced trainers. Of the D.XXIs the FRm is Mercury-powered and FRw Twin Wasp powered.

The order of battle of *Lentolaivue* 32 was then:

Squadron commander Captain Erkki Heinilä, Hyvinkää
- 1st Flight, Captain Paavo Berg, Hyvinkää (7 FRm)
- 2nd Flight, Captain Kullervo Lahtela, Hyvinkää (7 FRm)

LeR 3 was tasked with the protection of industry in the area Turku – Lahti – Helsinki – Tammisaari – Turku, plus preventing air raids on the most important port, Turku, and oc-

Brewster coded BW-369 of LLv 22 at Helsinki Malmi on 6 April 1940. With the formation of Lentorykmentti *3 on 27 March 1940, LLv 22 was to change its name to LLv 32 and become part of it. The Brewsters were exchanged with the Fokkers of LLv 24. All this was done by 19 April 1940. (Finnish Air Force)*

Fokkers of 1/LLv 32 at Utti in 3 July 1941. The closer one is FR-108 assigned to 2Lt Kyösti Karhila while the other, FR-90, was assigned to Cpl Mauno Kirjonen, both becoming aces in a couple of months. (SA-kuva)

casional protection of shipping in the Turku archipelago. The main task of LLv 32 was to protect the main railway going north from Helsinki.

The first observations of large bomber formations entering the airspace of southern Finland were made in Turku at on 25 June 1941 at 6 am. A couple of hours later 1Lt Veikko Evinen climbed for interception with seven aircraft of 1/LLv 32 and observed DB-3 bombers returning via Helsinki. He shot two down:

"At 0800-0810 hours. I engaged a nine-aircraft DB formation at 4,000 metres flying eastwards. I fired at the aircraft flying a little behind the others on the right flank and got the starboard engine in flames. The aircraft went into a dive, when I began to fire at one aircraft on the left flank and managed to put the port engine of it in flames.

The formation was then about 10 km east of Helsinki flying southwards. The first aircraft crashed about 2 km east of Malmi airport and the second in the sea south-east of the bay of Puotinkylä. A large hole in the rear part of the fuselage of my aircraft, probably caused by friendly flak. Bullet scratch in the propeller. My plane was FR-116."

On 28 June 1941 the 3rd Flight was re-established, tasked with type training on the Twin Wasp engined Fokker D.XXI.

On 1 July 1941 to the tasks of *Lentorykmentti* 3 was added the protection of the field army south-west of the river Vuoksi and the industrial areas at Vuoksenlaakso. This task was given to LLv 32, which was fortified by a Hurricane flight, *Osasto Kalaja*, of LLv 30.

On 3 July 1941 2Lt Esko Ruotsila's Hurricane pair of *Osasto Kalaja*/LLv 32 was over the Karelian Isthmus and intercept-ed three *Chaikas* of 7 IAP over Enso. In the ensuing combat two enemy aircraft were shot down, as the first Hurricane victories. A Fokker pair of 1/LLv 32 also claimed a Pe-2 bomber, which was quite an achievement for the much slower D.XXI.

On 4 July 1941 during continuous combat air patrols in the defence of Kotka and Kouvola, LLv 32 was engaged twice with the enemy, shooting two bombers down. One of these was an SB from 205 SBAP.

At 4 am on 11 July 1941 two MiG-3s of 5 IAP, KBF attacked Utti, shooting at a line of new Fokker D.XXIs of the 3rd Flight, setting fire to two and damaging four. The disastrous day came to an end by one Fokker hitting a roller on the runway and burning.

LeR 3 commander LtCol Einar Nuotio was working hard in order to get first-line fighters to his squadrons. This brought results on 14 July 1941, when the Curtiss Hawks of LLv 14 on the same base were switched with the Winter War vintage Fokker D.XXIs. Within a few days LLv 12 was also flying Curtiss Hawks. By now Maj Olavi Ehrnrooth had fully taken over the command of LLv 32. Under his leadership the planes were not distributed to the flights. The pilots were assigned to flights and those on duty just took any serviceable Curtiss.

The Hurricanes of *Osasto Kalaja*, subordinated to LLv 32, had scored three more aircraft by 15 July 1941, when the HC pair claimed more, as told by 2Lt Esko Ruotsila:

"At 1750-1925 hours. Air surveillance reported five fighters at Simola. We met south of Simola a three-plane patrol coming from the east, at 1,000 metres altitude. We attacked with the sun

behind our backs. After the pass we separated a little from each other and engaged enemy planes on every side of us. Finally I saw one evading in the Merijoki direction. I chased it to Nurmi level, when my engine started to act up.

I turned back to home. Then I saw Sgt Aikala chasing one enemy aircraft in the Vilajoki direction. I turned towards them, when the enemy took evasive action. Sgt Aikala's engine was also acting up. Both engines had the supercharger loose. The oils were leaking to almost dry. Simola observation post confirmed two aircraft having crashed. My plane was HC-452."

On 16 July 1941 the Curtiss planes of LLv 32 flew their first successful aerial combat, when Maj Olavi Ehrnrooth headed

a three-plane patrol and engaged in the Nuijamaanjärvi area four *Chaikas*, shooting down one of them.

On 21 July 1941 the squadron commander Maj Olavi Ehrnrooth got his next air victory, when the wreck was later found. The *Chaika* had belonged to 7 IAP and Ehrnrooth wrote in his combat report:

"*At 1005-1010. altitude 1,800 m. I was leading a pair (wingman 2Lt Karhila) during a fruitless search and then took a course away the front-line, flying at slow speed, when two I-153s approached from behind. When they were in the firing distance we pulled as tight turn towards them, commencing three opposite attacks, passing and tight turning again, scoring hits.*

Fokker D.XXI, FR-109 of E/LLv 32 flown by the squadron commander Capt Erkki Heinilä in July 1941, based at Utti. (Karolina Hoda)

Fokker pair FR-108 and FR-90 of 1/LLv 32 at Utti on 3 July 1941. LLv 32 flew continuous air combat patrols over south-eastern Finland during the first three weeks of the Continuation War. (SA-kuva)

The lack of 7.5 mm tracers affected my gunnery, The enemy dodged at a very close distance. My plane CU-557 had a couple of holes from explosive ammunition."

In the early morning of 22 July 1941 two pairs of Curtiss scrambled to meet *Chaikas* attacking Utti. Cpl Mauno Kirjonen entered in combat with them and claimed two shot down, writing in the report:

"At 0325-0335, altitude 200-700 m. After take-off I flew in the Haukkajärvi direction, meeting there one I-153 and got to fire at it obliquely from behind a good burst, when tracers began to come past both sides. I dodged and took a dive.

I pulled up above Haukkajärvi and at the same time one I-153 came towards me from above, we both fired and the Russian dodged below. Platings threw off my plane, the engine stopped and smoke entered in the cockpit. Because I was low I did not remain to wait for possible flames and bailed out.

After the first dodging after the first plane Cpl Heikinheimo had seen it crash east of the airfield in the Kouvola direction. Those on the ground saw the other dive into lake Haukkajärvi. My plane CU-501 fell on the field at Karhula and became a total write-off."

The *Chaikas* had belonged to 7 IAP, one piloted by Lt V.N. Shavrov, whose collision killed him and was entered in the records as a taran. But Kirjonen's report showed clearly that it was a shoot down

Karelian Isthmus

The offensive to take back the Karelian Isthmus commenced on 31 July 1941. The attack was at first aimed east of Viipuri and it was supported by LLv 32 from LeR 3, with army co-op aircraft of LLv 14 subjected to the land forces and the bombers of LLv 44. In order to increase the fighter power 3/LLv 24 was placed for the use of LeR 3.

On the ground the opposition was the 23rd Army on the Karelian Isthmus. Its air forces consisted of 5 SAD (mixed aviation division), which had two fighter regiments, 7 and 153 IAP plus two ground-attack regiments, 65 and 235 ShAP. These units

had originally about 100 I-153s, 30 I-16s and 100 MiG-3s plus 140 I-15bis, but the compliments were halved for different reasons by the Finnish offensive. Additionally 65 ShAP was transferred in August 1941 to 7th Army air forces in Karelia.

On 10 August 1941 1Lt Pentti Nurminen's swarm of 1/LLv 32 was engaged in a combat over Kirvu with an I-16 flight, dispatching two of the detachment. On the return flight several trucks, engines and artillery positions were strafed. Two days later Capt Kullervo Lahtela's swarm of 2/LLv 32 destroyed one I-153 from a ten-plane mixed formation in Antrea area.

On 13 August 1941 1Lt Pentti Nurminen's five aircraft of 1/LLv 32 met over Antrea four *Chaikas* and succeeded in destroying three of them in a fierce battle, with all friendly planes having plenty of hits. 2Lt Kyösti Karhila fought this way:

"At 1345-1400, altitude 2,000 m. On a combat air patrol we met three I-153s. I fired at them from opposite direction on three occasions. I pulled up gaining altitude and dived back, Cpl Kajanto on my wing, to attack the enemy patrol, which scattered. I saw Cpl Kajanto take one plane and I took another. After pulling up I attacked again and fired. This repeated 10-12 times until I ran out of ammunition. My victim was pouring light smoke.

Curtiss can combat the I-153 using pendulum tactics, dive, firing and pull up and again the same. No damage to my plane, CU-561."

Next day Capt Kullervo Lahtela's swarm of 2/LLv 32 entered combat over the Isthmus at Lenijärvi with a 16-aircraft strong fighter formation of 7 IAP. Out of the Russians two I-16s were sent down while Finnish losses were one Curtiss. Lahtela related this:

"At 0500-0600, altitude 200-500 m. On a combat air patrol mission south-east of Antrea the radio began transmitting. For better reception I flew to Kirvu at 500 m. At Leinjärvi I saw one enemy plane Vultee (in-line engine) throwing leaflets, I saw it straight below me but lost it from my sight twice.

I saw above me Cpl Virtanen circling with an I-16, which had yellow wing tips, no yellow belt, No. 10 on the rudder. I climbed above it and made three attacks. In the third attack I fired from

Curtiss Hawk 75A coded CUw-563 ready for a mission at Lappeenranta on 16 August 1941. The squadron, CO Maj Olavi Ehrnrooth, gives last minute instructions to the pilot, the 2nd Flight leader Capt Kullervo Lahtela. (SA-kuva)

LLv 32 pilots at Lappeenranta on 16 August 1941. Sitting from left: 1Lt V. Evinen, Capt A. Bremer, Capt H. Kalaja, Maj O. Ehrnrooth, Capt K. Lahtela, 1Lt P. Nurminen, 1Lt A. Euramo and 1Lt J. Arnkil. Standing from left Cpl M. Sarparanta, Cpl V. Virtanen, Sgt E. Visuri, Cpl N. Erkinheimo, 2Lt S. Alapuro, Cpl J. Laakso, Cpl J. Kajanto, Cpl M. Kirjonen, 2Lt K. Karhila, Cpl T. Herttua, Cpl E. Emaus, MSgt T. Heijari, SSgt M. Fräntilä, Sgt J. Tiivola, Sgt K. Perjo, WO E. Koskinen, Sgt P. Salminen, 2Lt J. Nyholm, MSgt V. Ikonen, 2Lt J. Hillo, Sgt A. Gerdt, SSgt N. Satomaa, 2Lt A. Lakio, Cpl E. Tähtö and Sgt Y. Pallasvuo. (SA-kuva)

1Lt Aimo Euramo (right) of 2/LLv 32 demonstrates his latest combat to his fellow pilots at Lappeenranta on 16 August 1941. From the left: 2Lt Mauri Aalto, Sgt Yrjö Pallasvuo and SSgt Mauno Fräntilä. (SA-kuva)

straight above when it took a steep turn to evade. The I-16 was then at 100 m altitude, where it turned on its back, dived in the forest and caught fire.

Due to poor visibility two of our four planes had separated, one of them returning to base. By reports of other participants the enemy formation consisted of 1 Vultee, 3 I-16s, 10 I-153s and 2 I-18s. My plane was CU-565."

On 18 August 1941 *Lentolaivue* 32 flew with Detachment Karhunen of LLv 24 top cover for the crossing of the river Vuoksi from the early morning. 1Lt Pentti Nurminen's swarm of the 1st Flight prevented a 13-plane detachment from entering the protection area by shooting down four I-153 fighters and scattering the rest of the formation. Cpl Mauno Kirjonen became the first ace of LLv 32 and told this:

"At 1400-1420 hours. We met one bomber with an escort of about one dozen I-153s. I got to shoot at first a short burst from ahead against an I-153, after which I pulled into a cloud. When I came out of the cloud I saw one I-153 diving towards the crossing area. I went after it and managed to shoot a long burst. The Russian began to pull up during the firing and disappeared in the cloud. I waited and after a while it came in front of me, when I was able to shoot from about 50 m from aside and behind. It went in a dive on our side and I saw a long flame coming from the starboard side. I had to pull in the cloud and lost visual contact. At the end of the fight I saw one aircraft catch fire and a man bail out from it. The aircraft crashed at the beach of Sintola next to the road. My plane was CU-566."

On 21 August 1941 Capt Aulis Bremer's three-plane patrol of 3/LLv 32 met above Rautu a *Chaika* flight and downed three aircraft of the detachment. Sgt Yrjö Pallasvuo opened his account:

"At 1415-1550 hours, altitude 1,500 m. We saw four I-153s bombing the railway at Paakkola in the Vuoksi direction. We attacked and I got into the opposite firing position with one I-153. I shot and when curving back I went in a spin. After recover-

ing and climbing back I went three times in opposite firing position and shot every time. On the last occasion another I-153 was shooting from ahead and aside, I flew in a cloud and when coming out saw the I-153 going to the east. I went after it and caught it between Metsäpirtti and Rautu, where I shot at it from behind and when it curved from straight above, when it burst into flames and crashed in the forest. The tyre of my left wheel was punctured. My plane was CU-566."

Next day LLv 32 got orders to fly top cover for the chase of the IV Army Corps in the Nuijamaanjärvi – Enso – Kaltavesi area. 1Lt Pentti Nurminen's swarm met over Pölläkkälä three I-153 fighters, sending all of them down.

On 23 August 1941 Capt Aulis Bremer's swarm of 3/LLv 32 engaged a five-plane *Chaika* flight in the Suulajärvi area and shot them all down. Bremer fought this way:

"At 0945-1030 hours. I attacked against the first I-153 with 2Lt Aalto and Sgt Pallasvuo. I fired at it from ahead, in a left bank and from straight behind. After the first one crashed on the north shore of Muolaanjärvi I attacked another. I fired at it from straight behind and below behind. Judging by the tracers the hits were right in the fuselage. Because the combat took place very low, far on the enemy side, I cannot specify the location of the aircraft. During the battle I observed a few flak explosions, which were relatively far away from the planes. My plane was CU-563."

On 24 August 1941 the IV Army Corps on the Karelian Isthmus commenced the crossing of Viipurinlahti and cut the road from Viipuri to Koivisto. Three Russian divisions, which were defending Viipuri, began their retreat too late and became encircled, leading to a surrender at Porlammi. Five days later the Finnish troops took Viipuri.

Stalemate

With mobile warfare at an end on the Karelian Isthmus, the largest combat of LLv 32 so far took place on 3 Septem-

Curtiss CUw-560 of LLv 32 taking off from Lappeenranta on 23 September 1941. The aircraft were not assigned to pilots, but 2Lt Kyösti Karhila flew this frequently and became an ace four days earlier flying this plane. (SA-kuva)

Hawker Hurricane I HC458 of Os Kalaja/LLv 32, flown by WO Aarne Arte in September 1941, based at Utti. (Karolina Hoda)

ber 1941. Capt Paavo Berg's 8-aircraft strong 1st Flight jumped a *Chaika* squadron of 65 ShAP over Siestarjoki and shot down seven aircraft in fifteen minutes. 2Lt Kyösti Karhila describes the encounter this way:

"At 1055-1105 hours. On a surveillance mission we met half a dozen I-153s. I saw one break off to the sea. I was about 1,000 m higher. I caught it easily and got by surprise behind its tail. I opened fire from 20 m off and the plane caught instantly fire and crashed in the sea. My plane was CU-566."

On 15 September 1941 Capt Kullervo Lahtela's swarm of 2/LLv 32 encountered three DB-3F bombers over Lempaalanjärvi on the Karelian Isthmus, attacking them. After a short battle all the Russians were shot down. 2Lt Sakari Alapuro wrote thus:

"At 1545.1725 hours. On an interception I saw in the Petäjänmäki area three DBs, which were heading north. I attacked one slightly from above and behind in a shallow bank. At the beginning of the firing I managed to silence the gunner and the port engine began to emit black smoke. Due to my higher speed I had to pull past it. I curved back again and fired at the starboard engine continuously, until it caught fire. At the same time the DB slipped in a cloud ahead. I banked away to the south of the cloud waiting the aircraft to return on its side, but when it did not come back, I flew towards Lempaala, where I saw a burning aircraft on a swamp west-south-west of Rahonjärvi. My plane was CU-570."

On 19 September 1941 the tasks of *Lentorykmentti 3* were specified:

1) with its main forces to protect troops on the Karelian Isthmus and crossing points over the Vuoksi river,
2) with some parts to protect Kouvola, Kotka and Viipuri areas and industrial plants at Vuoksenlaakso and
3) to reconnoitre for the army groupings on the Karelian Isthmus. LLv 32 received the western parts part of the first task.

Hurricane HC452 of LLv 32 at Lappeenranta in September 1941, after a short exchange to Tiiksjärvi with 1/LLv 10. It was assigned to 2Lt Esko Ruotsila, who claimed 2½ air victories with it.

On the same day Capt Paavo Berg's 8-plane 1/LLv 32 engaged in the Ohalatva area five MiG-3s and destroyed them all, three of them by Berg, who recounts:

"At 1320-1330 hours, at about 3,000 m altitude I saw among the clouds at first four enemy aircraft flying in the opposite direction. I took a diving turn and went after them. Just after coming out of the cloud I saw a bit lower one I-17 only 200 m away. I fired at it and it started smoking instantly and went down diving on its back.

I followed the others, which still numbered as four. I opened fire on the nearest, when it pulled down to the left pulling up again. I followed and fired when it started to smoke. At the same time two more (Alppinen and Kajanto) came to fire at it. I radioed that there are more on the left side and turned to the left to look for a new target.

To the left I saw right down to the surface two I-17s, which banked to the left and continued thereafter to the south. I caught quickly one of them, which was remaining behind. When I fired at it, it began to bank to the left, I pulled after it firing every now and then, getting closer all the time. When I got the bursts even a couple of times ahead of it, it suddenly flipped to the left crashing. The I-17 was easily caught by the CU. My plane was CU-563."

On 21 September 1941 Capt Kullervo Lahtela's 8-plane 2/LLv 32 was engaged on the Karelian Isthmus over Ollila in a combat with a dozen *Chaikas*, of which three were shot down. WO Eino Koskinen claimed one:

"At 1540-1800 hours. After arriving above Terijoki I saw two 4-plane patrols of I-153s. At the same time I observed one turning towards me, when I chose to attack from ahead and opened fire. The *Chaika* dodged down and I pulled up to the left, when observing another in front of me in a banking climb. I managed to shoot at this one from about 50 m distance when it suddenly flipped into a dive. It ended under my aircraft, when I lost visual

Hawker Hurricane I HC-451 of LLv 32 in October 1941, based at Immola. (Karolina Hołda)

Hurricane HC-451 of LLv 32 photographed during an overhaul at Immola on 21 October 1941. The acute shortage of spares kept the Hurricanes grounded for long periods. (SA-kuva)

contact and could not establish if it did crash. After this I did not see a single enemy aircraft. My plane was CU-558."

In order to destroy the Russian aircraft in Hanko base, given to the Russians in the Winter War peace treaty, an 8-aircraft Curtiss outfit was formed and known as Detachment H. It was headed by Capt Paavo Berg. The detachment moved to Nummela, closer to Hanko, on 29 October 1941.

On 1 November 1941 Capt Paavo Berg's pair of 1/LLv 32 tried to lure the Russians into combat over Hanko. A solitary I-16 was fired at by the Finns, received hits and crashed into the ground. In spite of the Finnish top cover pair the Russian top cover swarm downed the Finnish lead plane, killing Berg. Capt Kullervo Lahtela was posted to command Detachment H.

On 5 November 1941 Capt Kullervo Lahtela's swarm of 1/LLv 32 engaged over Tvärminne three I-16 fighters which chose to fight. Three more Russian aircraft took off from Hanko. One I-16 was sent down but also one Curtiss was downed.

LeR 3, based entirely on the Karelian Isthmus, for the coming winter season its main task was specified on 14 December 1941 as protection of the army groupings on the Karelian Isthmus and prevention of enemy aerial reconnaissance. A secondary task was the protection of Vuoksi industrial area and traffic centres. Reconnaissance to be carried out on each mission. LLv 32 flew on the west of the Isthmus, left border at Ohta – Punnusjärvi – Vuoksi.

The last air battle of the year on the Karelian Isthmus took place on 23 December 1941. 2Lt Mauri Aalto's five aircraft of 3/LLv 32 carried out a routine surveillance mission to the front-line. At Ohalatva two R-5s were engaged and both were shot down, these being actually Polikarpov U-2s of 174 ShAP.

Brief Analysis

During 1941 the fighters of LeR 3 had claimed 87 aerial victories. *Lentolaivue* 32 had flown 2,300 sorties, shooting down 65 enemy aircraft, losing eight planes and eight pilots in combat.

At the beginning of 1942 *Lentolaivue* 32 had 14 serviceable Curtiss and one in overhaul, plus two serviceable Hurricanes and one in overhaul.

The first battle of the year was fought on 8 January 1942, when 3/LLv 32 was engaged in combat on the Karelian Isthmus. 1Lt Pentti Nurminen's flight of six Curtiss met three I-15bis biplanes over Lumisuo, claiming all shot down.

On 19 March 1942 the Finns decided to take Suursaari back when the advance could be made on the ice. Detachment Pajari was formed for this purpose. LeR 3, LeR 4, LLv 6, 3/LLv 24 and an ambulance/liaison flight were ordered to provide support. Maj.Gen Aaro Pajari, commanding the forces, on 21 March 1942 ordered the 3500 man strong invasion force to attack:

"For the invasion LeR 3 was ordered with tasks 1) to protect the march of Detachment Pajari to Haapasaari and from there onwards, 2) to support the attack on Suursaari, 3) to hamper the retreat of the enemy from Suursaari by causing losses and 4) to carry out reconnaissance on every mission."

On 27 March 1942 the invasion troops began the approach to Suursaari. 57 aircraft were assembled for air cover, the main part from LeR 3: sixteen Fokkers of LLv 30 and thirteen Curtiss from LLv 32. Additionally LLv 6 provided five SB bombers and six I-153 fighters, LLv 24 six Brewsters and LLv 42 eleven Blenheims.

Capt Aulis Bremer's Curtiss swarm of 2/LLv 32 patrolled together with the Fokkers of 1/LLv 30 in the Suursaari area,

sharing one victory and sending down another two enemy fighters on their own account.

On 28 March 1942 in the morning Maj Olavi Ehrnrooth's six Curtiss of LLv 32 fought against a mixed formation and shot down two *Chaikas* and one *Rata*. 2Lt Jaakko Hillo had one confirmed and one damaged:

"At 0530-0730, altitude 0-500 m. On a top cover mission I observed east of Suursaari I-153 and I-16 aircraft surrounded by Wasp Fokkers. I fired at one I-153 from above and behind, when the Russians turned towards Lavansaari. While shooting at one evasive I-153 several times, one I-16 managed to shoot at me from behind, when I dodged and fired at it in a bank. The plane went down about 10 km south-east of Suursaari's southern tip.

After levelling from a steep bank I observed one I-153 surrounded by several Curtiss. After firing at it once I ran out of ammunition and broke off. I-16 had more than one heavy machine-gun and I-153 fired a couple of rocket projectiles. My plane was CU-553."

During the day the island was taken by the Finns and when LLv 32 was flying top cover for the occupation parade, 29 Soviet fighter approached in three formations. In a 20 minute combat the Finns claimed fifteen shot down without losses. 1Lt Pentti Nurminen downed two this way:

"At 1740-1800 hours. When I was leading a flight of six Curtiss in protection of Suursaari, I received a message 4/1 west from Seivästö. I took my flight west of Lavansaari. Then the radio informed that west of Suursaari first 8, then 11 and then 10 unknown aircraft at 3,000 metres. I climbed to 3,000 m above the clouds and then saw the streaks of the aircraft at about 2,000 m and right after the eight aircraft themselves, partly I-153s and part I-16s. I carried out an attack, when I got to shoot at one I-16 three good bursts from close range at the side and behind. The aircraft shed fragments and went down out of control. After this I fired at three I-153s from behind and aside. The I-153 which I shot from the side and behind descended smoking, but I had to pull up again due to other I-153s.

The Russians had no will for fighting and rather wanted to head to Leningrad and Karavaldai. I did not observe the Russians firing their mgs, instead they shot their rocket projectiles. I-16 is an inferior opponent compared with I-153. The Chaikas did not use a turn to opposite direction for dodging, but a climbing turn to port or starboard and then straight towards Leningrad direction (obviously extremely low on fuel). My plane was CU-571."

On 3 April 1942 Capt Kullervo Lahtela's eight Curtiss of 1/LLv 32 flew a reconnaissance mission to the Gulf of Finland, engaging in the Seiskari area a similar size *Chaika* outfit, of which five were sent down. The Russians shot down one Curtiss.

The last major air battle in the sector of LeR 3 on the Karelian Isthmus was fought on 7 April 1942, after which the air space became quiet for over three months. Capt Aulis Bremer's 2/LLv 32 engaged over Lumisuo a detachment of ten I-16s, sending six of them down. Bremer's report read:

"At 1520-1600 hours, altitude 500-3,000 m. I was on a reconnaissance mission (with 4 CU) to the front-line, north of Lempaalanjärvi I observed at first 3 I-16s curving at the edge of a cloud, about 1,000 m higher than us. I went up towards them, when I noticed that above the clouds there were at least ten more I-16s 1000-2000 m higher. A long-lasting combat against a superior enemy ensued. I finally got behind one I-16 at 50 metres distance and managed to shoot straight from the side and behind.

It began to smoke instantly and crashed at Pieni Suojala, northwest of Lumisuo.

I shot at another I-16, first from below and behind and after that a long and well aimed burst into the engine and cockpit from the opposite direction. After the pass I did not see the aircraft anywhere in spite of a search, so obviously it must have crashed.

As I later heard, it had indeed crashed. When the combat began we were at a low altitude due to the reconnaissance nature of our mission, so in the beginning we were in a considerably weaker position. Later this became fixed. Friendly flak disturbed the combat in the beginning by well aimed concentrated fire at our planes. In the opposite direction firing I got three hits, one which punctured the starboard tire. My plane was CU-556."

While the fronts remained calm for six months a re-organization was made in the air arm on 3 May 1942, based on territorial division. It received criticism from the fighter regiment leaders. They considered that in the new system the most important feature of the air arm, mobility, cannot be fully exploited.

Olonets

On the Olonets Isthmus *Lentorykmentti* 1 was reformed to operate for the Olonets Group. LtCol Viljo Rekola was appointed in command. LeR 1 was seconded two squadrons: LeLv 12 already reconnoitering in the sector and LeLv 32, which had earlier been a successful fighter outfit in the Karelian Isthmus. The operational area was specified as the Olonets Isthmus and the waterways on both sides, the right border was at Valamo-Uusi-Laatokka and the left one at Lohijärvi-Derevyannoye-River Vytegra.

Due to the spring thaw all elements of LeLv 32 had arrived to its new base at Nurmoila on the Olonets Isthmus by the end of May 1942.

The order of battle of *Lentolaivue* 32 was then:

Squadron commander, Major Olavi Ehrnrooth, Nurmoila (13 CU)

- 1st Flight, Captain Kullervo Lahtela, Nurmoila
- 2nd Flight, Captain Aulis Bremer, Nurmoila
- 3rd Flight, Captain Pauli Ervi, Nurmoila

South of the River Svir the Finns were faced by the 7th Detached Army and its air forces, consisted of 55 SAD. It had one bomber regiment (72 SBAP), one dive-bomber regiment (4 GPBAP), three fighter regiments (415, 427 and 524 IAP) and one reconnaissance squadron (119 RAE), This aviation division had 55 serviceable aircraft, of which only ten were fighters.

The frequent flying of the fast Russian reconnaissance aircraft via the western Olonets Isthmus needed to be controlled and the Curtiss fighters of LeLv 32 were ordered to interception duty. On 7 June 1942 MSgt Niku Satomaa was to first to get within firing range. In spite of scoring good hits on a Pe-2, its destruction was not observed and Satomaa was only credited with a damaged. The Pe-2s escaped often by pushing the throttles wide open, unless the CU-pilots had the advantage of surprise or altitude to cut the route. Later the same day, south at the mouth of River Svir, a CU-patrol claimed two MiG-3s sent down and opened the score of LeR 1. It began to accumulate fast.

On 15 June 1942 2Lt Sakari Alapuro's swarm of 1/LeLv 32 saw over Mergino air base four Yak-1 fighters in the landing circuit. The Curtiss quickly shot two down and exited the scene. Next day LeLv 32 took off for 33 sorties. 2Lt Kalevi Tervo's pair from 2/LeLv 32 engaged six LaGG-3 fighters over Mäkräjärvi. One was shot down and the others tried to break off. A second one was destroyed before the higher level speed of the Russians took them outside the firing range of the Finnish Curtiss pilots.

By 24 June 1942 a few Pe-2s were damaged over Olonets Isthmus before the first one was shot down, as told here by 1Lt Aimo Euramo of 2/LeLv 32:

"At 0550-0630 hours, altitude 800 m. Based on air surveillance report Capt Bremer ordered me and 2Lt Nyholm to scramble for intercept. Soon after this came the first bombardment, which

Curtiss CUw-560 of LLv 32 at Suulajärvi in late April 1942. This was the top scoring Hawk with 19 air victories to its credit.

was done by five Pe-2s. I climbed to about 1,000 m over the base and noticed suddenly two Pe-2s, which were passing low over the airfield towards the west. I went after them and with the sun behind me I got to about 300 m from the rearmost plane directly above. Then it observed me. I fired and the rear gunner of the enemy plane fired also 5-6 short bursts, after which there was no return fire. The aircraft began to gain distance. I followed for a while to Lake Ladoga shooting at the rearmost plane, when its port engine began to pour black smoke. The planes soon left me, when I gave up the chase returning to the base. My plane was CU-571."

Both the coast guard and air surveillance posts saw one Pe-2 sink in Lake Ladoga and Euramo was credited with an air victory. Later six Pe-2 planes escorted by two LaGG-3s bombed Nurmoila air base. 2Lt Kalevi Tervo's pair of 2/LLv 32 took off to intercept and shot one of the fighters down. The bombers were chased a long way and only damage was inflicted. Next day 2Lt Kalevi Tervo's pair of 2/LeLv 32 destroyed, in the vicinity of Mergino air base deep on the Russian side, one Pe-2 bomber.

The Soviets attacked Nurmoila with small formations throughout 5 July 1942, dropping burning naphtha and phosphorus. Several combats were fought during the alert missions. The air force headquarters credited 2Lt Kalevi Tervo of 2/LeLv 32 with two aerial victories:

"At 1130-1120, altitude 2000 m. I observed about 2,000 m lower one aircraft heading south. I pushed into a dive taking a course a little in front of it. When descending I noticed that it was a Pe-2. From about 75 m distance I shot the first burst into the fuselage and the second one to the port engine from about 30 m distance. In the middle of the burst thick smoke puffed out of the engine and it flashed into flames. The aircraft flipped then onto the port wing and crashed into swampy terrain below.

When I was about to return to my own side I suddenly saw tracers coming obliquely below and behind from the right. I quick-

ly turned my plane towards the fire and observed two LaGG-3s flying one behind the other. With a tight curve I got behind one of them. I fired a short burst at it, when the engine began to smoke and it began to descend towards the south-west. In the meantime the other tried to get behind my tail and fired a couple of bursts, which passed on the port side. Immediately after this I went to the deck and broke off the battle. I did not see any more the aircraft which I had fired at. My plane was CU-560."

2Lt Juho Nyholm's pair also scrambled over Nurmoila to meet the bombers. The wingman Sgt Niilo Erkinheimo reported:

"At 1630-1700 hours, altitude 1,500-2,500 m. I was on an intercept mission over Nurmoila and observed nine Pe-2s approaching from the west. I turned towards them and shot at one from ahead and above. I got then behind it and shot the left engine to smoke.

Then I became the target of three gunners and pulled to the right. I noticed then one plane below me and went after it. I fired from about 200 m a couple of bursts to the fuselage and then moved to the right engine, which began to smoke strongly. Then I shot at the left engine from 50 m and it began to emit black smoke, which a while later turned to white. I fired one more burst when the plane caught fire, One man bailed out.

Due to engine trouble I made a forced landing at Kuittinen on the belly, only the prop blades bent. My plane was CU-558."

On 7 July 1942 Capt Aulis Bremer's pair of 2/LeLv 32 took off for interception to the River Svir engaging two Pe-2 bombers escorted by one LaGG-3. The escort fighter and one bomber were shot down near Lotinanpelto. Bremer described his mission:

"At 1540-1720 hours, altitude 3,000 m. I was on an interception mission with WO Koskinen over the River Svir at about 3,000 m altitude. When I looked back I saw three small dots in the horizon. I turned towards them. When approaching I observed that they were two Pe-2s escorted by one LaGG-3 fighter.

Pilots of 1/LeLv 32 at Nurmoila on 21 June 1942. From the left: Sgt Niilo Erkinheimo, WO Viljo Ikonen, 2Lt Kyösti Karhila, flight leader Capt Kullervo Lahtela, SSgt Unto Alppinen, 1Lt Veikko Evinen and SSgt Paavo Aikala. (Finnish Air Force)

Pilots of 2/LeLv 32 at Nurmoila on 21 June 1942. From the left: Sgt Väinö Virtanen, 2Lt Kalevi Tervo, SSgt Paul Salminen, flight leader Capt Aulis Bremer, SSgt Jaakko Tiivola, 1Lt Aimo Euramo, 2Lt Mauri Aalto and 2Lt Juho Nyholm. (Finnish Air Force)

Pilots of 3/LeLv 32 at Nurmoila on 21 June 1942. From the left: MSgt Mauno Fräntilä, Sgt Aimo Gerdt, 1Lt Esko Ruotsila, SSgt Lauri Jutila, flight leader Capt Pentti Nurminen, SSgt Lauri Suominen, 2Lt Risto Pennola and SSgt Jaakko Kajanto. (Finnish Air Force)

Curtiss CU-503 of LLv 32 at Nurmoila in June 1942. This plane was credited with ten aerial victories. It was originally a Cyclone-engined Hawk 75A-4, but these engines were unreliable and were changed to Twin Wasps from December 1941 onwards.

When I arrived at about 1 km distance, they turned towards us. I took short cut and turned against the Pe-2s, when the LaGG-3 turned to me. I fired a short burst from ahead and below. Then I pulled a swift turn in order to get behind the LaGG-3. Now my plane went out of control for a moment. I managed to level it without losing much altitude. In the meantime it had attacked WO Koskinen, but he made an evasive manoeuvre. Now it was pushing towards me from the side, shooting at me all the time. But I was in no danger area since it was aiming too low and had too much deflection. I curved immediately behind it and got into a distance of only about 50 m behind it. I fired a good burst from left and behind, when it began to pour smoke. It made still a few shallow turns to both directions, but stayed solidly in my sight and I held continuous fire at it. It then went in straight dive to the Russian side over Lotinanpelto. I followed it all the time and kept firing until I ran out of ammunition. During the dive it stayed as well as possible in my sight, which was also proved by the tracers striking in the fuselage of the plane. South-west of Lotinanpelto it disappeared, crashing into the forest. My plane was CU-552."

On 19 July 1942 SSgt Paul Salminen's pair of 2/LeLv 32 was engaged during an escort mission in a combat with four fighters near Pasha airfield, sending three of them down. These proved to be 524 IAP fighters. Sgt Väinö Virtanen claimed two:

"At 1300-1430 hours, altitude 1,800-500 m. When protecting an artillery fire control plane, two I-16s and one LaGG-3 attacked from behind the fire control plane. I was 500 m above these, attacking after one I-16. I got within 50 m and started firing, when it banked to the right. Then I got to fire so that the rounds hit the engine and cockpit, when its nose went slightly up and flipped into a vertical dive and disappeared.

When Sgt Aaro Kiljunen of 1/LeLv 32 was hit in combat on 20 July 1942 flying CU-504, he made a good landing to Nurmoila, but right after that the plane caught fire and burned out. Even shovelling sand over the plane did not help.

Curtiss CUw-558 of LLv 32 on jacks at Nurmoila on 2 August 1942. The air force photographer took a series of pictures for the recognition manual. This plane scored 17 kills. (Finnish Air Force)

1/LeLv 32 pilot Sgt Nils Erkinheimo was hit in combat on 5 July 1942, making a good forced landing with CUw-558 at Kuittinen in Olonets. The plane was repaired by the unit and flew again after eight days.

I could not see the final fate of my victim as I was attacked by another I-16. I fired at it from opposite directions after which we circled for a while. After a tight curve I got behind it and shot a long burst, hitting the fuselage from 200 m. The plane made a couple of quick rolls, I continued firing when the plane went in a dive.

After this I saw SSgt Salminen circle with one I-16. I joined him. We chased it southwards shooting in turns. I ran out of ammunition, but SSgt Salminen continued, My plane was CU-552."

Next day 1Lt Veikko Evinen's swarm of 1/LeLv 32 escorted an FK fire control aircraft at Krestnojärvi, coming under attack by five fighters. One LaGG and one MiG were downed, but on landing at Nurmoila one Curtis caught fire since bullets had punctured the fuel tanks. The field personnel tried to ex-tinguish the fire by pulling the tail apart, but the fuselage was already gone.

On 25 July 1942 a Curtiss swarm of 1/LeLv 32 engaged near its base different enemies, three flying boats of 59 AE, KBF, sending all down. Capt Kullervo Lahtela described the melee:

"At 2120-2200 hours, altitude 0-200 m. When I took off for the interception I saw Olonets' flak firing and soon observed three MBR-2 flying boats flying from Olonets to Nurmoila at about 200 metres. When Sgt Erkinheimo and I attacked these from be-hind, the port wing plane broke off turning to the left. After that I saw WO Ikonen shooting and one MBR-2 went down in the forest about 5 km south-east of the base.

I went after the other one, which had turned to the left while WO Ikonen and Sgt Erkinheimo remained behind the one turning

Second lieutenants Yrjö Pallasvuo (left) and Kalevi Tervo of 2/LeLv 32 at Nurmoila on 24 August 1942, when both shot down one and damaged another MiG-3 over the River Svir between Mergino and Lotinanpelto.

to east. We shot in turns with Sgt Kiljunen at the plane having turned to the left, as long as the return fire stopped, the engine began smoking and the plane crashed in the forest. I landed with Sgt Kiljunen, since I did not see the third MBR-2, which was chased by two Curtiss. My plane was CU-552."

On 2 August 1942 1Lt Aimo Euramo's pair of 2/LeLv 32 was engaged over the River Svir with six LaGG-3s, destroying three of them. The remaining ones turned away to flee.

On 13 August 1942 2Lt Kalevi Tervo's swarm of 2/LeLv 32 was on an interception mission to Lotinanpelto, coming under attack by four MiG-3s. To the advantage for the Finns the Russians submitted to a turning battle and lost three fighters of their strength.

On 24 August 1942 Capt Aulis Bremer's swarm of 2/LeLv 32 struggled against eleven MiG-3s, sending three down and damaging another three between Mergino and Lotinanpelto. One of the combatants was 2Lt Yrjö Pallasvuo, who claimed one destroyed and another damaged:

"At 1545-1715 hours, altitude 5,000-10 m. When on a recce and escort mission I met about 20 km south of Lotinanpelto at about 5,000 m two MiG-3s, 100 m higher than me. The enemy flew to the north-east. I turned after them and got by surprise behind and below the wing plane to 50-75 m. I shot a long burst to the fuselage, after which the enemy banked to the port side, slid on the side some distance and then in a wide uncontrolled spiral went down. I kept my eyes in it and when a small dot it continued the spiral and went into the forest.

When the lead plane saw what happened it went into a dive via a half-roll, and I followed. The MiG-3 levelled after a 1,000 m dive, I stayed with it and shot a burst from behind when it pulled up to the left. The distance grew to 500 m. The Russian levelled after the pull-up and took a course towards Mergino. To disturb it I fired a burst towards it. I seemed to hit it, since its speed went down and I caught it easily. I managed to shoot from straight

behind from 100 m. The plane curved to the left and then I fired a long burst to the fuselage. Then it took a roll and went into a vertical dive towards the ground. I followed a while but then lost it from my sight.

During the last combat the machine-guns were not working properly and I had to take several cocking manoeuvres. I was in a shallow dive and tried the guns, but only the left fuselage gun worked a while and then all were quiet. Also the trigger had jammed. I pressed into a dive and went to the deck, where I thought my speed would take me away from possible enemies since I could not fight back.

When flying at tree top height, on right side of the engine was a sudden explosion and flash, The engine started to cough. I looked behind and saw a MiG-3 right behind me, shooting. The engine was running poorly and occasionally almost stopped, when I started to look for a place to force land. Luckily the engine continued to run, but coughing. I did not dare to make any curves and flew straight on waiting for the next Russian burst to hit. To my astonishment it did not come and I looked back. I saw about 500 m away from me on the right side a CU flying at the deck. I carefully curved and identified it as 2Lt Tervo's plane. We took a course to the north and together flew towards the River Svir. About 1 km before it we flew over a Russian flak battery, which shot fiercely at us. But by pressing to the surface of the nearby field we got undamaged to the friendly side over the river. 20 mm hit to the engine from right behind. Splinters to the propeller. My plane was CU-554."

After the FR reconnaissance planes of LeLv 12 had passed Mergino airfield, the fighter units always took off after the Fokkers. Since the number and battle willingness of the Russians had increased, a trap was arranged on 5 September 1942. The aim was to lure the enemy fighters after the Fokkers to the north side of the River Svir, but a Pe-2 passing the scene at random tempted the ambush planes too soon to chase after it, which meant that the combat had to be fought on the Russian side. Seven Curtiss of LeLv 32 and 35-40 Russian fighters participated in the battle. The Finns claimed the passing Pe-2 and ten fighters, damaging another seven, with no Finnish losses. MSgt Paul Salminen shot down three fighters:

"At 1120-1315 hours, altitude 0-3,500 m. I was on an interception mission with WO Koskinen. We engaged over Lotinanpelto one Pe-2 escorted by four LaGG-3s. I got after the Pe-2, but the mgs did not work. I made cocking manoeuvres and while doing so lost contact with the enemy planes. I flew south of the River Svir, where I met 2Lt Tervo. We caught in our vision the Pe-2, which was heading south. We chased it, but did not catch it.

2Lt Tervo dived onto the deck and then I saw four LaGG-3s and I-16s attack us from above. I began curving with these and got several times in a firing position behind LaGG-3s. The Russians were firing at me continuously, but very inaccurately. I noticed one LaGG-3 take a dive and I followed after it. I kept firing at it during the dive. It levelled on the deck and I managed to shoot at it from about 50 m distance. After a while the LaGG crashed in the forest 10 km north of Mergino.

I took a climb and engaged at about 2,000 metres altitude three LaGG-3s and two MiG-3s. I managed to take by surprise one LaGG-3. I shot at it from 50 m distance straight from behind. At the same time three planes were firing at me. I dodged and lost sight of the plane I had fired at.

I curved for a while, but when I could not do other than fire short bursts, I broke off by diving through a cloud to the deck. I flew to the River Svir and gained altitude. I saw below me one CU chased by three I-16s and one MiG-3. I dived and attacked the I-16 firing at the CU and it burst in flames with the first alvo. It crashed in the River Svir west of Lotinanpelto.

I began to chase another I-16. I got to fire at it from 30 m distance from behind. Its engine stopped and I was about to ram it. The starboard wing of the I-16 swept off my port side antenna wire. The plane crashed in the forest 5 km east of Lotinanpelto.

I began climbing and headed east and saw below me one CU, which was chased by an I-16. I dived and attacked the I-16, shooting first from behind and then in a bank. We got fierce flak fire from the ground and in a turn my engine was hit, beginning to emit black smoke. I had to abort the battle and returned to base. The I-16 continued its flight to the south. My plane was CU-558."

On 30 September 1942 a CU-swarm of 2/LeLv 32 managed to cut the route of a reconnaissance Pe-2. 2Lt Kalevi Tervo acted thus:

"At 1105-1200 hours, altitude 4,000-2,500 m. On an intercept mission I met one Pe-2 near Vonozero at about 4,000 m altitude. It came from north-east across my flight path. When I tried to cut its route, it curved first to the east and then straight to the north. I cut across after it, when it turned back to the south coming towards me about 300 m below. I fired first from ahead and then from the rear above. It went into a glide and lost speed so that I could easily follow it. I shot continuously to both sides from close range. The port engine was smoking strongly. When only one of my machine-guns worked, I stopped firing and just followed its flight. It went in a shallow dive into the forest, where it caught fire. My plane was CU-552."

The search on 12 October 1942 for an observation balloon in Olonets by LeLv 32 brought results. WO Viljo Ikonen described the shoot down with Sgt Lauri Mäittälä this way:

"At 1250-1335 hours, altitude 1,200 m. My task was to search for a Russian observation balloon over the mouth of the River Svir. I found it 2 km north of Ojatti at 1,200 m altitude. I managed to approach it in the cover of the clouds to 700 m distance being then 500 m higher. I fired twice. It caught fire on the second burst. I did not observe any bailout. 40 mm flak kept firing. The colour of the balloon was light grey, it blended well with the clouds behind it. My plane was CU-559."

On 7 November 1942 Capt Pentti Nurminen's CU-swarm of 3/LeLv 32 escorted an FK plane photographing the front. A MiG-3 swarm tried to get the FK, but the opposite happened and all the MiGs were downed. The combat report of SSgt Aimo Gerdt told:

"At 1215-1400 hours, altitude 500-3,000 m. On an escort mission I met on the east side of Lotinanpelto two MiG-3s. I got behind the lead plane when it was trying to escape in the Savijärvi direction. I shot it into flames from about 200 m distance and the pilot bailed out above Lyugovitsha. This occurred at 1,500 metres.

When I was heading towards Savijärvi at 1,000 metres a MiG-3 came towards me from above to port, trying to get behind me. I dodged down to port and so we took a 360° turn, after which I managed to shoot straight in the cockpit from behind on the port side. The Russian turned to its back and fiery sparks came behind the engine. Then it went down into the deck southeast of Lyugovitsha. I stayed easily behind it in a turn. In a climbing turn I could not follow up to the end. My plane was CU-503."

On 9 November 1942 1Lt Kalevi Tervo's swarm of 2/LeLv 32 reconnoitred the Lotinanpelto area observing a single Pe-2 bomber, which was shot down. Shortly after, it jumped a MiG-3 detachment downing two aircraft.

2/LeLv 32 leader Capt Aulis Bremer was hit in combat on 24 October 1942 and bellied CU-559 at the town of Olonets. This time the plane was sent to the factory for repairs and returned after six months.

The first six months on Olonets Isthmus were quite successful for *Lentolaivue* 32. It had destroyed 65 enemy aircraft without losses to itself.

On 1 January 1943 *Lentolaivue* 32 had nine Curtiss in working order and two in overhaul. The River Svir front was quiet, but still something happened.

The new *Lentolaivue* 34 was established on 23 January 1943. While waiting for the arrival of their new fighters, Messerschmitt Bf 109G-2s, LeLv 32 commander Maj Olavi Ehrnrooth took charge of LeLv 34. He also got the authority to hand-pick the flying personnel from other squadrons and most lost their best pilots, including LeLv 32 giving 14 pilots, leading to a re-organizing of the squadron.

On 9 February 1943 one Pe-2 photo plane was caught at noon over Olonets. 2Lt Kyösti Karhila of 1/LeLv 32 surprised it from above with the sun behind him, shooting it down near Shotkusha station. Two days later Karhila scored again:

"At 1005-1110 hours, altitude 0-4,000 m. On an intercept mission with 1Lt Arnkil, WO Ikonen and Sgt Kiljunen, I observed above Lyugovitsha at about 3,500 m one aircraft performing vertical turns. I radioed this to WO Ikonen, who was with Sgt Kiljinen higher and closer to the plane. I saw WO Ikonen attack the aircraft. This one pressed underneath him, then made a half roll and continued straight on into a vertical dive. Now the pilot bailed out of the plane, but I did not observe a parachute opening. The aircraft crashed into the forest about 5 km SE of Saarentaka station.

We assembled now and gained altitude over Lotinanpelto. Then I observed two aircraft on the south side of the power station, I informed the others by radio to go and take a look at them. I found out that they were two FR planes, but I also observed an aircraft in the south a little higher. I announced this on the radio and commenced the attack. When I got closer I discovered that there were more aircraft, one high in the sun and two lower in a pair, the aircraft being of the LaGG-3 type.

I pushed towards the pair, which after noticing me made a half roll. I went instantly after the wingman, who then pulled up. I managed to make a short cut then in the pull up and ended about 50 m from it and fired. The tracers hit the plane's engine as far as I could see. Now the LaGG-3 began to make a steep left curve. I got easily a good deflection and fired. Now the plane immediately straightened out and pulled up. I managed to shoot from about 50 m into the oil cooler of the LaGG-3. The plane instantly caught fire and went down burning in an inverted dive. I could not establish the crash site since the Russian lead plane was a little lower. This occurred over Bolato at 3,000-4,000 m.

I went immediately after the other LaGG-3. When I fired at it from about 100 m distance from behind and above, it pulled up to the left and began curving. I managed to shoot twice, the plane began to smoke. It went in a dive to the south and succeeded in leaving me. Then Sgt Kiljunen came behind me and because of greater altitude he got more speed, passed me and fired.

Then I observed at the port side below one LaGG-3 also heading south. I went after it and fired from about 100 m distance to the closest possible distance. The tracers sank in the fuselage. Now I noticed that the flaps dropped, I pulled away and saw Sgt Kiljunen also fire at the same aircraft and also pull away. I fired once more at the plane in question from the right side above and behind. The aircraft had only 50 m altitude then. It went down at an about 30 degrees angle and hit the ice in the north end of Lake Savijärvi. It raised a cloud of snow and when it settled I observed that the plane had travelled about 30 m and turned about 135 degrees to left. The port wing appeared broken, since the wings had a large V-angle, the starboard wing being considerably higher.

Now the previous LaGG-3 returned to check the fate of his comrade. I fired at it in a bank, but did not get a proper aim. The plane went to the south again in level flight, we followed it, but could not get a proper contact with it. The I observed that the flak was firing a few hundred meters to the right and there was Novinka air base. It was heavy flak, which shot at the shortest possible fuse, the explosions puffing halfway between the guns and my plane, no problem. I yelled in the radio 'get home and fast!'

I observed two large aircraft on the airfield, camouflaged in green. I failed to see the previous LaGG-3. I went down to the

CU-555 returned from factory repair on 22 November 1942 to LeLv 32 at Nurmoila, seen here a few weeks later, with touched up paintwork.

deck like Sgt Kiljunen and flew homewards. Nobody was following. Suddenly I observed a gaudy black and white striped U-2 flying on the deck towards me. I pulled up and shot from left and behind above at the U-2. The tracers hit the fuselage in the cockpit. The plane went straight in the forest in the north end of lake Malkjärvi, the port wings hit a tree and broke off and the plane nosed over. We continued the journey on the deck to our side and home. My plane was CU-560."

On 4 March 1943 a swarm of 1/LeLv 32 entered combat on the return flight, as described by Sgt Aaro Kiljunen:

"At 0940-1150, altitude 2,500-0 m. When I returned from an FK escort mission with 1Lt Taina, 2Lt Karhila and SSgt Erkinheimo, we observed with 1Lt Taina one Pe-2 and one MiG-3 at about 4000 m heading south. We chased them to the east side of Mergino but could not catch them. When we turned back I lost 1Lt Taina from my sight and shortly after that three MiG-3s came towards me slightly below me. I managed to shoot at one from straight ahead and slightly above. I saw that the tracers sank in the engine and the canopy, when I got the impression that it had received enough. With the sun behind my back, the aircraft did

not spot me since it did not fire back. When I pulled up to the left after this I noticed that the aforementioned plane going towards the ground at about 60 degree angle and the other two pulling up towards the sun. After this I could only dodge continuously, since they we using the so-called pendulum tactics attacking in turns and always pulling up towards the sun. I had to retreat to Segesjärvi level until the aircraft left me alone. I was now low on fuel and I had to return directly to the airfield. My plane was CU-562."

On 18 March 1943 LeLv 32 received from the depot the first refurbished war booty LaGG-3 fighter, which was intended to be used in the capture of the fast Pe-2 bombers. The LG planes were plagued with technical problems and the Pe-2s were not much bothered.

On 5 May 1943 a CU swarm of 3/LeLv 32 fought against a superior number of LaGG-3s, as told here by 1Lt Veijo Taina:

"At 1235-1415 hours, altitude 4,000 m. We were chasing two fighters. After crossing the lines I was about 500 m higher than the others. I had warned the boys of another possible six fighters, of which I had been informed by radio. When I looked back

LaGG-3 LG-3 of LeLv 32 flown by WO Eino Koskinen in March 1943, based at Nurmoila. (Karolina Hołda)

2/LeLv 32 leader Capt Aulis Bremer about to go for a test flight with LG-3 from Nurmoila on 19 March 1943. Contrary to the unit's other planes, the LaGG-3s were assigned pilots, this one WO Eino Koskinen. (Finnish Air Force)

I observed that two enemy aircraft had attacked our last planes. I made a steep bank and dive to left. One enemy fighter had began to perform a climb and banked to left. While it was banking I opened fire at it from the bottom side, it now flipped over on its back and commenced a dive, which ended on the ground. I saw it crash in the forest at the edge on a swamp. Looking for the location on my map it was about 9 km east of Sekeenkylä village.

The other three planes of the swarm continued their flight to the east. I lost them for a while from my sight until I observed two friendly aircraft and above them 4-6 enemy fighters, two of which were up at about 6,000 m altitude. I saw one of ours in a curve battle with the enemy and I tried to disrupt the enemy fighters with long distance firing, but obviously it had no effect. After these observations I had several clashes with the enemy fighters, but all without any results. Until one LaGG-3 suddenly dived to attack me from the starboard side. I thought that it was a CU plane and did not evade at first, but at the distance of about 30 m I realized that it was a LaGG-3. A short curve battle came out, the LaGG-3 broke momentarily off and came towards me slightly from the side. I opened fire and the burst hit its port wing, which literally snapped off. The plane flipped in a split second to a faster than usual spin and hit the ground in a twist at Saarentaka station. I followed it down to about 100 m altitude. My plane was CU-503."

More Curtiss Fighters

Fifteen more Curtiss Hawks were purchased from the German war booty depots and sent to the factory for refurbishing. The first completely overhauled Curtiss arrived with LeLv 32 on 10 August 1943. During the rest of the summer, seven more planes appeared, thus improving the plane situation of the squadron somewhat. On the average there were ten to twelve serviceable fighters.

On 8 September 1943 the 3rd Flight succeeded in shooting down a Pe-2 reconnaissance plane. 2Lt Pentti Virtalahti wrote:

"At 0840-0930 hours, altitude 5,500-2,000 m. While stalking a Pe-2 above Pisi at about 5,500 metres, I noticed a single plane east of our field. Being alone, I guessed it was a Soviet plane and headed for Kinkiyeva right away. As I was informed of it by radio I was close enough to feel it already. Over Kinkiyeva I got behind it at about 200 m and fired at the right engine. After a few short bursts it began to smoke and the plane accelerated so that I was left about 1 km. After a brief chase I gained on it again. By now the right engine smoked heavily. I took aim on the left engine and gave it a long burst; then I had to pull up as the Pe-2 was by now much slower than I was. After the pull-up I saw it diving on one side with the left engine in flames. From this attitude it went in a vertical dive and fell directly into a small lake. As I looked around then, I saw a single parachute float down. At first the Pe-2 shot back hard with ventral and dorsal guns. In the second stage it fired no more. No hits on my plane CU-553."

On 2 November 1943 the 2nd Flight sent six CU planes to cover a Blenheim photo recce plane south of the river Svir. The air battle of the return flight is described by 2Lt Jarl Lemström:

"At 1010-1020, altitude 4,000-1,000 m. On return from the escort we noticed three LaGG-3s to the front and left of us some 500 m below. I singled out the last of them and did a half roll to reach it. After it tried to evade with a tight turn and a hammerhead turn I managed to get in a shooting position. After three short bursts in the engine the LaGG-3 began to smoke. It stopped soon and the LaGG-3 pulled up. Due to my speed I could follow, the LaGG-3 started a tight turn again and turned on its back. Now I got to fire again and after a couple of bursts it ignited, the right wing burst, and the plane crashed vertically. Before impact the pilot bailed out and fell down some 3 km east of the plane. The man looked dead when I greeted him. Soviet 20 mm flak shot at us all through the battle. My plane was CU-578."

At the beginning of 1944 Lentolaivue 32 possessed 18 serviceable Curtiss while another three were in overhaul. On 14 February 1944 the unit became Hävittäjälentolaivue 32 (HLeLv 32).

On 16 February 1944 the Curtiss flight of 1Lt Reino Hakulinen of 3/HLeLv 32 reconnoitred the Ojatti-Alehovtsina region and met a Pe-2 at Kuuttilahti, accompanied by two LaGG-3s. One swarm tied up the fighters and the other dis-

Curtiss CU-551 of LeLv 32 in new Warpaint parked at Nurmoila on 1 July 1943. At this point the complement of the squadron was at its lowest, with only six Hawks and one LaGG-3 in working order.

Curtiss CU-505 of LeLv 32 running up at Nurmoila in August 1943. The splinter boxes were meticulously built of logs. The walls were supported by sand banks giving good protection to all but direct hits.

1/LeLv 32 at Nurmoila in October 1943. From the left: 1Lt Martti Joutsen, 1Lt Mikko Sarparanta, 1Lt Olavi Auvinen, flight leader Capt Jorma Visapää, 2Lt Hans Niemeier, 1Lt Jaakko Hillo, WO Viljo Ikonen and 1Lt Olli Mäenpää. To the right are the flight's mechanics.

patched the bomber. A pair of captured LaGG-3s joined in the combat, WO Eino Koskinen of 2/HLeLv 32 scoring the solitary air victory of LG planes, reporting this way:

"*At 1045-1135 hours, altitude 4,000-30 m. While on interception I met one Pe-2 and two LaGG-3s north-west of Kinkiyeva. I was some 200 m below them, and I took on the Pe-2, but the LaGG-3s noticed me and one of them turned into me. We flew at each other, but when the other got on my tail I had to duck, managing to lose it. Now the other came at me again and I shot at him face on. The plane I fired at dived away and fled the fight.*

The other came up behind me again. After a few turns we went into a curve battle, during which I fired and hit, and the right landing gear dropped out. The Russian tried a few vicious turns and finally went down in a dive. I went after it, and at the

Svir monastery I caught it. I fired, got some smoke out, and while it was puffing heavy smoke, we flew in tandem on the deck. Finally he went up to 50 m, and then on into the forest, on its left wing. The Russians fought bravely, even after only one remained. It was tough when there were two. My plane LG-1 was intact."

On 24 February 1944 1Lt Mikko Sarparanta of 3/HLeLv 32 took his Curtiss swarm to reconnoitre the Ojatti area, finding dummy planes in covered shelters, and a crash-landed LaGG-3 at the centre of the field. It was shot up. La-5s were seen in the air for the first time on the return leg.

26 March 1944 showed things to come. A photo reconnaissance Blenheim was escorted by twelve Curtiss from 2/HLeLv 32 to Lotinanpelto, led by Capt Aaro Virkkunen, where an aerial battle ensued with two LaGG-3s and three

1Lt Jaakko Hillo of 1/LeLv 32 flies Curtiss CU-580 on an escort mission over the River Svir on 16 October 1943. Hillo accumulated a total of 7½ air victories. (SA-kuva)

Curtiss Hawk 75A CU-551 of LeLv 32 flown by 3rd Flight leader 1Lt Veikko Evinen in October 1932, based at Nurmoila. (Karolina Hołda)

The escort mission of 16 October 1943 was also photographed in colour. The four Hawks of LeLv 32 involved were CU-580, CU-505, CU-574 and CU-581. (SA-kuva)

La-5s. One Curtiss was shot down, the pilot bailing out. On the same day at Nurmoila, two La-5s surprised a Curtiss engaged in ground target practising, and turned it into a sieve.

On 8 May 1944 1Lt Pentti Virtalahti's Curtiss swarm from 3/HLeLv 32 met two La-5 fighters over the western Karelian Isthmus and shot one down.

On 12 May 1944 six *Mersus* of 1/HLeLv 24 flew to Nurmoila, led by 1Lt Lauri Nissinen. The main task was to oppose the La-5s fighters, which were harassing the almost obsolete Curtiss Hawks at will. The assignment ended on 3 June 1944 and the *Mersus* claimed five La-5s shot down.

On 16 May 1944 1Lt Nissinen led five MTs of 1/HLeLv 24 to intercept the enemy over the town of Olonets. Over Mäkriä the detachment found two La-5s and attacked. Both Russians were shot down. Next day the three *Mersus* of 1/IILeLv 24 led

by 1Lt Nissinen met four La-5s above the Svir power station. In the short battle that resulted, one enemy fighter was shot down.

On 18 May 1944 a Curtiss swarm from 3/HLeLv 32, led by 1Lt Veikko Evinen, flew escort for Blenheims over Lotinanpelto when they met a flight of La-5s. In the ensuing battle one enemy fighter was shot down.

On 28 May 1944 a large battle was fought between Mergino and Svir. More than 20 Red fighters and six Curtiss of 3/HLeLv 32, as well as two Messerschmitts of 1/HLeLv 24, were in the fray. One Curtiss was shot down, balanced by six La-5s. 1Lt Lauri Nissinen reported:

"At 09.45 I took off with two MT for intercept. North-west of Savijärvi we met our CUs in battle with La-5s. Above the battle at 4,000 m there were La-5s and we went for these. I took a close shot at the top cover leader, which shed pieces and went down vertically.

Pilots of 2/HLeLv 32 in front of CU-560 at Mensuvaara in late July 1944. From the left: SSgt Erkki Emaus, SSgt Urho Pohto, MSgt Jaakko Tiivola, flight leader 1Lt Reino Hakulinen, 1Lt Ossi Marttila, 1Lt Jorma Pesola and 1Lt Kyösti Kitunen.

LaGG-3 coded LG-3 of HLeLv 32 at Mensuvaara in July 1944. It was still used for interception missions, but was becoming obsolete.

Curtiss P-40M KH-51 of HLeLv 32 in July 1944, based at Mensuvaara. (Karolina Hołda)

During the battle I took shots at many planes but the battle was too intense for me to check results. At 10.30 I took on five La-5s over Lotinanpelto airfield at some 2,000m and gave three bursts into the last La-5. It dived with a lot of smoke coming out. I curved around for a while with the others, but fuel starvation forced me to break off. My plane was MT-235."

On 6 June 1944 1Lt Esko Eerola and his CU swarm of 1/HLeLv 32 scrambled for an interception over Lake Ladoga. The Curtiss met a plane they took for a B-26 Marauder with Russian insignia, and they shot it down. Later it was learnt that it was a Douglas Boston.

Major Offensive

On 21 June 1944 the Red army started a major offensive in Olonets, by advancing over the River Svir. The air support came from the 7th Air Army, which possessed 220 fighters, 90 assault planes and 160 bombers with another 110 bombers for the break through. To oppose this air armada *Hävittäjälentolaivue* 32 threw into the battle fifteen Curtiss Hawks.

Early in the morning of 23 June 1944 the Russians invaded the Finnish rear with a coastal brigade at the mouth of River Tuulos on Lake Ladoga, and the Finns had to retreat before this overwhelming force.

Next day 3/HLeLv 32 attacked amphibious craft of the invasion fleet with eight planes, mission report stating this:

"Planes and crews: CU-581 Capt Evinen, CU-585 1Lt Alapuro, CU-573 1Lt Virtalahti, CU-553 1Lt Hietamäki, CU-562 1Lt Klemola, CU-584 1Lt Eerola, CU-574 SSgt Kiljunen, CU-578 Sgt Ojapalo. Low-level attack on amphibious craft emerging north-west from Ontranlahti. Attack at 14.30-15.30. No vessels or boats at Ontranlahti. Attack made against vessels approaching from Tuulosjoki bridgehead. Vessels probably had armour plating, as machine gun fire had no effect. Weather 5/10, clouds at 200 m. 20 mm and machine-gun fire shot down 1Lt Virtalahti and all planes were holed. Capt Evinen crash-landed at Vitele on the return leg (died next day)."

On 28 June 1944 the Finnish VI Army Corps initiated delaying tactics at Olonets and arrived behind the River Vitele line. Petrozavodsk was evacuated by Finnish troops at the same time and the troops went to the Vieljärvi region. In the evening 1Lt Alapuro and his swarm of 3/HLeLv 32 reconnoitred the Suurmäki-Säntämä area, and when they met two

HLeLv 32 also had one Curtiss P-40M Warhawk with serial KH-51, as it was thought to be a Kittyhawk. It was used only on familiarization flights, twenty in total. It is seen here at Mensuvaara soon after arrival on 3 July 1944.

LaGG-3s. They forced one to crash in the forest. Alapuro reported:

"At 1820-2000, altitude 100-10 m. On a recce mission east of Vitele I observed two LaGG-3s heading north. The Russians were below us. I dived after them. The LaGG-3s noticed us and banked west, one beginning to gain altitude and the other continuing straight ahead. I cut the curve and caught the latter LaGG-3. I fired at it from 300 m, when it took a tight turn to left trying to shoot against me. After the pass the LaGG-3 pulled a little up and tried on several times to turn behind me with slow speed.

On one occasion when the LaGG-3 turned behind me, I tightened the turn just so much that the LaGG-3 could not get enough deflection. When my speed was 200 km/h, I shut the throttle and pulled the control column. Then the speed of the LaGG-3 decreased too much and it stalled from 10 m altitude to the forest and caught fire.

The LaGG-3 pilot was tough and aggressively came on, but was too eager in choosing a combat manner which less suited the LaGG-3. My plane was CU-584."

On 1 July 1944 1Lt Jaakko Hillo of 1/HLeLv 32 and his Curtiss pair attacked a ground attack plane formation at Olonets and shot down one Il-2. Three days later, when 1Lt Sakari Alapuro of 1/HLeLv 32 was on a recce mission with six planes, they met a large Russian formation arriving from Lake Ladoga. The Curtiss attacked and shot down one Boston and two La-5s of 415 IAP. The fighters were claimed by 1Lt Martti Joutsen:

"At 1330-1520 hours. Nine Bostons escorted by twelve fighters came towards us on a recce mission. We attacked them, I got behind one fighter and fired from 20 metres. It pulled up and dived into Lake Ladoga.

On return a large number of La-5s and LaGG-3s attacked us over Salmi. After 1Lt Alapuro told us to pull up to the cloud line, I saw 500-1,000 m lower two La-5s. I dived to attack and began shooting from a longer distance. One went into a dive and increase somewhat the distance. It levelled at the deck and continued straight ahead. I dived an a 45 degree angle cutting the corner and got to fire from close above. The enemy went into the forest.

Then 2+2 La-5s attacked, harassing me for about 20 minutes. When fighting at the deck the enemies attacked at the same altitude in 100-200 m distances. When one pulled up the other started shooting. CU is hopelessly slow and poor in climb compared to the La-5. My plane was CU-586."

The VI Army Corps retreated on the coast of Lake Ladoga and reached the Pitkäranta-Loimola line (also known as the U line) on 10 July 1944, and settled down for defence. The U line withstood all Russian attacks and the fighting subsided in a week.

On 15 July 1944 six Curtiss of HLeLv 32 and four Brewsters of HLeLv 26 escorted the 33-plane bomber formation of LeR 4 as they attacked artillery positions on the coast of Lake Ladoga. Enemy fighters intervened and shot down one Blenheim, but not without loss: the Brewsters took down one Yak-9 and the Curtiss one La-5.

2/HLeLv 32 achieved the last victory of the squadron on 27 July 1944. On reconnaissance to Hyrsylä, a six plane Curtiss detachment sent down a U-2 liaison plane. The detachment leader 1Lt Jorma Pesola related thus:

"At 0410-0540 hours, altitude 50 m. When on a recce mission observed west of Piitsjoki station one U-2 flying on the deck. I was then at 1,000 m. I banked towards the plane and shot at it from above and behind a long burst, from 300 to 20 m. The plane crashed among the tree stumps. I fired also at the wreck on the ground. My plane was CU-556."

The last mission of HLeLv 32 was flown on 31 August 1944 by a Curtiss pair taking off for interception.

The commander of the Air Force ordered the air regiments to tell squadrons to cease fighting at on 4 September 1944 7 am. A ceasefire commenced and two weeks later it was confirmed by The Moscow Armistice.

Brief Analysis

Hävittäjälentolaivue 32 had flown 8,000 sorties and claimed 202 enemy aircraft destroyed, 190 thereof by the Curtiss. It had lost thirteen fighters in combat with ten pilots killed and one POW.

Battle-worn Curtiss CU-559 of HLeLv 32 at Mensuvaara in August 1944. Though becoming obsolete the Hawk was a rugged plane and due its manoeuvrability could hold its own in combat.

Lentolaivue 30

When a large number of fighters were received during the Winter War and an unlimited licence obtained for the State Aircraft factory to produce Fokker D.XXIs, a second fighter regiment was established on 27 March 1940, *Lentorykmentti 3*. Three squadrons were subjected to it: *Lentolaivue 30*, *Lentolaivue 32* and *Lentolaivue 34*, the last being an advanced training unit.

The personnel of *Lentolaivue 30* came from the disbanded LLv 10. Capt Kaarlo Lejon was posted in command. The intention was to equip the squadron with Caudron Renault CR.714 lightweight fighters, of which France had promised 80, but in the end only six arrived.

On 9 May 1940 the factory received an order to produce 50 Fokker D.XXI aircraft fitted with Twin Wasp Jr engines, assembling these within a year. When waiting for these *Lentolaivue 30* received a dozen Gloster Gauntlet fighter trainers.

LLv 30's base was set at Pori on the west coast. Its first combat planes were received on 29 August 1940, when LLv 28 delivered its nine Hurricanes, which equipped the 1st Flight.

When the factory had assembled the six CR.714s, an attempt was made to fly them from Tampere to Pori. Only four made it and two of these were damaged on landing. LLv 30 accepted the two intact ones on 11 September 1940, but they were grounded permanently the day before.

Caudron Renault CR.714 CA-556 of LLv 30 in September 1940, based at Pori. (Karolina Hołda)

Caudron Renault CR.714 coded CA-551 in a shelter at Pori in early September 1940. Four planes were flown to Pori, two of which were damaged on landing. The type was grounded permanently on 10 September 1940.

The Fokker D.XXIs began to arrive from the factory from 19 March 1941 onwards, LLv 30 taking delivery of 25 by the next conflict.

Continuation War

When Germany informed the Finnish military leaders of their offensive on the Soviet Union, a full-scale mobilization was initiated in Finland on 17 June 1941. The German attack began on 22 June 1941 and three days later the Red air forces bombed Finland, starting the Continuation War.

The Finnish fighter force was then positioned in two air regiments, *Lentorykmentti* 2 commanded by Col Richard Lorentz and *Lentorykmentti* 3 led by LtCol Einar Nuotio. The latter had three squadrons: *Lentolaivue* 30 with Fokker D.XXIs and Hurricanes, *Lentolaivue* 32 with Fokker D.XXIs and *Lentolaivue* 34 with trainers.

The order of battle of *Lentolaivue* 30 was then:
Squadron commander Captain Lauri Bremer, Pori
- 1st Flight, Captain Heikki Kalaja, Pori (7 HC)
- 2nd Flight, 1st Lieutenant Veikko Karu, Pori (9 FRw)
- 3rd Flight, 1st Lieutenant Erkki Ilveskorpi, Pori (10 FRw)

The task of LLv 30 was to protect the urban areas and industry in south-western Finland, preventing air raids on Turku and occasional protection of shipping in the Turku archipelago.

On 1 July 1941 the whole 1st Flight was transferred to LLv 32, which operated east of LLv 30's sector. Two days later the main parts of LLv 30 flew to Hyvinkää, still tasked with the protection of areas in southern and south-western Finland, the

Pilots of 3/LLv 30 in front of Fokker FR-156 at Turku in late June 1941. From the left: flight leader 1Lt Erkki Ilveskorpi, 2Lt Heikki Keso, 2Lt Veijo Taina, 2Lt Mikko Salomaa, Sgt Urho Lehto, 1Lt Martti Kalima, SSgt Pentti Rekola and 2Lt Kullervo Virtanen.

Hurricane coded HC458 of 1/LLv 30 at Hollola on 29 June 1941. In front of the plane is the flight leader Capt Heikki Kalaja. Two days later the whole unit was annexed to LLv 32 at Utti. (SA-kuva)

Hawker Hurricane I HC456 of Os Kalaja/LLv 32 flown by the flight leader Capt Heikki Kalaja in August 1941, based at Lappeenranta. (Karolina Hołda)

most visible elements becoming the interception of the enemy over the Gulf of Finland and attacks against light surface vessels.

LLv 30 was flying constant combat air patrols and on 4 July 1941 made the first contact. 1Lt Martti Kalima's pair of 3/LLv 30 engaged two SB bombers over the south-western coast at Dragsfjärd. Kalima spent all his ammunition on these two, but both exited with one engine smoking. Two days later the 3/LLv 30 swarm of 1Lt Martti Kalima scrambled at dawn and scored the unit's first air victory by shooting down an SB bomber of 117 RAE (reconnaissance aviation squadron) on a mission to the Turku area. Kalima wrote in the combat report:

"At 0320-0350 hours, altitude 2500 m. A three plane SB patrol flew towards the north-west over Stortervo. When I observed the planes I was 700 m from them and 200 m higher. The enemy was flying straight towards us. The clouds were at 2700 m. We took a half-roll and dived on them, receiving strong defensive machine-gun fire. I got to 50 m behind and below one, fired at the gunner and then pulled up having additional speed, and then back.

My wingman 2Lt Salomaa had spent all his cartridges on the gunner of the starboard plane, which did not shoot any more

when I made the second attack. In the third attack from below the left flank plane took a strong swing and went in a vertical dive, both engines smoking.

I tried to attack the remaining two bombers, but they managed to evade in the clouds. They had made a steep turn inside the cloud, because when came out of the cloud a while later, I saw two SB planes flying south at 2300 m. The distance was then 2-3 kms and I was heading west.

After the first attack they dropped their bombs. My plane FR-148 had five bullet holes. 2Lt Salomaa's plane was intact."

On 8 July 1941 the pair of 1Lt Veikko Karu of 2/LLv 30 engaged two flying boats of 15 AP, KBF (Baltic Fleet aviation regiment), shooting down both into the sea near Naissaari on the Estonian coast. Karu told of this event:

"At 06320-0740 hours. After 2Lt Mattila observed two enemy (MBR-2) pairs about 3000 m below us we dived to attack them. I targeted the rightmost plane. My machine-guns did not at first work, so I was forced to pull up behind the enemy tail (flying right at the deck). During this move I saw 2Lt Mattila shoot, the plane catching fire and diving into the sea.

Fokker FR-125 of 2/LLv 30 at Hyvinkää on 18 July 1941. This plane was assigned to future ace 2Lt Ture Mattila, though on this occasion 2Lt Oiva Louko sits in the cockpit, while the mechanics turn the inertial starter crank. (SA-kuva)

Four Fokkers of 2/LLv 30 at Hyvinkää in late July 1941. Closest is FR-124 assigned to 1Lt Veikko Sauru. The victory bar on the rudder top dates to 19 July 1941, when an MBR-2 flying boat was destroyed. The next planes are FR-122, FR-123 and FR-129.

After charging the guns I attacked again from behind and shot from 30-40 m distance. Instantly after this the plane dropped vertically in the sea. I felt like having hit the pilot.

The plane shot by Mattila dropped something with a parachute, looking like a mine. My plane was FR-129."

On 14 July 1941 2Lt Ture Mattila's pair of 2/LLv 30 was guided by radio to intercept two MBR-2 flying boats of 44 OAE, KBF (Baltic Fleet detached aviation squadron). Both were caught and shot down outside Tallinn on the Estonian coast. Mattila destroyed one:

"At 1445-1525 hours, altitude 1-5 m. After observing two enemy planes south of Pellinki, I began the chase. When I caught then at the Estonian coast at 3-4 m altitude I fired three bursts to the rightmost plane from straight behind, when the plane caught fire. I also observed that the plane which Sgt Nuorala had fired at was burning. My plane FR-127 had 13 bullet holes all around the plane."

Nuorala destroyed the other and his plane flipped over on landing, caused by combat damage. Nuorala also reported briefly:

"At 1445-1525 hours, altitude 1-5 m. I followed the lead plane on the left wing. When the lead plane attacked, I shot at the leftmost plane. After four bursts it caught fire. I observed also that 2Lt Mattila's victim was burning. Hits on my plane caused damage: left tyre punctured, windshield broken, hit in the fuel lines. My plane FR-121 was damaged on landing."

3/LLv 30, which had remained in the defence of south-western Finland, engaged on 22 July 1941 a fighter swarm north of Turku. 1Lt Martti Kalima heading the swarm and shot two down as follows:

"At 1800-1845 hours. We took by surprise north of the airfield a four-plane (2 I-16 + 2 I-153) patrol, of which one I-153 was left behind. We attacked from above behind against the latter. I shot one burst from a long distance and we began to chase it. North of Pensari Island when we were about to enter into effective

These four pilots of 2/LLv 30 shot down, on 5 August 1941, four MBR-2 flying boats. From the left: Sgt Aaro Nuorala, 1Lt Teuvo Ruohola, 2Lt Oiva Louko and Sgt Aarne Tirkkonen. Behind the armourer is Ruohola's Fokker FR-119.

Fokker D.XXI FR-157 of E/LLv 30 flown by the squadron commander Maj Lauri Bremer in September 1941, based at Utti. (Karolina Hołda)

firing range, from south behind the island five I-153s and four I-16s appeared on the scene.

We could not avoid any more the clash. A blind fight commenced and it was fought under 100 metres altitude. At one stage an I-16 had managed to get behind one of ours. I shot at it in a bank from 50 metres distance. It flipped into a vertical dive and fell in the sea. During the combat all four aircraft of our patrol had to fire at the enemy planes.

We saw at least four of them break off the combat and head south. 2Lt Salomaa and I fired at one I-153, which went into a dive at Pensari Island. On the island a strong fire was lit.

The Russians avoided coming down to the surface. Best evasive manoeuvre was a fast level bank at the surface. The antenna was shot off my plane. One of the bullets had travelled through the main spar of the port wing and the aileron actuator and the trailing edge. Five hits in the wing tip. Starboard tyre was punctured. The landing succeeded. My plane was FR-148."

On 5 August 1941 the 2nd Flight flew several combat air patrols with Fokker pairs. The Russians lost a total of five MBR flying boats over the Gulf of Finland in three encounters.

On 25 August 1941 Capt Veikko Karu's pair of 2/LLv 30 attacked two flying boats near Pien-Tytärsaari in the eastern Gulf of Finland destroying them both. Karu told about his scoring this way:

"At 1400-1410 hours. When I was on a combat air patrol south of Pien-Tytärsaari I observed at 1335 hours two enemy planes. I gave a signal to my wingman (pilot 2Lt Mattila) and began a chase. I caught the enemies, who had descended right to the surface in just about five minutes. I made the attack from straight behind against the starboard plane. I shot at the engine and port wing root. When I got right behind the tail I observed a puff of smoke from the enemy's engine and received a shower of fuel on my wind shield.

At the same time my machine-guns jammed. I pulled up to the right and began further off making charging moves. When pulling up I took a glimpse to the wingman, whom I observed to pull steeply to the left behind the other plane and head north. I understood that he had to give up the fight for some reason (2Lt Mattila was wounded).

After getting my starboard guns to work again I observed that the flying boats continued in the previous direction, the starboard

one slightly smoking. I attacked this one straight from the side, but because my shooting seemed to have no effect and I received strong counter fire, I moved the attack from straight behind. I shot a burst from a very short distance to the engine, when it and the port wing caught fire. Immediately after this I shot a burst into the cockpit, when the plane crashed ahead of me in the sea.

I moved to the next. I shot a long burst from very close range into the engine. The shooting appeared to have no effect at first even though the bullets went exactly in the target. All of a sudden the enemy burst into flames, and I almost collided with it. When I pulled over the enemy plane I felt a strong shock in my aircraft and it flipped to the port side. After having levelled my aircraft in the normal position I saw only loose pieces on the sea surface and further back behind me a black column of smoke was rising from the previous aircraft, which I saw on the return flight to north of Suursaari.

I did not observe the enemy to do any other evasive manoeuvres but a couple of shallow banks. I used about 500 rounds. 27 holes in my own aircraft. My plane was FR-129."

Eastern Karelia

On 20 August 1941 *Lentolaivue* 10 was established on a temporary basis at Onttola. Ten days later LLv 10 commenced operations at Tiiksjärvi, being tasked with reconnaissance for the 14th Division advancing in the Rukajärvi direction. A side task was the intercept in the area. For this purpose 3/LLv 30 flew to Tiiksjärvi, forming 1/LLv 10. The first six Fokker D.XXIs arrived on 21 September 1941 and the rest within nine days.

On 23 September 1941 1/LLv 10 was twice engaged in combat with *Chaikas* at Rukajärvi. In the first clash 1Lt Aaro Virkkunen's swarm sent one out of eight Russian aircraft down and in the second Capt Erkki Ilveskorpi's five aircraft fought with a dozen, setting one on fire.

On 27 September 1941 1Lt Martti Kalima's 1/LLv 10 was on an interception at Ontajoki, engaging in a combat with I-153 fighters. Squadron mission report told:

"On an alert mission at 0925-1025 hours we flew towards the front with eight aircraft. The lower patrol, where belonged 1Lt Kalima (FR-150), 2Lt Salomaa (FR-154), 2Lt Virtanen (FR-155), Sgt Nuorala (FR-148) and Sgt Kilpinen (FR-141), flew at about 100 m altitude. The upper patrol, 1Lt Lehtonen (FR-156), 2Lt Ukkonen (FR-153) and 2Lt Leino (FR-138) about 200 m higher.

The top patrol leader 1Lt Lehtonen observed on the sun side five Russian fighters of I-153 type picking up at the front-line. We were about 5 km from them. 1Lt Lehtonen tried to come and inform the lower patrol his observations, but before this 2Lt Salomaa flying the left wing of the lower patrol had already observed the enemies and informed the lead plane 1Lt Kalima this.

We turned towards them and tried to obtain more altitude since the Russian were already at 500 m. Only four enemies had remained, one had broken off earlier and the others were going east in a shallow climb. When noticing us they turned towards us and dived in a line towards us with a 150 m altitude advantage. Our own aircraft were in an inclined echelon caused by the bank and that the top patrol had just tried to inform us of the Russians.

We pulled at them and started firing and after the pass we turned after them, after which began a blind combat, where the Russians as more manoeuvrable in spite their inferior numbers always got behind somebody's tail, but then we were saved by good comrade spirit. This was exhibited in us trying to fire at a Russian plane, which had got behind the tail of some of us. At the very beginning 1Lt Lehtonen got an exploding bullet to the cockpit through the wind shield and was forced to break off, since the splinters of the bullet hit the eyes so badly that his sight was blurred.

2Lt Ukkonen shot at one I-153, which came towards him so long that he had time to make only a small dodge, but the Russian flew straight on. The result was that 2Lt Ukkonen's landing gear swept off the port upper wing of the Russian, and the I-153 crashed; 2Lt Ukkonen had great difficulties in regaining control of his aircraft and was forced to break off the fight. The aircraft caught fire and 2Lt Ukkonen had to bail out 20 km from the frontline on our side.

An I-153 was behind the tail of Sgt Kilpinen firing at him. 1Lt Kalima got behind the Russian and gave a burst. Then the Russian quit firing and pulled into a shallow climbing turn, when 1Lt Kalima shot from a bank a burst and the Russian aircraft flashed up in flames and crashed in a pond.

Liberated from the Russian Sgt Kilpinen pulled a bank, when he got from behind one banking Russian into his sight, fired a burst and the Russian caught fire and crashed. 2Lt Salomaa got behind the last I-153 and shot a burst, when the Russian took a shallow dive to escape and while smoking broke off due to his higher speed."

Ukkonen's Fokker collided with the *Chaika* of 152 IAP piloted by Lt M.N. Piskunov.

On 1 November 1941 *Lentolaivue 10* was annexed to *Lentolaivue 14*, which had arrived to Tiiksjärvi four days earlier. 1/LLv 10 became formally again 3/LLv 30, but in practice was named *Lentue Kään* after its leader, Capt Pekka Kään.

With the Finnish advance to Karelia, *Lentolaivue 30* on the home front received new tasks on 1 September 1941. They were to protect the Kouvola-Kotka-Viipuri-Ylävuoksi area and occasionally Helsinki, Tammisaari and Turku areas with small forces. For this purpose the main parts of squadron flew the next day to Utti.

Gulf of Finland

By the autumn the flying units, or what was left of them, of the Baltic Fleet air forces had retreated into the shelter of the huge air and naval base of Kronstadt, outside Leningrad.

With the lack of aerial activity *Lentolaivue* 30 operations (aside the main task of reconnaissance on the Karelian Isthmus, which became a stalemate on 2 September 1941) turned into harassment of light surface vessels in the Gulf of Finland.

On 17 October 1941 Capt Veikko Karu's 6-plane 2/LLv 30 group observed, on combat air patrol, a transport vessel in the Gulf of Finland and sank it near Södeskär. On a later mission Karu's pair shot one motor torpedo boat into flames on the south side of Suursaari.

On 29 October 1941 Capt Veikko Karu's pair of 2/LLv 30, engaged in a reconnaissance mission to the Seivästö area, came

Fokker FR-144 of 1/LLv 10 (ex-3/LLv 30) at Tiiksjärvi in eastern Karelia on 4 October 1941. It was assigned to the flight leader Capt Pekka Kään. He chose the letter A as a tactical number. (SA-kuva)

Fokker FR-156 of 1/LLv 10 ready for a mission at Tiiks-järvi on 25 September 1941. The assigned pilot, 2Lt Mikko Salomaa, climbs into the cockpit. This was the 10th plane of the flight. (SA-kuva)

across a transport vessel and guard ship, of which the former was shot in flames and the latter began to smoke. On next mission Karu's pair set fire to the guard ship. Later Karu's six-plane flight fired at a motor torpedo boat outside Garavaldai and east of Krasnaya Gorka, which exploded.

During November and December 1941 the pattern repeated itself.

On 10 November 1941 Capt Antti Naakka's pair of 1/LLv 30 met, on a regular combat air patrol to the Gulf of Finland, an MBR-2 flying boat, shooting it down. Next day Capt Veikko Karu's swarm of 2/LLv 30 shot near Lavansaari a motor torpedo boat into flames during the first mission to the Gulf of Finland and sank a guard boat in the same area on the next mission.

On 17 December 1941 Capt Veikko Karu's 2/LLv 30 encountered an MBR flying boat escorted by three I-153s in the vicinity of Narvi lighthouse. Four Fokkers attacked the escort fighters, which escaped. Two Fokkers then shot the flying boat down.

Brief Analysis

During 1941 the fighters of *Lentorykmentti* 3 had destroyed 87 enemy aircraft. Of these LLv 32 had dispatched 65 planes, but lost eight planes and four pilots. LLv 30 had flow over 2,400 sorties and shot down 22 aircraft, losing four planes and three pilots in combat. In addition LLv 30 had sunk 14 and damaged 12 light surface vessels.

1942

At the beginning of 1942 the nucleus of *Lentolaivue* 30 had seventeen serviceable Fokker D.XXIs, while the still formally 3/LLv 30 had six further north.

The order of battle of *Lentolaivue* 30 was then:
Squadron cvommander Major Lauri Bremer, Utti
- 1st Flight, Captain Antti Naakka, Utti (8 FRw)
- 2nd Flight, Captain Veikko Karu, Suulajärvi (9 FRw)
- 3rd Flight, Captain Pekka Käär, Tiiksjärvi (6 FRw)

The first combat of the year occurred on 19 January 1942. The swarm of Capt Veikko Karu of 2/LLv 30 claimed

three MBR-2 flying boats on a CAP mission to the Suursaari-Lavansaari area.

On 11 March 1942 LLv 30 received new tasks. The main task was close-range reconnaissance on the Karelian Isthmus and Lake Ladoga next to it, plus the ice-covered Gulf of Finland. The secondary task was interception over the eastern Gulf of Finland and the protection of the railway network and industry in the Virolahti-Luumäki-Saimaa canal area. The task of the 3rd Flight (*Lentue Käär*) at Tiiksjärvi remained the same.

On 19 March 1942 the Finns decided to take Suursaari back when the advance could be made on the ice. Detachment Pajari was formed for this purpose. LeR 3, LeR 4, LLv 6, 3/LLv 24 and an ambulance/liaison flight were ordered to support. MajGen Aaro Pajari, commanding the 3,500 man strong invasion force of Suursaari, issued on 21 March 1942 the order for attack:

"For the invasion LeR 3 was ordered with tasks 1) to protect the march of Detachment Pajari to Haapasaari and from there onwards, 2) to support the attack to Suursaari, 3) to hamper the retreat of the enemy from Suursaari by causing losses and 4) to carry out reconnaissance on every mission."

On 27 March 1942 the invasion troops began the approach to Suursaari. For air cover 57 aircraft were assembled, the main part from LeR 3: sixteen Fokkers of LLv 30 and thirteen Curtiss from LLv 32. Additionally LLv 6 provided five SB bombers and six I-153 fighters, LLv 24 six Brewsters and LLv 42 eleven Blenheims.

Capt Antti Naakka's six Fokkers of 1/LLv 30 flew a combat air patrol mission to Suursaari and engaged a mixed formation, sharing with a swarm of LLv 32 Curtiss one I-153 sent down. Two more were damaged. One hour later Capt Veikko Karu of 2/LLv 30 was patrolling in the same area and downed one I-153 of the encountered formation.

2/LLv 30 flew a combat air patrol to Suursaari at dawn on 28 March 1942, with Capt Veikko Karu's seven Fokkers engaging a formation of about twenty fighters flying in two groups. During a ten minute turning fight the Finns sent down three aircraft, when the enemy planes split and began to break off. The Fokkers made a fruitless chase. Karu claimed one fighter destroyed and another damaged:

"At 0542-0645, altitude 1500-0 m. When we met east of Suursaari the enemy formation flying at three altitudes, I attacked them. I hit first the I-16 patrol highest up. A curve battle ensued, during which one I-16 managed to get behind my tail. It got to fire only once and then the bullets went below me. After curving a while I got behind the tail of an I-16 and fired a burst from straight behind. It began a climbing curve to the left, when I managed to fire a long burst to the engine in the inner curve. Then the engine puffed smoke and the plane flipped uncontrolled to the left. We were then at 200-300 m altitude.

Then instantly three I-153s attacked me, not allowing me to follow the fate of my victim. I managed to shoot at one I-153, but then the Chaikas broke off by going to the deck.

I observed below and ahead of me one I-16 going eastwards. I fired at it in the inner curve and got a long burst to the engine. My guns jammed and I had to pull up. After charging the guns they worked again. I saw the I-16 I had fired going smoking in a shallow dive towards the north tip on Lavansaari. My plane was FR-136."

On 3 April 1942 Capt Antti Naakka's eight Fokkers of 1/LLv 30 scattered a 100 man detachment on the way from Tytärsaari to Lavansaari and 1Lt Veikko Sauru's five Fokkers of 2/LLv 30 strafed a company on the ice between Seiskari and Garavaldai. Later in the day the same Sauru detachment was attacked by a dozen *Chaikas* on another mission to the Seiskari area. The Finns claimed two enemy fighters shot down, both claimed by SSgt Otto Karme.

Pilots of 2/LLv 30 at Suulajärvi on 26 January 1942. From the left: flight leader Capt Veikko Karu, 1Lt Veikko Sauru, Sgt Helge Krohn, 2Lt Reino Linko, kers Pentti Väisänen and 2Lt Tauno Saalasti. (SA-kuva)

Fokker FR-157 of E/LLv 30 assigned to the squadron commander Maj Lauri Bremer, parked here on the ice outside Hamina, shortly before its flip over in a landing on 24 March 1942.

Fokker FR-121 of 1/LeLv 30 at Suulajärvi in June 1942. The assigned pilot was 1Lt Olavi Lilja. An air base guard with a steel helmet poses in the cockpit.

On 15 April 1942 the Fokkers of 2/LLv 30 shot up seven trucks with full fuel loads on an ice road from Garavaldai to Seiskari, and also sent down one out of two *Chaikas* appearing in the scene. Capt Veikko Karu claimed the fighter:

"*At 0820-0830 hours, altitude 1000 m. When I was flying a CAP mission with six planes over lake Lempaalanjärvi at 1500 m altitude I observed slightly below me an aircraft heading to south over Lumisuo. It saw me, too and began to gain altitude and when observing that cannot break off turned right towards me. Then another I-153 flashed under me heading south. I continued my attack against the first one. We were over Perämäki at that moment.*

The Russian began shooting slightly earlier, but after getting a good aim and having shot a long burst in its engine, the firing ceased instantly after the first bullets had hit it. Then it suddenly flipped on its port wing and light smoke puffed out of the engine. I had time to shoot into the cockpit when it went on the side in front of me. When the Russian flashed pass me I saw it smoking pretty well. The Russian was making a left curve. Due to the excessive speed I had gained in the dive I could not follow it even if I pulled so that I blacked out.

When I saw it again, it was diving at about a 45° angle south at about 300 m altitude. I decided to go after it, but then I saw above me to the south four enemy aircraft (I did not observe the type). I began to gain altitude. The enemy planes broke off however and I flew to our own side to carry on the mission. No hits on my plane. The short burst which the enemy managed to fire went over me. My plane was FR-129."

This proved to be the last fruitful encounter for the next two years. Most of the time was spent in uneventful reconnais-

sance over the Karelian Isthmus and dull maritime reconnaissance over the eastern Gulf of Finland.

On 28 October 1942 the air force C-in-C, Lieutenant General Jarl Lundqvist, paid a visit to LeLv 30's base at Suulajärvi, informing them that they will be equipped with captured I-153 fighters brought from Germany, and that the IT-flight of LeLv 6 will be attached. Both squadrons will be part of the new *Lentorykmentti 5* with LtCol Knut Ilanko in command. The regiment would work for the naval forces commanded by LtGen Väinö Valve.

On 6 November 1942 the Mannerheim Cross was awarded to the 2/LeLv 30 leader Captain Veikko Karu, who was credited with seven air victories and sharing the sinking of thirteen surface vessels.

LeLv 30, which was transferred to Römpötti on the Karelian Isthmus, was seconded to *Lentorykmentti 5* on 16 November 1942. The task of LeLv 30 was also the submarine search and maritime reconnaissance east of Lavansaari. The times as a fighter outfit were over, for now.

1943

At the beginning of 1943 *Lentolaivue 30* possessed five Fokker D.XXIs in the 1st Flight and three I-153s in the 2nd Flight. A few encounters took place during the year, but none led to a combat producing results.

When LeLv 30 handed over its Fokkers of the 1st Flight on 27 February 1943 to LeLv 12, it was left with the *Chaikas* of the 2nd Flight until the factory could refurbish the war booty I-153s bought from Germany.

Polikarpov I-153 IT-11 of 1/LeLv 30, flown by 1Lt Esko Lappalainen in April 1943, based at Römpötti. (Karolina Hołda)

On 24 March 1943 the IT planes of LeLv 30 encountered enemy aircraft more frequently above the Gulf of Finland . The reconnaissance mission of the 2nd Flight began at 7 am, told here by 2Lt Olavi Puro:

"While leading a three-plane IT patrol on a recce mission we met north of Lavansaari two Russian I-153 fighters. In the ensuing combat the oil tank of IT-13 (pilot Sgt Jänkävaara) was shot up, which caused the plane to begin smoking and it was force landed on the pack ice zone located halfway between Peninsaari and Narvi. The pilot was safe and started heading towards Tiurinsaari.

Of the combat it can be specially mentioned that the Russians were very skilful as pilots, but were shooting very long bursts, which were poorly aimed. The guns of my plane jammed right in the first burst and in spite of several cocking manoeuvres they remained out of order. The Russians tried to follow the smoking plane of Sgt Jänkävaara, but Sgt Durchman and I prevented it. After curving for a while the Russians turned towards Lavansaari and landed on the island.

Heavy and 40 mm flak kept firing as much as they could during the whole battle. We returned to base, re-fuelled and then took off to set fire to IT-13. When Sgt Jänkävaara had reached the edge of the pack ice, I landed and brought him on my wing to a safer area by taxiing closer to Tiurinsaari. My plane, IT-18, was intact."

Chaika IT-15 of 2/LeLv 30 being re-fuelled at Utti on 28 September 1943, on the way to the factory for through overhaul, which took five months.

1944

At the beginning of 1944 *Lentolaivue 30* had three serviceable I-153s in both flights. This had been the typical complement through the previous year. On 14 February 1944 LeLv 30 became fighter squadron *Hävittäjälentolaivue 30* (HLeLv 30), according to an existing plan on transforming it to a night fighter unit.

The re-organization began instantly, establishing a temporary 3rd Flight receiving the *Chaikas*. The 1st Flight moved to Utti at once, where it was issued FIAT G.50 fighters and began type training.

On 6 March 1944 Malmi-based 2/HLeLv 34 was attached to HLeLv 30 with all its planes and pilots, forming 2/HLeLv 30. The task of the squadron was to protect Helsinki and intercept the enemy over the central Gulf of Finland and, in Utti, to train new pilots to be ready for the *Mersus*, using FIATs.

By 1 May 1944 the compliment of 2/HLeLv 30 had risen to ten Messerschmitts and a week later 1/HLeLv 30 received three *Mersus* at Utti and began to train pilots on them.

On 22 May 1944 Capt Veikko Karu was assigned as the commander of HLeLv 30. Karu was a Mannerheim Cross holder and he had arranged the LeR 3 night fighter course a year earlier. He had been given type training in the MT, which made him qualified for night fighter operations.

HLeLv 30, protecting Helsinki, scrambled every time that Russian reconnaissance planes crossed the Gulf of Finland. On 24 May 1944 WO Oiva Tuominen of the 2nd Flight did not manage even to get a damaged plane in the records, as he reported:

"At 1645-1735 hours. I took off on an intercept, and I found a Pe-2 some 20 km south-east of Stenskär. I tried deflection shooting first, after which it went into a cloud pillar. As it emerged I shot from above and behind at the left engine which caught fire. I gave it another burst and the fire spread into the fuselage. I did not see it any more after it went into a cloud some 30 km south-east of Tytärsaari. Heavy snow and bad visibility. My plane was MT-407."

On 28 May 1944 the scramble produced results, as told by 1Lt Kyösti Karhila of 2/HLeLv 30:

"At 0420-0535, altitude 7000 m. On an intercept mission I observed the condensation stream created by a plane flying about 4000 m higher than me. I gained altitude following the moisture stream. The stream broke, but I followed in its direction and after a while I saw the plane far left above me, at about 7000 m altitude. The plane banked to the south, I caught it and recognized it as a Pe-2.

I fired only the cannon as a trial from 600 m distance from behind left. Now about 1 metre piece departed from the wing tip, the plane banked into a vertical dive continuing to the ground, shedding some parts while in a dive. The crash site was at Kuusalu (Estonia), 100 m north of a road passing on the north side of a lake about 10 km north-east. My plane was MT-403."

On 2 June 1944 HLeLv 30 terminated its FIAT training and the planes were sent to the Air Fighting School. *Mersus*

LeLv 30 commander Maj Eino Luukkanen (l) and 1Lt Tauno Saalasti (r) at Römpötti in March 1943. Behind is one of four I-153s of the 2nd Flight.

FIAT FA-25 of 1/HLeLv 30 at Utti in March 1944. The FIATs acted as intermediary types when training pilots to fly the Messerschmitt Bf 109G.

replaced the FIATs at Utti, and thus the squadron had only 109s in both flights.

In order to speed up the night fighter training the air force commander gave an order on 9 June 1944 to send a twenty man detachment to Germany. The pilots were picked up from various squadrons (but not *Mersu* squadrons) and Capt Martti Kalima was posted to lead the outfit. The men of *Gruppe 34/44* left in a few days. The plan was to place these men after training with HLeLv 30. An attempt was made to acquire Bf 110G-4 night fighters for two squadrons, but it failed and the *Mersus* already in the country received modified instrumentation instead.

The major Russian offensive started in the Karelian Isthmus on 9 June 1944. Next day LeR 5 received an order to transfer four MTs to Kymi as soon as possible and assume the responsibility for protecting Kotka. At the same time the night fighter training in Utti was suspended for the time being. The *Mersu* swarm arrived two days later at Kymi.

On 15 June 1944 the HLeLv 30 planes at Kymi, practically its 2nd Flight, was transferred with its pilots to Immola and annexed to HLeLv 34. Kotka was to be protected by anti-aircraft artillery alone. HLeLv 30 now had only four MT planes which remained to cover Helsinki, though this area would remain quiet for more than a month.

HLeLv 30, which had been set to defend the capital Helsinki, made intercept missions, when 20 July 1944 saw again the approach of Russian planes in the western airspace of the Gulf of Finland. A few clashes followed.

On 5 August 1944 1Lt Esko Lappalainen of 1/HLeLv 30 took off with his pair on an interception over the Gulf of Finland at 1815 hours, engaging an La-5 fighter of 3 GIAP, KBF:

"*After hearing on the radio that the Russians were some 10 km away, I went to full throttle, and after a couple minutes I could see ahead and below 1/1. I notified "Pig" control station and my wingman and I went to attack. It rained and the enemy plane flew at the edge of the cloud, as I dived below it. Before I could fire the enemy began to turn right, so I had to fire in a climbing turn. I saw hits at the left wing-fuselage joint where my cannon shells exploded. Small parts came off the enemy and after I hit it, it turned left and entered the cloud. I did not see it after that. My plane was MT-485.*"

Night fighter capability development was re-initiated on 28 August 1944 with LeR 3 and LeR 5, but these were suspended a week later when the cessation of hostilities between Finland and Soviet Union took place.

The final mission of HLeLv 30 was an interception on 3 September 1944 to the Gulf of Finland, flown by a pair of Messerschmitts.

The Air Force commander ordered the air regiments to tell squadrons to stop fighting on 4 September 1944 at 7 am. A ceasefire commenced and two weeks later it was confirmed by the Moscow Armistice.

Hävittäjälentolaivue 30 had flown in the Continuation War a little short of 6,000 sorties, 4,000 of them by October 1942. The Fokkers had claimed 33 enemy aircraft shot down and the Messerschmitts three more. The squadron had lost eight fighters and eight pilots in combat.

The last wartime equipment of HLeLv 30 was the Bf 109G-6. Here is MT-408 soon after arrival at Helsinki Malmi, which occurred on 26 March 1944. It was assigned to the 2nd Flight and 1Lt Tauno Kallio.

Officers of HLeLv 30 at Utti in early June 1944. From the left: 2Lt Ilmari Toiviainen, 2Lt Toivo Uusiaho, 1Lt Esko Hämäläinen, 1Lt Reino Hyvärinen, 2Lt Birger Vesoma and 1Lt Arvo Weissmann.

When the six air stations were re-formed into five aviation regiments on 1 January 1938 the sixth, *Lentoasema 2* was reduced to a detached squadron, *Erillinen Lentolaivue* (Er.LLv) flying Blackburn Ripon floatplanes.

On 31 May 1941 the name of Er.LLv (detached squadron) was changed to *Lentolaivue 6*. The squadron was subordinated to the naval forces and it was tasked with maritime reconnaissance, submarine search and attacks against Soviet shipping. LLv 6 consisted of a reconnaissance flight, bomber flight and fighter flight. In the mobilization of 18 June 1941 the reconnaissance flight was annexed to the bomber flight. The 3rd, fighter, flight flew captured Polikarpov I-153 fighters.

The order of battle of *Lentolaivue 6* was then:
Squadron commander, Major Knut Ilanko, Turku
- 2nd Flight, Captain Birger Ek, Turku (3 VP, 2 KO)
- 3rd Flight, Captain Lauri Karjalainen, Turku (5 VH)

In the morning of 25 June 1941 the Soviet air forces bombed Turku with three waves of 40 bombers. The Finnish fighters that climbed for interception did not catch the enemy. The Continuation War had begun.

Early next day another eighteen aircraft bombed Turku. 3/LLv 6 leader Capt Lauri Karjalainen's pair took off and caught the invading SB bombers and shot one down in the sea. WO Tarmo Hämelä claimed his aerial victory thus:

"*At 0300-0400 hours. I was landing at the airfield, when I observed a 9-plane detachment over Turku. I climbed to 3,000 metres. I caught the Russians at about the same altitude. Howev-er, I could not get closer than 200-300 metres, because my landing gear was down and the pressure out to retract them. I fired at the closest aircraft with short bursts for about a minute, after which it started to smoke and the formation evaded into a cloud at 2,000 metres. Pansio fire control post observed the aircraft to crash into the sea. The Russian fired from every aircraft all the time when an opportunity arose. They scored no hits. My plane was VH-14.*"

The airfield at Turku became unsuitable for operations after the bombardment and the 3rd Flight flew to Nummela. From here the task of the flight was to reconnoitre the Hanko area and carry out strafing missions with bombs and machine-guns against the port and air base. The missions were usually flown in pairs.

During the summer and autumn the reconnaissance and strafing missions to Hanko continued. Due to the small quantity of aircraft only the reconnaissance was satisfactorily performed.

On 3 December 1941 reconnaissance mission, the Hanko area was confirmed to be empty of the Russians. The 3rd Flight flew on 12 December 1941 to Helsinki Malmi. It had three airworthy I-153s.

On 13 January 1942 the 3rd Flight flew to an ice base outside Hamina and was tasked with the support of the infantry occupying the outer islands in the Gulf of Finland.

For next two months the 3rd Flight flew reconnaissance and strafing missions to Suursaari and Tytärsaari island, usually in pairs. Additionally a few times top cover was flown for the squadron's SB bombers.

Captured Polikarpov I-153 coded VH-12 of 3/LLv 6, photographed at Turku on 13 June 1941. The VH combination in the serial comes from the Finnish words Venäläinen Hävittäjä *(Russian Fighter). (Finnish Air Force)*

Polikarpov I-153, VH-101 of 3/LLv 6 flown by Capt Lauri Karjalainen in June 1941, based at Turku. (Karolina Hołda)

On 27 March 1942 began the occupation of Suursaari. For two days the flight attacked with four aircraft swarms the Russian positions on the island. Next day the occupation was completed.

On 3 April 1942 an infantry company was observed on the ice heading from Tytärsaari to Lavansaari. The detachment was spread around in the attack of 1Lt Paltila's four aircraft.

On the following day's reconnaissance mission 1Lt Paltila observed a large camp area on the ice near Lavansaari. The swarm made three firing passes and created a lot of confusion in the camp.

On 23 April 1942 the flight flew to Malmi during the spring thaw and on 27 May all aircraft were transferred to Römpötti on the Karelian Isthmus. The task was reconnaissance in the eastern Gulf of Finland.

On 4 June 1942 an order was issued to change the VH prefix of the serial to IT. This was done a week later in the flight.

Chaika coded VH-12 of 3/LLv 6 running up on the ice outside Hamina in late March 1942, during the re-occupation of Suursaari (Gogland) in the middle of the Gulf of Finland. Behind is VH-18 of the same unit.

3/LeLv 6 plane VH-16 parked at Helsinki Malmi in May 1942, after the unit abbreviation became LeLv on 3 May 1942. This flight flew to the Karelian Isthmus on 27 May 1942 and was based at Römpötti.

Polikarpov I-153, IT-16 of 3/LeLv 6 flown by the flight leader Capt Per-Eric Ahonius in July 1942, based at Römpötti. (Karolina Hołda)

Capt Per-Eric Ahonius, the 3/LeLv 6 leader, leans on his Chaika IT-16 at Römpötti on 9 July 1942. Four days earlier he had shot down two I-16s, as shown on the rudder tip. The VH combination in the serial was changed to IT on 4 June 1942. (SA-kuva)

On 22 June 1942 Capt Per-Eric Ahonius' reconnaissance pair observed near Seiskari island a few motor patrol boats, two of which were shot into flames.

On 4 July 1942 Capt Per-Eric Ahonius' pair took by surprise two Russian I-16 fighters at Seiskari and shot both down. His combat report reported:

"*At 22.50 I observed a dust cloud appearing on the shore of Seiskari and flew there. Three apparently I-16s were taking of from the sandy beach. I shot at the middle aircraft, which instantly caught fire and remained on the shore burning. The first machine had become airborne. I began now firing at it and when it curved to the port, it caught soon fire, when I was able to shoot from the side at 50 m distance and it went into the sea. The third one about to take off either stopped there or managed to slip and hide to starboard. My plane was IT-16.*"

East of Lavansaari a motor patrol boat was destroyed in a machine-gun attack. Next day Capt Ahonius' pair near Shebeleva lighthouse shot up another motor patrol boat.

In the following months the sea area was reconnoitred daily during a so-called morning round with a pair. The flight had usually three I-153s in flying condition.

On 4 October 1942 1Lt Mainio Paltila led three *Chaikas* on a reconnaissance mission to Lavansaari, where a combat was fought with five Russian *Chaikas* of 71 IAP, KBF. Both parties lost one aircraft shot down. The Finnish victor was 1Lt Olavi Puro, relating this:

"*At 0810-0925. We were on a regular reconnaissance mission with three aircraft, 1Lt Paltila leading the patrol. When going around Lavansaari we went down to the deck due to strong flak. On the west side of the island three Russian* Chaikas *attacked us from above and after a while two more arrived on the scene.*

A fierce curving battle turned out and the Russians favoured shooting from opposite directions. Early in the combat I observed MSgt Salminen's plane IT-17 turn on its back at about 150 m altitude and then go in a vertical dive into the sea 2-3 km south of the island. I fired at several aircraft from straight ahead and in a turn without any significant results. Finally I got rid of the carousel and was able to climb to 300-400 metres altitude, from where I could begin diving and hitting the Russians.

Now I could surprise one Chaika approaching from the sun and shot a burst to its engine from 50 metres. A strong black-grey cloud of smoke puffed from the plane and it headed in a shallow

Chaika *IT-20 of 3/LeLv 6 parked at Römpötti in September 1942. The regular pilot was the new flight leader 1Lt Mainio Paltila. On 16 November 1942 the whole flight was attached to LeLv 30 and the squadron no longer had a fighter element.*

dive towards Lavansaari airfield, turning on its back and going into the sea about 1.5 km from the shore. I fired at a few planes still, but then one took me by surprise from the sun hitting my aileron, which was damaged to such an extent that I was forced to break off. My plane was IT-18, which had 7 hits. Ammunition spent 1500 rounds."

On 12 November 2Lt Olavi Puro was on a submarine search jumping on a solitary Pe-2 bomber, which was shot down near Peninsaari island. Puro wrote this in his combat report:

"At 1030-1140, altitude 250-300 m. I led a two-plane patrol on a recce mission. After the reconnaissance and when returning at about 300 m just under the clouds north of Peninsaari, I observed a single Pe-2 coming towards us about 30 metres lower than me.

I attacked it from ahead and shot from a close range a burst into its starboard engine and wing. Then I turned swiftly after it, when I observed the starboard engine pouring smoke. I managed to fire from behind another burst into it, but could not make it to the close range any more. This burst hit from about 100 me-

tres the starboard engine, when it began to smoke much stronger leaving behind a black trail of smoke having a length of about 50 metres, and the plane began to swerve to the port. At this point the rear gunner fired with the dorsal gun, but the bursts went behind and aside.

Barely being able to gain altitude the plane managed to climb in the clouds then engine even stronger smoking heading towards Lavansaari. We did not see it any further, even when Sgt Durchman and I circled over the area for a while. Ammunition spent 450 rounds. My plane IT-18."

The Pe-2 bomber had belonged to 73 BAP and it was later seen to crash in the sea.

On 16 November 1942 the 3rd Flight was annexed to *Lentolaivue 30* forming its 2nd Flight. After this there was no fighter element in *Lentolaivue 6*.

The 3rd Flight of *Lentolaivue 6* claimed five aerial victories with the I-153 and destroyed four light surface vessels. Four *Chaikas* were lost, two in combat, one due to a technical fault and one in a flying accident. Three pilots were killed.

3/LeLv 6 in a group picture at Römpötti on 1 November 1942. The four pilots are in the centre, with the flight leader 1Lt Mainio Paltila in the dark blue tunic. Left of him is WO Kaarlo Lahtonen and to the right 2Lt Olavi Puro and Sgt Matti Durchman, the latter two later becoming aces. (Finnish Air Force)

Lentolaivue 34 appeared in the Finish Air Force on three occasions. The first was on 10 October 1934, when the squadrons were numbered by the order of the day of the air force commander. Now the squadron of *Lentoasema* 4 based at Turkinsaari, outside Viipuri, became *Lentolaivue* 34, flying Blackburn Ripon floatplanes. In the re-organization of 1 January 1938 LLv 34 became a reconnaissance squadron, re-numbered *Lentolaivue* 14 and subordinated to *Lentorykmentti* 1.

When large numbers of aircraft arrived in Finland during and after the Winter War, a second fighter regiment, *Lentorykmentti* 3, was established on 29 March 1940. It comprised two fighter squadrons, LLv 30 and LLv 32 and a training squadron, which was named on 24 July 1940 as *Lentolaivue* 34. The intention was to equip it with Fokker D.XXI fighters assembled by the factory, but while waiting for these the unit used a variety of advanced trainers. By an earlier air force HQ order *Lentolaivue* 34 was annexed to advanced training squadron T-LLv 35 on 25 June 1941.

The new *Lentolaivue* 34 was established at Immola on 23 January 1943 with Major Olavi Ehrnrooth in command. It was part of *Lentorykmentti* 3 commanded by LtCol Einar Nuotio. The intention was to make this an elite outfit. While waiting for the promised Messerschmitt Bf 109G-2 fighters to arrive, the air force commander gave Maj Ehrnrooth a free hand to pick the pilots from volunteers of other squadrons. With bitter minds other squadron commanders gave away their best pilots. In return they also "gave away" part of the weaker technical personnel, villains and those with disciplinary problems.

Messerschmitts

LeLv 34 received its first aircraft on 13 March 1943, when the squadron's pilots brought sixteen Messerschmitts from Germany to Helsinki Malmi. The fighter was given the nickname *Mersu*.

The first successful encounter took place on 24 March 1943. Capt Pauli Ervi and WO Ilmari Juutilainen of 1/LeLv 34 scrambled after a Pe-2 photo plane while it crossed the Gulf of Finland. At 14.30 hours Juutilainen shot it down as the first air victory of the squadron and the Messerschmitts:

"Based on an air surveillance message we took off for intercept with Capt Ervi. The Pe-2 was approaching Helsinki from the south and I was guided to the chase by the flak observers, who saw us both. In square 117 a 5 I met the enemy at 2,500 metres. Obviously it did not see me. From the sun I shot at the fuselage (from aside, above and behind) and also from behind and above. Pieces came off the plane and it went vertically into the sea. MT has magnificent armament, the shells worked well. My plane was MT-212."

On 27 March 1943 Maj Olavi Ehrnrooth commanding LeLv 34 was killed in a flying accident. His post was taken two days later by the previous LeLv 30 commander, Maj Eino Luukkanen.

The next victory came on 13 April 1943 when the *Mersu* pair of 3/LeLv 34's 1Lt Paavo Myllylä scrambled to the Lavansaari area, where it shot down a solitary Pe-2 bomber. Five days later LeLv 34 claimed its third air victory when the *Mersu* pair of 3/LeLv 34's 1Lt Reino Valli destroyed a single La-5 fighter nearby Lavansaari. After this the score began to rise steadily.

Bf 109G-2 coded MT-201 of LeLv 34 taxis at Helsinki Malmi in March 1943. It was assigned to the squadron commander, Maj Olavi Ehrnrooth, and after his death on 27 March 1943 to Maj Eino Luukkanen.

Mersu *MT-207 of 1/LeLv 34* at Helsinki Malmi in April 1943. It was assigned to 1Lt Veikko Evinen, who later became an ace with six victories.

Mersu *MT-207 of 1/LeLv 34* in the process of painting the national insignia and serial number in front of the main hangar at Helsinki Malmi airport on 19 March 1943.

On 19 April 1943 3/LeLv 34 leader Capt Olli Puhakka took three MT planes for an intercept south of Kotka. At Lavansaari the detachment entered a combat with three Pe-2s escorted by six LaGG-3 and three La-5 fighters. One bomber and one La-5 were claimed shot down. 4 GIAP, KBF, reported having lost one La-5 and one Pe-2 of 73 BAP made a forced landing on the ice outside Kronstadt.

On 21 April 1943 1Lt Aimo Euramo of 2/LeLv 34 scrambled over Kotka for an intercept. Three Pe-2 bombers try to evade, flying towards Estonia, but one was caught send sent down near the coastline.

On 23 April 1943 WO Oiva Tuominen's *Mersu* pair of 1/LeLv 34 took off for an intercept over Helsinki, engaging a solitary Pe-2 aircraft, which was chased and shot down near Seiskari. It had belonged to 15 ORAP, KBF.

On 2 May 1943 WO Oiva Tuominen's pair of 1/LeLv 34 sent down a solitary Boston bomber of 15 ORAP, KBF close to the Estonian coast. Three *Mersus* scrambled later to the Lavansaari direction, where they were engaged in combat with approximately thirty Soviet aircraft. In the ensuing clash the Russians lost two La-5 fighters. Next day Capt Olli Puhakka's pair of 3/LeLv 34 sent down near Lavansaari a single Boston, also belonging to 15 ORAP, KBF.

On 4 May 1943 the Messerschmitts of LeLv 34 were airborne on several sorties and shot down six aircraft in three aerial combats. WO Yrjö Turkka's pair of 3/LeLv 34 intercepted in the Lavansaari area an La-5 squadron. Three enemy aircraft were shot down, two of them by MSgt Lauri Jutila:

"At 1020-1140 hours. I encountered with WO Turkka about ten La-5s between Suursaari and Lavansaari at 5,000 m altitude. I at-

Mersu *MT-213 of 3/LeLv 34 parked at Utti on 24 April 1943. At this point the assigned pilot was 1Lt Reino Valli. (SA-kuva)*

Pilots on duty in front of 2/LeLv 34 machine MT-201 at Utti on 2 May 1943. From the left: Sgt Aaro Nuorala, 1Lt Kyösti Karhila, SSgt Aimo Gerdt and Sgt Osmo Länsivaara. All were or became later aces.

tacked one two-plane patrol from the side and above. I fired at the wing plane, which then went in a vertical dive and fell into the sea near the shore of Lavansaari. After this I attacked another pair from behind and above. I fired at the wing plane, when it began to pour smoke after receiving numerous hits. Because of my greater speed I passed it and lost it from my sight. No damage to my plane. The enemy aircraft supported each other well. My plane was MT-209."

Both 3 GIAP, KBF and 4 GIAP, KBF lost La-5s, three in total. 1Lt Kyösti Karhila of 2/LeLv 34 tells how he engaged enemy aircraft the first time piloting a *Mersu*:

"At 1105-1225 hours. On an intercept with 2Lt Pallasvuo I encountered two LaGG-3s south of Seivästö, I got behind the wing plane and fired from about 50 m distance, cannon shells hitting the plane, it shed large pieces of plating, caught fire and crashed in the sea. Now the other one went in a dive, I caught it

and fired. The Russian made small evasive manoeuvres, but when my shells hit it, it shed large pieces too, caught fire and crashed straight and vertically in the sea. The cannon is a splendid weapon. My plane MT-214."

On 5 May 1943 1Lt Esko Ruotsila's swarm of 2/LeLv 34 headed towards Haapasaari, where two Bostons were approaching escorted by three La-5s and five LaGG-3s. The Finns split into pairs and claimed both bombers and two fighters shot down. SSgt Niilo Erkinheimo reported this:

"At 1630-1720 hours. I encountered 8-10 km north of Haapasaari one Boston III and 4-5 LaGG-3s at 6,000 m altitude heading east. I attacked the bomber from behind at the right side. I fired a burst, the port engine began to smoke. I made five passes from behind and after these the whole plane was on fire and ditched.

Pilots of 2/LeLv 34 and 3/LeLv 34 sun bathing at Utti on 2 May 1943. From the left: 1Lt Lasse Lehtonen, 1Lt Kyösti Karhila, 2Lt Mauno Kirjonen, 1Lt Reino Valli, 3rd Flight leader Capt Olli Puhakka, intelligence officer Capt Timo Tanskanen, a mechanic, Sgt Osmo Länsivaara and lying down 2Lt Yrjö Pallasvuo. Behind is 2nd Flight leader Capt Kullervo Lahtela's Mersu MT-203.

Mersus of 1/LeLv 34 at Helsinki Malmi in mid-April 1943. From the left: MT-215, 210, 207, 212 and 205.

During the last pass I got an La-5 after me, I dodged it by pulling up to the left. The La-5 lost its speed and it flipped into a dive. I turned after it and at the deck got about 100 m behind it, when I gave a burst. The plane dodged up to the left. I fired another burst, when it began to pour strong smoke and it went into a vertical dive. The altitude was now at about 800 m. One man had bailed out and came down where I shot the La-5. No damage in my plane MT-203."

On 16 May 1943 *Lentolaivue 34* received 14 more Messerschmitts. Early in the morning MSgt Urho Lehtovaara's pair of 3/LeLv 34 encountered near Peninsaari three LaGG-3s, destroying two of them, Lehtovaara wrote:

"At 0450-0545 hours. When I was on an intercept mission over the Gulf of Finland I saw a ship convoy 10 km north-west of Seiskari heading west, escorted by fighters, I approached the convoy from north-west at 2,500 m altitude and observed north of the convoy three LaGG-3s slightly below me. I attacked the

starboard wing plane of the formation and fired at it from the left side and behind. The aircraft caught fire and went in a shallow dive into the sea. I gained instantly more altitude and attacked another plane, firing at it from the right side above and behind. It banked quickly to the left and went in a spin into the sea about 20 km north-east of Peninsaari. I was forced to break off after having spent all my ammunition. My plane was MT-213."

Another pair of the 3rd Flight sent down one of the encountered two *Chaikas* in the Lavansaari area. 1Lt Esko Ruotsila's swarm from 2/LeLv 34 met a solitary LaGG-3 and sent it into the sea nearby Seiskari. The latter was actually a Yak-7 of 13 IAP, KBF.

On 17 May 1943 LeLv 34 *Mersu* pairs of 2Lt Yrjö Pallasvuo and MSgt Lauri Jutila were in the Lavansaari area. Sgt Klaus Alakoski wrote in his combat report:

"At 19.10-20.15 hours. On an intercept mission I saw between Lavansaari and Peninsaari one I-153 plane going in a shal-

low glide towards Lavansaari. I was coming from Seiskari at 1,000 m altitude towards Lavansaari. I pressed into a dive and got to open fire from about 300 m distance. The cannon shells hit the fuselage, tearing off large pieces. After this the plane flipped over and went vertically to the sea about 8 km east of Lavansaari. At the shooting the altitude was 300 m.

Soon after this the ground station radioed that two I-153s were approaching Lavansaari from the north. I climbed to meet these and saw them over the north tip of Lavansaari. I gave a small burst, after which I pulled up and got behind them. The planes were heading to Seiskari. I opened fire from a long distance, but in the middle of the shooting I observed that they were friendly. I quit firing instantly after which they broke off. When Russian I-153s are in the air at the same time, it is difficult to separate them from friendly aircraft, especially over Lavansaari. Besides, the friendly plane message had stated that four IT planes were around. My plane was MT-213."

On 18 May 1943 Capt Olli Puhakka's swarm of 3/LeLv 34 attacked a fighter squadron in the Lavansaari area and sent down two *Chaikas* and one *Rata*. Next day LeLv 34 commander Maj Eino Luukkanen's swarm attacked Soviet fighters near Lavansaari, but these were able to avoid greater losses by their manoeuvrability. One I-153 out of six may have been destroyed. After this the aerial activity over the Gulf of Finland accelerated considerably and LeLv 34 was engaged in combat almost every day, mostly above Lavansaari, where the Russians had a significant air base.

On the morning of 20 May 1943 the search patrol of 2/LeLv 34 fought with a *Chaika* pair. The lead plane received hits and crashed later in the sea. Five MT planes of 3rd Flight leader Capt Olli Puhakka were involved in combat near Lavansaari with 23 Russian planes, sending down one I-16 and a Yak-1. Puhakka's report stated:

"At 1330-1445 hours, altitude 50-1,000 m. I saw three I-16s flying low over a ship convoy on the north-west side of Seiskari. I managed to take the last one by surprise and shot it down next to a ship with a few rounds. After this I observed plenty of aircraft south-east of Seiskari heading to the east. I caught one Yakovlev fighter and after shooting three bursts into it, it dived in flames into the sea. Over the Kronstadt airfield I tried a curve battle with two Yakovlev fighters, but I was low on fuel and I had to interrupt the combat as futile. I landed on Suulajärvi airfield. My plane was MT-229."

In the evening in an air battle in the vicinity of Peninsaari, 1Lt Paavo Myllylä's swarm of 3/LeLv 34 destroyed one *Chaika* out of a twelve-plane *Chaika* squadron.

On 21 May 1943 WO Yrjö Turkka's *Mersu* pair of 3/LeLv 34 flew an intercept mission to Lavansaari engaging an I-153 squadron. Three German Messerschmitts also arrived on the combat scene. In the Finnish firing the Russians lost two *Chaikas* while the Germans claimed one more. A while later 1Lt Veikko Evinen's swarm of the 2nd Flight encountered in the same area four I-153s. One was sent down and SSgt Anatoli Sitvinikov collided with MT-228, killing its pilot 1Lt Tauno Saalasti.

In the evening Maj Eino Luukkanen led eleven MT planes of the squadron to the Seiskari-Lavansaari area, where a large air battle emerged against four Il-2s, 17 LaGG-3s and La-5s. The Russians received reinforcements from the east, but the Finns were also assisted by 16 Brewsters and 12 German Focke-Wulfs. The MT pilots downed two Il-2s, two LaGG-3s and two La-5s. Sgt Klaus Alakoski reported:

"At 1840-1955 hours. When I was on an intercept mission I observed two La-5s flying from Tytärsaari to the east. After the first pass the aircraft pulled in the clouds. I climbed over the cloud, but could not find these aircraft. When I came under the cloud I saw LaGG-3s west of Lavansaari. I attacked from the rear and above and got to fire at one, which began to take a shallow climb. My burst was hitting well and the machine flipped into a vertical dive, going straight into the sea about 5 km west of Lavansaari.

I pulled up to the cloud base and got to fire from there at another LaGG-3 from straight above. The plane nosed up and went into a dive. Probably the pilot was killed at once since the plane went out of control to the sea about 2 km south of the south tip of Lavansaari. The combat took place at 3,000 m altitude. Then the cannon and one of the machine-guns jammed. I continued to fight with two La-5s firing with one small gun. I was forced to break off the battle since I was getting low on fuel. My plane was MT-220."

On 22 May 1943 Maj Luukkanen's five MT planes of LeLv 34 sent two Lavochkins out of nine to the sea near Lavansaari and damaged a further three. Next day Capt Kullervo Lahtela's pair of 2/LeLv 34 destroyed one of four *Ratas* in the Lavansaari area.

On 2 June 1943 near Lavansaari, 1Lt Väinö Pokela of 1/LeLv 34 led six *Mersus* to combat a large Soviet formation, which consisted of 28 planes in all. Two Pe-2s, two La-5s and one LaGG-3 were destroyed, but one MT was hit by the debris of a 73 BAP, KBF Pe-2 and made a forced landing in the sea. WO Oiva Tuominen reported:

"At 1415-1445 hours. I scrambled for Kotka. South of Haapasaari I saw 8 Pe-2 and La-5 aircraft. We instantly attacked the bombers with a swarm. I fired one burst at one in the fuselage, but did not observe any results. I shot another in the starboard engine, which stopped and smoked heavily. I shot the third in the cockpit when the canopy broke. After that from about 30 m to the starboard engine, when the wing snapped off. The plane exploded and crashed in the sea. From this I got pieces in my propeller so that the blades were bent and I had to make a forced landing on the beach of Someri. My plane was MT-212."

On 5 June 1943 a *Mersu* pair of 1/LeLv 34 was also in the same region, where WO Ilmari Juutilainen shot three aircraft down. His combat report stated:

"At 1515-1625 hours. We were on a scramble mission with 1Lt Tervo on the east side of Seiskari. I saw two planes fly towards Baternaya bay. I followed them and radioed 1Lt Tervo about this. I caught them at 1,500 metres under the clouds east-north-east of Uusi-Kernovo. I shot at a LaGG-3 flying on the wing, it caught fire and crashed 8 km east of Uusi-Kernovo.

I continued my firing towards a Tomahawk, but it pulled up, as did I. When it tried to escape by diving I shot three times in the fuselage, and shedding pieces around it crashed in a swamp on its back, 12 km east-north-east of Uusi-Kernovo.

I observed more aircraft in the east so I turned away. I got now four I-16s behind me from the clouds. I pulled over the clouds when Kovashi flak was shooting. One I-16 came up, I fired, it dived under the clouds, me after. I shot again and still a third time when it pulled up in front of me. I scored a hit in the starboard wing, which broke off. The pilot bailed out with a square parachute while the plane crashed in a swamp 4 km south of the centre of Karavaldai, burning cheerfully."

1/LeLv 34 pilot WO Oiva Tuominen climbs into Mersu *MT-215 at Utti on 2 June 1943. Tuominen was the first air force Mannerheim Cross holder and had a total of 47 air victories.*

Four pilots of 1/LeLv 34 in front of Mersu *MT-222 at Utti on 2 June 1943. From the left: 1Lt Lauri Pekuri, SSgt Eino Peltola, Sgt Lauri Mäittälä and SSgt Erik Lyly. All but Mäittälä gained ace status.*

Pilots of 1/LeLv 34 in front of Mersu MT-212 at Utti on 2 June 1943. From the left: flight leader Capt Pauli Ervi, 1Lt Väinö Pokela, MSgt Mauno Fräntilä, WO Oiva Tuominen, 1Lt Kalevi Tervo, SSgt Gösta Lönnfors and Sgt Urho Lehto.

1Lt Väinö Pokela of 1/LeLv 34 on returning from a mission to Utti on 2 June 1943. The plane is MT-214 belonging to the 2nd Flight, in which Pokela shot down one Pe-2 bomber of 73 BAP, KBF.

Four 1/LeLv 34 pilots watch others doing aerobatics over Utti on 2 June 1943. From the left: SSgt Eino Peltola, Sgt Lauri Mäittälä, 1Lt Lauri Pekuri and SSgt Erik Lyly. Behind are MT-212, 222 and 207.

The flak was fierce when I flew away, inside the encirclement was peaceful. The MT's cannon is a splendid weapon. No damage to my MT-222."

After this the Gulf of Finland went quiet. The clashes became rare since the new base at Seiskari protected the formations returning from the German front. In spite of several efforts the Russians did not enter into combat and chose to remain under the strong flak defences of the island, where it was strictly forbidden for the Finns to fly.

On 24 June 1943 Maj Luukkanen's patrol of LeLv 34 encountered over the Karelian Isthmus two Bostons escorted by a LaGG-3 flight. The fighters knew their job since the bombers escaped and one LaGG-3 was sent down. Later a *Mersu* pair of 1/LeLv 34 downed one *Chaika*, which had just taken off from Lavansaari.

The main opponent for the Messerschmitts was the Baltic Fleet air forces, which on 1 July 1943 possessed 100 fighters, 50 bombers, 70 assault planes and 20 others.

The order of battle of *Lentolaivue* 34 was then:
Squadron commander Major Eino Luukkanen, Utti
- 1st Flight, 1st Lieutenant Lauri Pekuri, Utti (6MT)
- 2nd Flight, Captain Kullervo Lahtela, Utti (6 MT)
- 3rd Flight, Captain Olli Puhakka, Malmi (6 MT)

On 10 July 1943 1Lt Kalevi Tervo scrambled his swarm of 1/LeLv 34 to an interception over the Lavansaari-Seiskari area, where they met a recently airborne formation with fourteen I-153s, two Pe-2s, and two Lightnings. Two MT planes remained in high cover and two more attacked, shooting down one Lightning and four *Chaikas*. The Lightning was more likely a captured German Fw 189. WO Ilmari Juutilainen shot down three:

"At 1535-1650 hours. We were returning from an alarm mission between Lavansaari and Someri, when we were notified of ten I-153s en route east. West of Seiskari I noticed three I-153s and shot down the leader before they made it to the island.

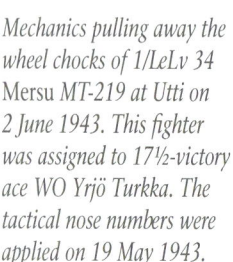

Mechanics pulling away the wheel chocks of 1/LeLv 34 Mersu MT-219 at Utti on 2 June 1943. This fighter was assigned to 17½-victory ace WO Yrjö Turkka. The tactical nose numbers were applied on 19 May 1943.

Four duty pilots of 1/LeLv 34 at Utti on 2 June 1943. From the left: WO Oiva Tuominen, 1Lt Kalevi Tervo, SSgt Gösta Lönnfors and Sgt Urho Lehto. On Tuominen's next mission he had to ditch his MT-212 and swim ashore.

As we continued east I found some fifteen I-153s, 2 Pe-2s and 2 P-38Ds. I circled to attack them out of the sun, and thus surprised two I-153s with 1Lt Tervo; they fell as one.

The bombers on the deck, covered by the fighters, were hard to get at because the fighters used plenty of rocket projectiles flying some 200 m above the bombers. One P-38 flew to one side for a second and I shot it down, aiming my cannon and machine gun fire at the cockpit.

As soon as we attacked the enemy began to circle around the spot, and we were sorry to leave it there due to dwindling fuel. The I-153s were painted bright black, red and green. The Pe-2s and the P-38D were dark grey. My plane was MT-217."

On 16 July 1943 the new commander of LeR 3, LtCol Gustaf Magnusson, redefined squadron tasks. The area of action for LeR 3 was still from Valamo to Helsinki, and its mission was combat, air cover, and reconnaissance. LeLv 34 operated between Lavansaari-Virolahti and Porkkala-Naissaari, its tasks being the protection of Helsinki and Kotka.

On 19 July 1943 WO Ilmari Juutilainen of 1/LeLv 34 scrambled to intercept a lone Pe-2 which appeared above Suulajärvi. It escaped but over Koivisto WO Juutilainen ran into three LaGG-3s and shot down two. 1Lt Kyösti Karhila and his patrol of 2/LeLv 34 shot down a lone Pe-2 over Seiskari. In the evening, a *Mersu* patrol of 2/LeLv 34 led by WO Oiva Tuominen, shot down one of three LaGG-3 planes over Someri.

On 20 July 1943 Maj Eino Luukkanen led his patrol over Suursaari to attack four Il-2s and eight LaGG-3s, shooting down two fighters:

"At 0500-0550 hours. I took off with two planes to intercept four Il-2s which were flying around Suursaari. While airborne, I heard over the radio that eight LaGG-3s were also in the air, so I called in three planes by radio. While over Kotka, I saw flak plumes west of Suursaari, and these enabled me to find four Il-2 planes right away. Sgt Leskinen and I circled around the plumes to avoid the flak and then we attacked the Il-2s from above and behind. As I was firing, the plane dropped its bombs and just then the whole area was full of flak plumes (several hundred of them). Now I spotted ships to the left and front, and the Il-2 bombs fell some 2 km from the nearest ship.

The flak forced me to abort and I left to the west of the ships, where I found eight LaGG-3s below me. I radioed Sgt Leskinen to come in, but he heard nothing on the radio. I fired at five LaGG-3s from the side and behind, and one of them fell from 400 m, igniting before impact and leaving a large oil slick.

Another LaGG-3 went into a spin as I fired at 200 m height, and it went in the sea 25 km south-west of the Suursaari southern tip at 5.21. I broke off, having run out of cannon shells. At the south tip of Suursaari I saw the Il-2 planes retreating east at 100 m. I fired at one with the small guns, to no avail, so I left and saw Sgt Leskinen firing at the same plane. The three planes I had

WO Oiva Tuominen shot down on 2 June 1943 a Pe-2 of 73 BAP, KBF. Its debris hit the propeller of Tuominen's Mersu MT-212 and he had to ditch it at Someri. Here the plane has been pulled out of the sea and taken to Hamina harbour. The salt water had already ruined the plane.

Mersu MT-215 of 1/LeLv 34 parked at Helsinki Malmi in late May 1943. It was assigned to 11-victory ace SSgt Eino Peltola.

called in saw the enemy planes only at Seiskari. Bad coordination of LaGG-3 pilots. My plane was MT-201."

WO Oiva Tuominen of 2/LeLv 34 and his patrol flew an intercept mission to Seiskari, scoring hits on one Yak, which then flipped over into the sea after the end of the runway. It had belonged to 13 IAP, KBF.

On 24 July 1943 WO Oiva Tuominen and his patrol of 2/LeLv 34 met in the Suursaari-Seiskari area nine LaGG-3s flying escort for four Il-2s. The Soviets lost two fighters to Tuominen:

"At 0755-00855 hours. Scrambled for interception. I saw four Il-2s and three LaGG-3s between Suursaari and Lavansaari. I caught the LaGG-3s south of Lavansaari, and the Il-2s went to land at Lavansaari. I did a curve battle with the LaGG-3s, now and then entering clouds, when we got to south of Seiskari. I shot down the first one just as it lowered its landing gear and it crashed in the sea.

Six more LaGG-3s appeared. I shot one into the sea from a curve south of Peninsaari. I shot at one more, but I could not follow it when I had to go into a cloud. I broke off when they began to fly over Seiskari. The Soviets flew in circles with one of them attempting an attack. Combat altitude 800 m. My plane was MT-220."

On 26 July 1943 1Lt Veikko Evinen of 2/LeLv 34 flew his three-plane MT patrol to Haapasaari where they met a 12-plane Il-2 formation, of which one was damaged and fell into the sea later. 1Lt Paavo Myllylä of 3/LeLv 34 fought with his patrol in the Someri area against a flight of La-5s flying cover for Il-2s. One of each type was shot down. 1/LeLv 34 flew single-plane patrols over the Karelian Isthmus, reporting one Boston and one Il-2 shot down. Three days later a swarm from 2/LeLv 34, led by its commander Capt Kullervo Lahtela, destroyed one La-5 from a detachment of ten in the Haapasaari-Seiskari area.

On 1 August 1943 1Lt Paavo Myllylä and his patrol from 3/LeLv 34 met four unescorted Pe-2 bombers at Tytärsaari. As

confirmed by the Russians, only one 73 BAP, KBF Pe-2 escaped, as reported by Myllylä:

"At 1605-1710 hours. I took off for interception with MSgt Lehtovaara at 16.05. At the Juminda cape, I spotted 4 Pe-2s eastbound at 3,500 m. I reached them before Pien-Tytärsaari and shot down one Pe-2 at some 15 km NW of Tytärsaari. I shot down another with MSgt Lehtovaara some 25 km NE of Tytärsaari. MSgt Lehtovaara shot down a third some 15 km NE of Tytärsaari. We fired at the fourth plane killing its machine gunner and causing the hoods to fly off, but we ran out of ammo and the plane managed to land at Lavansaari. My plane was MT-209."

On 2 August 1943 the main part of LeLv 34 was flown to Kymi, and its area of responsibility was extended east to the Viipuri-Oranienbaum line. Because the plane situation of LeLv 30 was still dismal, due mostly to accidents, LeLv 34 was ordered to assist in the reconnaissance of the eastern Gulf of Finland, especially in conjunction with its intercept and alert missions.

On 7 August 1943 1Lt Kalevi Tervo and his patrol of 1/LeLv 34 was sent on an interception over the Gulf of Finland. Over Peninsaari a lone Pe-2 was spotted, and later three LaGG-3s. Tervo reported on the attack:

"At 1510-1555 hours. While on interception north of Seiskari, I saw a Pe-2 roughly at Peninsaari, at 500 m, eastbound. I was at some 2,000 m. As I dove at the Russian, it turned north again towards Lavansaari. I caught it some 5 km before Lavansaari. As I began firing I found my cannon inoperable. I shot a long burst into the cockpit, stopping only at some 20 m distance. As I pulled up and right, the Pe-2 continued in a shallow glide towards Lavansaari. I did not attack again, as this would have meant entering the flak. I looked on as the Pe-2 glided in, crashed on the runway and caught fire.

As I gained altitude north of Lavansaari I saw three La-5s slightly above me. As they attacked my patrol mate Sgt Mäittälä

Mechanic Esko Rinne stands in front of 1/LeLv 34 Mersu MT-207 at Suulajärvi in August 1943. This fighter was assigned to 21-victory ace 1Lt Kalevi Tervo.

Mersu MT-222 of 1/LeLv 34 at Suulajärvi in August 1943. It was assigned to WO Ilmari Juutilainen for one year, who claimed with it 15 kills out of his 94.

I entered the fray. Sgt Mäittälä had to break off in a dive. I remained in the fight, trying to get above the LaGG-3s. After some turns I managed to shoot at a La-5 from above as it pushed below me. I stopped firing very close, when I was vertical going down. As I pulled up, I saw the LaGG dive in the sea, the pilot in his parachute.

I did some more curving around with another La-5, but when it found itself threatened it dove away to the safety of Lavansaari but I did not follow. Then I saw three motor torpedo boats racing towards the crash site some 2 km NW of Peninsaari. After running around for a while they fished out the pilot and then left for

Seiskari. LaGG-3 pilots were pretty good, used the plane's agility to their advantage. My plane was MT-205."

On 10 August 1943 WO Ilmari Juutilainen and his patrol from 1/LeLv 34 fought at Seiskari with LaGG-3s, shooting down two. Later the patrol of 1the 1st Flight led by 1Lt Reino Valli acted as top cover for the Brewster flight, and it was attacked by three La-5 planes. One of these was shot down. A week later Capt Lahtela and his patrol of 2/LeLv 34 was sent on interception to the Someri-Lavansaari area, where air combat ensued with four Il-2s and six Yak-7s. Two of the *Sturmoviks* of 7 GshAP, KBF went into the sea.

On 20 August 1943 the swarm of 1Lt Kyösti Karhila of 3/LeLv 34 met a sixteen plane detachment over Seiskari, of which a Boston and a La-5 were shot down. In the afternoon, three *Mersus* acted as top cover for 16 Brewsters as they fought fifteen Lavochkins west of Kronstadt. The Finns had superior numbers for a change, and four LaGG-3s and three La-5s were shot down. Capt Olli Puhakka shot down one of the latter. On his next flight, he shot down two more planes and he wrote this report:

"At 1630-1745 hours. Sgt Nuorala and I ran into two Il-2s east of Tytärsaari. I fired at one from straight behind, probably badly wounding both pilot and gunner. The plane flew on belching heavy smoke, then crashing into the sea in flames after 5 to 10 km. In the meantime we were jumped by four LaGG-3s. We fought them for maybe 20 minutes. Two of them left after being hit. During the brawl I saw one crash on Lavansaari. My plane was MT-216."

The last to fight was 1Lt Väinö Pokela's patrol from 1/LeLv 34 who found a mixed enemy formation over the Gulf of Finland, and shot down one IL-2 and one LaGG-3. The Finns lost one of the two Messerschmitts.

On 29 August 1943 Capt Olli Puhakka and his swarm of 3/LeLv 34 met and shot down two Yak-7 fighters south of Suursaari. These had belonged to 12 KAE, KBF. Two days later the *Mersu* swarm of 1/LeLv 34 led by 1lt Lauri Pekuri shot down an La-5 pair at Seiskari.

On 7 September 1943 Maj Eino Luukkanen led a swarm of LeLv 34 Messerschmitts to intercept a 25 plane formation around Tytärsaari, shooting down two Yak-7s of 13 IAP, KBF.

On 8 September 1943 Maj Eino Luukkanen led six MT planes of LeLv 34 around Tytärsaari to battle with two Il-2s and their escort of four LaGG-3s. All enemy planes were shot down without losses. 35 ShAP, KBF admitted the loss of two Il-2s and 12 KAE, KBF the loss of two Yak-7s.

The last great aerial battles before the winter were fought on 23 September 1943 above the eastern Gulf of Finland. After 13.00 four Brewsters of 3/LeLv 24, led by 1Lt Martti Salovaara, and two *Mersu* patrols of 1/LeLv 34, fought twenty Soviet fighters close to the Shepelevskiy lighthouse and reported three Yakovlevs and five Lavochkins shot down. WO Ilmari Juutilainen reported shooting down three fighters, but only two were credited to him:

Polikarpov I-15bis, IH-2 of LeLv 34 in July 1943, based at Utti. (Karolina Hołda)

The target and target-tow plane of LeLv 34 was this captured Polikarpov I-15bis coded IH-2. It is seen here at Utti before flipping over in a landing on 2 August 1943.

Mersu MT-207 of 1/LeLv 34 faced engine trouble and WO Yrjö Turkka bellied the plane near Lintula on the Karelian isthmus on 11 December 1943. The repair in the unit took 17 days.

"At 1310-1350 hours. I took off with Sgt Lyly on interception to the Gulf of Finland. 1Lt Pekuri and SSgt Peltola were there already. As we got to the area, I found enemy planes NW of Shepelevskiy. The BWs were already attacking. I joined the first BW patrol in the attack. In the ensuing curve battle with many La-5s, LaGG-3s and Yak-1s I shot down one Yak-1 which fell burning into the sea outside Kopornoye bay.

One La-5 followed SSgt Peltola at 50 m, I shot it into flames, and it went in a few kilometres NW of the former. After an extended curve battle I managed a shot at yet another La-5, which burning and smoking fiercely fell in the sea between Ino and Kopornoye bay. It had a large white fuselage sign '64'. My plane was MT-222."

LeLv 34 recorded its 100th aerial victory on 4 October 1943 when 2Lt Yrjö Pallasvuo of the 3rd Flight shot down a pair of planes, actually Yak-9s of 15 ORAP, KBF, on reconnaissance mission, reporting thus:

"At 0945-1055 hours. While on interception, I met two LaGG-3s east of Loviisa at some 3,500 m. A curve battle ensued during which I hit a LaGG-3 from straight behind in the fuselage. It let out a huge white smoke cloud and it went in a vertical dive. I did not follow but started to look for the other LaGG-3. I met it south of Haapasaari at some 3,000 m, shot twice at it from behind, and its radiator flashed flames. Soon it went into the sea between Someri and Lavansaari. It had a new national roundel: a narrow white band around the red star. My plane was MT-233."

On 10 October 1943 the 3rd Flight sent MSgt Antti Tani and his patrol on alarm to the Utti-Luumäki area, where it shot down a solitary Boston. Capt Kullervo Lahtela, the commander of 2/LeLv 34, and his patrol shot down a single Pe-2 near Porkkala. Four days later 1Lt Aimo Euramo's Mersu patrol of 3/LeLv 34 shot down a solitary Boston near Luumäki.

In the morning of 27 October 1943, LeLv 34 commander Maj Eino Luukkanen led his three Mersus to attack an Il-2 detachment covered by six fighters, shooting down one ground

attack plane. Around noon the MT swarm of Maj Luukkanen went on interception to Lavansaari, meeting a 13-plane mixed gaggle of enemy planes and shot down one Il-2 and one Yak-1. WO Ilmari Juutilainen and his patrol of the 1st Flight took off to chase a pair of La-5s who flew over Suulajärvi, managing to catch and destroy one. As it returned, the patrol found four La-5s at Ino and shot down two. Again, Juutilainen reported three shot down but was credited with only two:

"At 1440-1535 hours. Two La-5s flew over our base where we had two BWs. I was sent with Sgt Mäittälä to chase them. We caught up with them but they escaped in a cloud. We met them again at the coast by Vammelsuu and I began to shoot at the lead plane. The last shot was taken about one km from Kronstadt, and the plane pulled vertical. We turned away and saw it striking the ground some 200 m from the SW corner of the Kronstadt airfield, leaving a 500 m high column of smoke.

Some time later we met four La-5s at Ino, at a height of some 2,000 m. In the fight we had in the fog above Tolbuhin, I fired at the lead plane which shed its right wing and exploded, causing a huge black cloud to engulf it.

I also shot one La-5 into flames and saw it crash in the sea east of Tolbuhin. I saw one speedboat leave Kronstadt and head for the crash sites. The Russians were eager to fight. After we had shot down all of them, no more planes appeared. Flak did not fire though it could have reached us at some point. My plane was MT-222."

On 4 November 1943 three Mersus of LeLv 34, led by Maj Eino Luukkanen, battled 15 Pe-2s escorted by 19 fighters in the Someri-Haapasaari area. Despite the numerical odds, two La-5s were shot down and two damaged. Luukkanen reported:

"As 15 enemy bombers and 19 La-5s flew west at Someri we went on to intercept just to be sure with all three serviceable planes, though it was likely the attack would be on vessels at sea, as it had been all day. At take-off I saw many flak plumes over Kotka, and then I saw 19 fighters above me. We had no time to assemble, and

Mersu MT-231 of 2/LeLv 34 running up at Helsinki Malmi on 8 January 1944. MSgt Mauno Fräntilä then flew an uneventful interception mission.

Mersu MT-213 of 1/HLeLv 34 lands at Suulajärvi. WO Ilmari Juutilainen brought the plane there on 6 April 1944 and flew it until 1 May 1944, when it was handed over to HLeLv 24.

I saw no planes of our own until I broke off from the action. South of Kotka I attacked the La-5s because I saw no bombers. I fired at three separate planes, one of which gave out smoke and at 13.00 went from 300 m through a half roll into a dive; I do not think it recovered before impact. I gave a long burst to another plane from behind at 200-50 metres, and my cannon shells burst the canopy and the rudder broke off, remaining tied to the plane with a wire. The plane lost control at 13.07 going into a tailspin from 1,500 m. I could not follow as a La-5 fired at me from behind. I had to break off due to exhausted ammo. I did not think the battle was too hard despite the numbers, because this enabled me to move around in the enemy formation and to be noticed only when firing. Enemy planes had a hard-to-see red star. Planes painted with black and green stripes seen from the top. The armour of La-5 seems effective. The green planes had a white spinner. My plane was MT-216."

On 19 November 1943 the swarm of Capt Olli Puhakka of 3/LeLv 34 met a 10-plane detachment of Soviets at Suursaari, shooting down one Yak-7. 1Lt Paavo Myllylä's patrol from the 3rd Flight fought five Sturmoviks escorted by four Yaks and managed to destroy one fighter. 1Lt Reino Valli led five Mersus of the 1st Flight on two interception flights to the eastern Gulf of Finland, first shooting down a solitary Pe-2 and then, on another flight, meeting a detachment of four Il-2s and four La-5s. They shot down one Il-2 and two fighters.

Bf 109G-2 MT-213 of 1/HLeLv 34 flown by WO Ilmari Juutilainen in April 1944, based at Suulajärvi. (Karolina Hołda)

Top *Mersu* scorers of *Lentolaivue 34* in 1943:

Rank	Name	Flight	Victories	Plan	Total victories
WO	Juutilainen, Ilmari	1	18	MT-222	51 + 2
MSgt	Lehtovaara, Urho	3	12½	MT-218	22½ + 1
Maj	Luukkanen, Eino	E	11	MT-201	25½ + 2½
Capt	Puhakka, Olli	3	11	MT-204	23 + 7
WP	Tuominen, Oiva	2	10	MT-220	32 + 9
SSgt	Alakoski, Klaus	3	10	MT-209	11
1Lt	Karhila, Kyösti	2	8½	MT-224	18½
1Lt	Tervo, Kalevi	1	6	MT-207	23
1Lt	Myllylä, Paavo	3	5½	MT-229	7
MSgt	Tani, Antti	1	5½	MT-223	10½

The total column includes also the previous victories of the pilot, with Winter War score after the + sign

On 1 January 1944 *Lentolaivue* 34 possessed 16 serviceable *Mersus* and one in overhaul. As during previous winters the air space over the front was quiet.

The first contact of the year was made on 6 February 1944 when 1/LeLv 34 sent 1Lt Lauri Pekuri and his swarm for an interception mission over the Karelian Isthmus, and they entered a battle over Siestarjoki. Eight Yak-7s were met and two were shot down. The Finns identified the types wrongly, as 3 GIAP, KBF actually lost two La-5s.

On 14 February 1944 LeLv 34 became *Hävittäjälentolaivue* 34 (HLeLv 34).

Initiated at mid-January 1944, the main offensive of the Red Army against the German North Army Group broke the siege of Leningrad by the end of February 1944. In this new situation, the commander of LeR 3 redefined on 20 February 1944 the areas and tasks of the squadrons thus: HLeLv 34 functioned with Porkkala-Naissaari as its right boundary and as its left Viipuri-Seiskari, with flying south of the line Naissaari – Prangli – Vaindlo – Tytärsaari – Lavansaari – Seiskari forbidden. HLeLv 34 was also to protect Helsinki and Kotka; all other tasks were subject to an order by the regiment commander.

By the beginning of March 1944, the Red Banner Baltic Fleet air forces changed their tactics when flying against the mine fields in the Gulf of Finland. They began to attack the northern maintenance depot in the city of Kotka, and in some cases, Hamina 20 km to the east. These attacks also affected the shipping of war materiel bound for other fronts.

The first major attack on Kotka was on 6 March 1944. At 13.45 the commander of HLeLv 34, Maj Eino Luukkanen,

took off to intercept a 40-plane formation, which was met at Narvi, and the Finns shot down five Pe-2s and two La-5s. The Baltic Fleet 12 GBAP is known to have lost at least four Pe-2s. Luukkanen wrote:

"The first air raid report came from Someri, and based on that I took off with all five available planes. I saw 27 Pe-2s and 12 La-5s above me, south of Kotka. I went for the bombers because the fighters were quite far off to the side. I got the first Pe-2 burning with a short burst of the cannon, and two men bailed out. The plane went through the ice at Kilpisaari.

Of the other Pe-2 I got the right engine to quit at 20 km NW of Narvi, and it went down in a dive through a half flat spin, but I was unable to follow it as I was jumped by eight La-5s. I managed a shot at one from below and behind, with pieces flying off below the engine, and it went through the ice 10 km NW of Narvi.

As I shot at the next La-5, I heard a loud bang in my plane, for one La-5 had crept up behind me unnoticed, and it hit me with four cannon shells, in the fuel tank, the radio, and the back armour. I broke off and performed a good belly-landing at the base. The Pe-2s flew in two formations, and the latter one turned back before Kotka when they saw us coming. My plane was MT-201."

Two and a half hours later, 1Lt Paavo Myllylä of 3/HLeLv 34 rushed his six Messerschmitts to a meeting with a smaller detachment, and sent four enemy planes into the sea outside of Kotka. In the evening 1Lt Väinö Pokela and his three MT planes of 1/HLeLv 34 met many Pe-2 planes with fighter cover at Seivästö. Two bombers and one escort fighter were shot down. The other triple victory of the day went to WO Ilmari Juutilainen:

"At 1710-1805 hours, altitude 1,000-10 m. As we patrolled the Gulf of Finland with the other BW planes, I noticed a formation heading east at Seiskari. I called the others, shot a burst to show the direction, and went on to attack. Due to a misunderstanding, the others did not see the target. On the way I was joined by two enemy fighters which I mistook for MT planes when I checked them up behind me.

I went for the 16 Pe-2s, shooting down two myself at 10 and 15 km west of the Shelepevskiy lighthouse. Now I found I had strange company. I pulled up, and the fighters went for Seiskari, but I followed them. My cannon was out, so I shot some 500 bullets into the nearer plane, and it crashed onto the ice. Both fighters were completely white. My plane was MT-222."

Also on 6 March 1944 2/HLeLv 34 was attached to HLeLv 30 with all its planes and pilots. The latter's task was to cover Helsinki and intercept the enemy over the Gulf of Finland, releasing HLeLv 34 from these duties.

Three distinguished commanders at Immola on 15 June 1944. From the left: 1/HLeLv 34 leader Capt Lauri Pekuri, HLeLv 26 commander Maj Lauri Larjo and HLeLv 34 commander Maj Eino Luukkanen. (SA-kuva)

HLeLv 34 commander Maj Eino Luukkanen was awarded the Mannerheim Cross on 18 June 1944, when his score stood at 39. He sits here in his assigned Mersu MT-417 at Immola three days earlier. (SA-kuva)

On 7 April 1944 HLeLv 34 received the first new Bf 109G-6 planes from Germany and began to hand over its older Bf 109G-2 planes to HLeLv 24.

On 8 May 1944 the commander of HLeLv 34 Maj Eino Luukkanen took nine Messerschmitts to attack Russian bombers over Kotka. Of the 29 adversaries, one Pe-2 and three Yak-9s were shot down.

By 17 May 1944 HLeLv 34 had so far tried every time to climb above any intruders as they attacked Kotka or Hamina, but now it was time to try something else. The Russians were taken by surprise when eleven MTs led by Maj Luukkanen hit a formation of 27 Pe-2 straight from the climb at 10.40, taking out eight bombers before the fifteen escort fighters flying above could do anything. In the fighter battle that ensued one MT was traded for three Yaks. Capt Olli Puhakka wrote this of his battle:

"Right after the start, I saw more than ten bombers in formation approaching Kotka with an escort of about 20 fighters. Before reaching the interception area, the planes turned east, probably after seeing so many fighters taking off. Because I had no time to get above the fighters, I decided to get at the bombers from below. This worked well as the planes dived at the Hamina harbour. As soon as they pulled up and levelled off I caught the Pe-2 formation.

The first plane caught fire and went into the sea right outside Hamina. With the same speed I attacked another plane which also flamed and dived. I had no time to see where it fell. Large parts including the hood came off, and I saw a man bail out, but had no time to see the parachute. Soon I was at the third one. It shed pieces too, went into a shallow dive, and soon crashed into the sea at the outermost small islands. One man exited this plane too, but I saw no chute. My plane was MT-419."

Maj Eino Luukkanen on scramble duty sitting in MT-417 at Immola on 15 June 1944. He took a flak hit to this plane on 19 June 1944 and made a forced landing between the lines at Summa, walking away. (SA-kuva)

WO Ilmari Juutilainen visited Lappeenranta on 28 June 1944 flying MT-451 of 1/HLeLv 34. On this date he was awarded the Mannerheim Cross for the second time for 75 air victories. (SA-kuva)

On 19 May 1944 Capt Olli Puhakka of 3/HLeLv 34 scrambled to Kotka at 04.35 with nine planes, to meet twenty Pe-2s escorted by 15 Yak-9s. In a chase that went all the way to Lavansaari, two bombers and four fighters were shot down. 1Lt Paavo Myllylä described the combat:

"Altitude 2,000-100 m. I saw several Pe-2s escorted by Yak-9s approach Kotka from the south-east at 3,000 metres. I managed to shoot at one Pe-2 over the south side of Kotka, it caught fire instantly between the starboard engine and fuselage. I was not able to follow the aircraft as I got several Yaks in my back. But Capt Puhakka saw the plane come down 5-10 km south-west of Rankinsaari. Then I fought with the Yaks and shot one into the sea 5-10 kms north of Lavansaari. My plane MT-406."

This was to be the last major attack of the Russians on Kotka for now, since the Karelian Isthmus was to be the point of interest for the Russian Army and its supporting air units in the near future.

At the end of May 1944, masses of Russian troops, tanks, artillery pieces and other materiel were in position facing the Finnish front on the Karelian Isthmus, north-west of Leningrad. Fighter reconnaissance had no trouble seeing this, but getting the HQ to believe the reports was harder.

Great Attack

After the success achieved against Germany in the spring of 1944, the Soviet Union began the fourth of ten strategic efforts, in the Karelian Isthmus; this was to be the only one which did not reach its goals. The offensive started on 9 June 1944, and on the next day the first Finnish line of defence was breached, forcing the Finns to retreat.

The massive attack on the Karelian Front was supported by the Soviet 13th Air Army with more than 1,100 planes, and the left flank was protected by 220 planes from the Baltic Fleet

Juutilainen buckles up in front of MT-451 of 1/HLeLv 34 at Lappeenranta on 28 June 1944. (SA-kuva)

Air Force. These warplanes, totalling over 1,300, were concentrated on a 20 kilometre strip of land at the eastern side of the Gulf of Finland.

Against this massive air armada aimed at the Karelian Isthmus, *Lentorykmentti* 3 could pitch the 14 Messerschmitts of HLeLv 24, the 16 *Mersus* of HLeLv 34, and the 18 Brewsters of HLeLv 26. LeR 3 was tasked with reconnaissance of the entire Karelian Army area, the blocking of Russian air attacks over the whole of the Karelian Isthmus, and the escort of Finnish bombers over the Isthmus and the nearby seas.

On the first day of the assault the Russians sent up 1,150 sorties. In the inclement weather, all LeR 3 squadrons took to the air in small detachments, and in six separate battles the Finns sent down nine enemy planes.

HLeLv 34 was able to fly only eight sorties due to the low clouds. A three plane detachment temporarily based at Suulajärvi and led by 1Lt Paavo Myllylä shot down two La-5s at Haapasaari.

Next day the 13 Air Army flew 800 sorties, and the intercepting LeR 3 interceptors took down sixteen fighters in five separate engagements.

Mersu MT-425 of 1/HLeLv 34 returns from a mission over Saimaa on 20 June 1944. SSgt Osmo Länsivaara was at the control on this occasion, claiming one Yak-9 shot down as his fourth victory. And there was one more to come.

Three *Mersus* of 3/HLeLv 34 led by 1Lt Myllylä were at Terijoki when they met and fought with ten bombers and some 60 fighters; Russian losses were one Tu-2, two La-5s and one Airacobra. Later when on reconnaissance to the front lines, a swarm from the 1st Flight shot down one Il-2.

On 14 June 1944 the swarm of HLeLv 34 led by Maj Luukkanen flew top cover to the unloading of ground troops at Perkjärvi station when it was attacked by dozens of enemy fighters. In the battle, three La-5s and two Airacobras were dispatched and the Finns lost one *Mersu*. Luukkanen's short report was:

"At 1050-1150 hours. On an escort mission with 14 MT we met close to the front-line 10 enemy fighters, I shot two La-5s into flames, one to Kuuterselkä and the other to Terijoki. Then I observed at the coast line at 700 m altitude a silver-colour observation balloon. I attacked from the sea with the sun behind me and shot the balloon down with one short cannon burst. The previous fighters were flying top cover to the balloon. Extremely strong flak. My plane was MT-422."

During the day, Capt Puhakka's 3rd Flight shot down two *Sturmoviks*, two La-5s and one Airacobra in two separate en-gagements over the front lines. Later at Kuuterselkä, Maj Luukkanen led his 10 MTs to battle with dozens of enemy fighters and managed to down one La-5 and one Airacobra.

3/HLeLv 34 was flying cover for infantry on 17 June 1944 in the Kivennapa-Valkjärvi area, when six *Sturmoviks* appeared with five fighters. The MT pilots shot down one Il-2, one Yak-9 and two Airacobras. WO Urho Lehtovaara wrote a short report:

"At 0520-0635 hours, altitude 100-2,000 m. South of Uusikirkko, I met many ACs. 2Lt Pallasvuo fired at one AC and apparently hit it, because it broke off. I made a subsequent attack on it and it went in a bog eight km north-west of Ino. Some Yak-9s and La-5s appeared on the scene. I got to fire at one Yak-9 from low and behind, and it crashed 4 km west of Ino. My plane was MT-417."

Capt Puhakka's swarm of the 3rd Flight shot down a solitary U-2 and one Pe-2 over the front-line. Puhakka related:

"At 0820-0930. According to the order I flew with four planes to Kämärä and the south. We got an announcement of five Il-2s and soon I saw them under me. The other planes separated due to poor radio contact and also that they had seen elsewhere other enemy planes.

Bf 109G-6, MT-435 of 1/HLeLv 34 flown by SSgt Urho Lehto in June 1944, based at Lappeenranta. (Karolina Hołda)

MT-445 of 3/HLeLv 34 parked among the trees at Lappeenranta on 22 June 1944. This fighter was assigned to 28-victory ace SSgt Klaus Alakoski.

I shot at two Il-2s so that pieces flew around, but could not get them down. Then a lone U-2 came towards me at the deck. I shot, breaking the tail and top wing, the plane crashed at the edge of a swamp south of Jäppilä.

When flying towards Johannes I observed a lone Pe-2 heading to Summa. I caught it soon and fired a burst from the tail to the right engine. The plane caught fire between the fuselage and engine and crashed on the beach of Summajoki, I saw two men bail out. My plane was MT-419."

Later Maj Luukkanen took five MTs to the Kämärä-Leipäsuo area to fight strafing Il-2s and shot down two. 1Lt Myllylä and his patrol added one ground attack plane to the tally.

On 18 June 1944 HLeLv 34 commander Maj Eino Luukkanen was awarded the Mannerheim Cross. His aerial victories stood then at 39.

A swarm of 1/HLeLv 34 led by 1Lt Pokela was on morning reconnaissance, when it chanced upon a 25 plane strong fighter regiment, of which they shot down one La-5. Capt Puhakka led six *Mersus* of the 3rd Flight to meet fifteen Il-2s attacking troops at Virolahti. One *Sturmovik* was shot down before the escort of fifteen fighters managed to tie down the Finns. Two Airacobras and one Yak-9 were destroyed in this event.

On 19 June 1944 new Bf 109G-6 deliveries from Germany began to ease the plane situation. HLeLv 24 and HLeLv 34 both came close to having 25 planes on duty.

Russian flak hit Maj Luukkanen's *Mersu* and he performed a successful forced landing between the lines. Capt Puhakka and his five MTs on a reconnaissance mission to Muolaanjärvi shot down three Il-2s and one Airacobra of a fifteen plane detachment.

In the evening 3/HLeLv 34 sent Capt Puhakka with eight planes to Viipuri and Capt Wind of 3/HLeLv 24 followed him with ten more MTs, running into a fight with numerous Russian air regiments. With no losses of their own, the Finns dispatched six Pe-2s, three Airacobras, two Il-4s and two La-5s, 13 victories in all. WO Urho Lehtovaara shot down four and reported:

"At 2000-2055 hours. I took off on a recce mission to the Vammelsuu-Haapakangas-Kyyrölä direction. Right after take-off I saw some 30 enemy planes at Säkkijärvi, and a few moments later, some 20 more planes appeared from there, heading north. I also saw enemy formations approaching from south-east, and I had to fight the enemy fighters with SSgt Nuorala over Viipurinlahti.

I took a surprise shot at two ACs from behind and below, and one of these fell 4 km west of Koivisto and the other at 8 km south-east of Koivisto. The fight depleted my fuel and I could not complete the recce, so I returned.

Arriving south of Viipuri I saw many Pe-2s in the south-east, covered by some ten fighters. I attacked them and managed a stern shot at two Pe-2s which caught fire immediately. One fell 2 km south of Kämärä and the other 3 km south. My plane was MT-406."

Lehrovaara's wingman SSgt Aaro Nuorala claimed three enemy planes shot down, stating:

"I took off at 20.00 with WO Lehtovaara for a recce mission. When being west of Viipuri I saw between Säkkijärvi and Simola an enemy bomber formation, flying in three waves escorted by fighters towards Simola. At Vahviala came one 10-plane formation, also with fighter escort heading north-west. I radioed the

base about this and continued the mission. Over Viipurinlahti we were engaged in a combat with enemy fighters. I fired at one La-5, which crashed at Härkölä by Koivirta stream.

Over Kämärä we met Pe-2 planes, I fired at one from the right behind, it caught fire and fell north-east of Huumola. I fired at another Pe-2. At Koskijärvi at fell to the left and went in a steep dive south-east. It was smoking and small flames came out.

I could not follow its fate as I was attacked by several fighters and had to dodge in the clouds. I met other enemy planes I fired at but did not get any results. I broke off having run out of cannon ammunition and mgs jammed. My plane was MT-416."

In the evening of 20 June 1944, the Russians captured Viipuri, regrouped, and directed the attack towards the narrow terrain of Tali-Ihantala; this route had the only available tank track to the north-west. This was where the Soviet attack was finally stopped for good.

This day was the peak of air combats. Before noon, the *Mersus* sent down 35 enemy planes in three large engagements. During the five more fought in the rest of the day, MT pilots downed 16 more enemy planes. The total catch of the day was thus 51 victories of which 31 went to HLeLv 24 and 20 went to HLeLv 34, all without losses to the Finns. WO Ilmari Juutilainen of 1/HLeLv 34 shot down four for breakfast:

"At 0710-0800 hours, altitude 2,000-100 m. Flying in formation, we met the Russkies below us between Ristiniemi and Ruonti. We made a formation attack on the enemy, I managed to shoot at one Yak-9, and it went into the sea smoking.

At this point I saw ten or eleven DB-3Fs coming from Ristiniemi in Virolahti, and there were Il-2s too below and closer to Ristiniemi. In a steep dive I shot down the leading plane of a five plane formation.

1/HLeLv 34 leader 1Lt Väinö Pokela with his Mersu at Taipalsaari in August 1944. He became an ace with five confirmed and two damaged victories.

Late model Bf 109G-6 coded MT-469 at Utti on 30 June 1944. On the same day it was transferred to 1/HLeLv 34 and assigned to MSgt Antti Tani, a 20½-victory ace.

There was another formation behind it (5-6 planes), and I shot into flames the right engine of one DB-3F on the right side. Someone came up from behind me, and put the other engine on fire. The plane went in 8 km south of Ristiniemi.

After this I had to fight two Yak-9s. I shot one down. Another MT appeared and between us we tried for a long time to get the other Yak. We stopped that between Narvi and Seiskari and broke off, when we saw more planes coming in. My plane was MT-426."

22 June 1944 was a busy day for HLeLv 34 with 46 sorties. WO Juutilainen and his swarm of the 1st Flight found a La-5 pair around Viipuri and shot down both of them. Flak shot down one MT of the 1st Flight when it escorted Junkers bombers. WO Lehtovaara's swarm of 3rd Flight shot down two La-5s at Tali.

On 23 June 1944 Maj Luukkanen of HLeLv 34 led seven MTs to attack a mixed enemy formation arriving from Viipuri and shot down three bombers, one ground attack plane and one fighter.

Because having both *Mersu* squadrons in one base at Lappeenranta was a risk, the commander of LeR 3 had them dispersed in case of an air attack. HLeLv 34 moved out to nearby Taipalsaari in the evening.

On 26 June 1944 a swarm of 3/HLeLv 34 fought ten Yaks over Tali and lost one plane. An hour later, Capt Puhakka and his seven *Mersus* escorted Blenheim bombers to Tali and repelled Russian attacks on the bombers. Three Russian fighters were shot down, and on the return flight, three Il-2s of a detachment they happened to meet were shot down. WO Ilmari Juutilainen scored three again:

"At 1255-1400 hours, altitude 1,000-10 m. The Blenheim escort was over and on the way home, when the Russki Airacobras, La-5s and Mustangs arrived. I shot at one Airacobra from aside and behind, it went into the forest shedding pieces. I managed to shoot at one Mustang from above and behind, it caught fire and fell burning north of Tamminiemi, Near it were several tanks, a big one burning. A pause followed and as if ordered the Il-2s arrived. We at-

Half of the pilots of 1/HLeLv 34 at Taipalsaari in August 1944. From the left: 1Lt Ilmari Joensuu, 1Lt Mauno Kirjonen, flight leader 1Lt Väinö Pokela, WO Ilmari Juutilainen, 2Lt Pentti Palm and SSgt Urho Lehto.

tacked them and shot at one from aside and behind, when it caught fire and crashed on a rock at Juustila. My plane was MT-422."

On 28 June 1944 Capt Puhakka and his 3/HLeLv 34 flight were on interception at Tali when they met 60 to 70 Il-2s along with 40 fighters. The Russians lost one *Sturmovik* and one Airacobra. Maj Luukkanen and his swarm reconnoitred the Vammelsuu area and shot down an artillery spotter U-2 and one of its escort fighters.

In the evening the whole squadron flew a 13-plane intercept mission to Tali. Of an enemy formation of more than a hundred planes the Finns shot down four Yaks and two *Sturmoviks*. The last three were claimed by SSgt Klaus Alakoski:

"At 1955-2105 hours. On an intercept mission I met at Viipuri enemy fighters. I engaged then at once. Then the battle shifted eastwards all the time. Near Kämäränjärvi I got to hit well one Yak-9, when it was pulling a bank towards me. I gave a short burst with three cannons. The burst hit obliquely under the fuselage, when the plane went on its back, going vertically into the forest north of Kämäränjärvi.

I met the Il-2s, which were heading towards Viipuri, with a fighter escort above. I attacked them, giving a burst from left above to the fuselage, Then plane fell instantly to its right wing and into a dive crashing in the forest about 10 km east of Viipuri.

I attacked immediately another Il-2, fired a long burst slightly obliquely behind, when a large puff of smoke came out, the plane went down in the city of Viipuri in a shallow dive. My plane was MT-445."

Commander of 3/HLeLv 34, Capt Hans Wind, and WO Ilmari Juutilainen of 1/HLeLv 34 were awarded the Mannerheim Cross for the second time, the only double knights created during the Continuation War. Both had 75 aerial victories at the time.

Next day the leader of 3/HLeLv 34, Capt Olli Puhakka, took his eleven fighters to meet the enemy over Tali. There was a big engagement with about 100 Pe-2s, 40 Il-2s and 40 to 50 fighters. Three Pe-2s and one Yak-9 were dispatched.

On the last day of June 1944 a most productive month was wrapped up with an "ace in a mission" sortie. It was the last of only two such events during the wars. WO Juutilainen reported:

"At 1050-1215 hours. We were blocking the enemy with eight planes. I took my planes to Tali and there we met enemy fighters. Above Viipuri I shot an AC in the back, and it lost most of its tail assembly. It crashed on the Punaisenlähteentori Square on the Torkkelinpuisto Park side. It emitted fuel spray or something resembling that as it went.

The battle moved in the direction of Säiniö. Close to Säiniö I hit one AC from behind and above, and it started to burn and fell between Säiniö and Karhusuo, burning even on the ground.

From the east I saw more than 50 bombers arriving with escort. I called the boys together to Juustila and Tali and there we went for the Russians. At Juustila I shot down a Yak-9 which went in burning, and another Yak-9 lost a wing and fell there too.

For some time we escorted Stukas, and then more Russian bombers and fighters got there. Of these I did not get any results. I saw Il-2 planes without escort. I went to attack these with a steep side attack and sent one Il-2 in flames into Juustila.

At the same time I was surrounded by La-5s. After wrangling one for 5 minutes I shot it into flames at 2,000 m. It crashed in

Pilots of 3/HLeLv 34 at Taipalsaari in July 1944. From the left: WO Urho Lehtovaara, MSgt Ilmari Pöysti, Sgt Matti Durchman, MSgt Lasse Aaltonen, flight leader Capt Olli Puhakka, 1Lt Henrik Salomies, 1Lt Lasse Lehtonen, 1Lt Paavo Reinikainen, 2Lt Jaakko Nurmi, SSgt Aaro Nuorala, SSgt Osmo Länsivaara, MSgt Onni Paronen and Sgt Viljo Leskinen.

flames some 3 to 4 km north of the Il-2 by a road. I ran out of ammo and had to break off. Fuel would have allowed me to stay for maybe 10 minutes. My plane was MT-457."

Air Force HQ confirmed six aerial victories for Juutilainen, two of them witnessed and four without witnesses. On this mission flown between 10.45 and 12.00, the eight MTs of HLeLv 34 and eight MTs of HLeLv 24 met 200 to 300 Russian planes and dispatched fifteen of them.

In the evening, HLeLv 34's Maj Luukkanen escorted 40 bombers to Portinhoikka with 18 MTs. Try as they did, the Russians did not get at the bombers, and the MTs shot down two Airacobras and one La-5. Luukkanen wrote in his combat report:

"At 2005-2110 hours. On an escort mission with 18 MTs for BL-, DN-, DB – and JK – bombers at Tali, I saw at 4,500 m altitude several enemy fighters. After shooting at one La-5 it began to smoke. I kept firing while the plane dived down banking. After the crash in the forest the plane caught fire for a moment.

After gaining altitude I saw behind a JK bomber an AC shooting from 200 m distance. I attacked from the side when the AC tried to escape in a dive. I got to fire from above and behind, when the tail lost pieces and it went from 600 m in a spin to a field 4 km east of Juustila. I assume that the JK gunner saw the case. My plane was MT-415."

The intensity of the battle of Tali-Ihantala reached its high point as June turned into July; the Russian advance was finally stopped in the IV Army Corps sector at the Viipurinlahti – Ihantalanjärvi – Vuoksi line. The Russians kept attacking the Finnish defence positions for another week but could not break the lines.

For HLeLv 34 the first day of July 1944 was the busiest of the month with 50 sorties. Seven MTs of the 1st Flight flew with Maj Eino Luukkanen to protect ships en route to Teikarsaari. A few Il-2s and Yak-9 were strafing the ships. The flight went to assist and turned the attackers away, shooting down one fighter in the process. A swarm of the 3rd Flight was at Viipuri when they shot down two out of six Pe-2s.

Maj Luukkanen and his swarm were escorting bombers when they met Il-2 formations at Koivisto. All of the enemy planes were shot down. MSgt Antti Tani was flying a 3-cannon *Mersu* and this is what he wrote:

"At 1410-1530 hours. While on escort south-east of Juustila, I met Il-2s who were strafing and bombing ground troops. I attacked one 6-plane formation, managing to down one of them. After that I proceeded to the next one, of which pieces came off and it began to smoke. Then it crashed. The next formation had nine planes and they turned south-east very fast. I fired at one and managed to break its tail assembly. I had no time to see it go in, but after pulling up I saw a fire in the forest. My plane was MT-453."

WO Juutilainen was the other pilot who worked on the ground attack planes and saw all Tani's victories while scoring two of his own. In the evening 1Lt Paavo Myllylä of the 3rd Flight led a 10-plane formation to cover ground troops. They met some Pe-2s and Yak-9s and shot down three of them. Myllylä claimed two this way:

"At 1845-1045 hours, altitude 3,000-2,000 m. While on a recce mission the lead post altered our mission to intercept. We me at Taavetti about 15 Pe-2s, which were escorted by about 10

Three pilots of 1/HLeLv 34 at Taipalsaari at the end of the Continuation War. From the left: 1Lt Kauko Risku, WO Ilmari Juutilainen and MSgt Antti Tani.

Yak-9s, I shot one Yak-9 into flames and it fell burning to the ground near Ylämäki. After this I shot at one Pe-2 near Säkkijärvi. When I last saw it it was flying towards Koivisto heavily smoking. I could not follow it as I got several fighters behind my back and had to fight them, without getting any worthwhile results. My plane was MT-432."

On 2 July 1944 ten *Mersus* of HLeLv 34 took off on interception in the morning to Juustila and Ihantala where they met Airacobras. One was shot down in exchange for one MT. Swarms of the 1st and the 3rd Flights scrambled to attack a Soviet formation that was bombing Lappeenranta. In the chase that resulted, the Russians lost seven Il-2s, three Pe-2s and one Yak-9. WO Urho Lehtovaara reported this way:

"At 2005-2055 hours. After a scramble I met at 4,000 m north-west of Viipuri several Pe-2s escorted by several La-5s. Below the formation at 1,000 m I saw a lone Pe-2 with starboard engine stopped. I approached from above and behind and fired at the port engine. The plane caught fire and went into the sea about 1 km south of Viipuri.

I approached now the bomber formation from below and behind and managed to shoot at two bombers, which both went into flames and crashed about 3 km south-east of Viipuri.

Then the escort fighters observed me and attacked, when I had to break off. My plane was MT-448."

Next day fourteen MTs of HLeLv 34's 1st and 3rd Flights were in the Viipuri-Juustila area when they shot down three DB-3Fs of a flight and two Yak-9s out of a formation a couple dozen planes strong. The flight lost one MT. Maj Luukkanen led six MTs to reconnoitre the front lines at Ihantala. A Russian formation of forty planes was patrolling the area and the Finns had to fight the lot. Three

Il-2s, one Pe-2 and one Yak-9 were shot down. Luukkanen reported of this event:

"At 1105-1220 hours. On a recce mission with six MT planes we met nine Il-2s firing at our positions at the front-line. I shot at two, of which the latter crashed in the forest south of Tammisuo. After exiting via south of Viipuri I saw six Pe-2s west of Lihaniemi at 1500 m altitude. After shooting one engine of a Pe-2 into flames I got three Yak-9s behind my tail, forcing me to break off, and not being able to follow the fate of my victim. But the air surveillance post at Särkijärvi informed that in the mentioned location one Pe-2 was seen crashing in flames at 11.50. In addition I saw one Il-2 in addition to mine coming down north of Tammisuo. My plane was MT-415."

On 3 July 1944 LeR 3 commander LtCol Gustaf Magnusson advised the pilots of both HLeLv 24 and 34 that from now on the shooting down enemy aircraft has to be put aside. The most important task is to protect the bombers on their missions. The *Mersu* pilots took this very seriously, as under their escort missions not a single bomber was lost to fighters over the Karelian Isthmus.

As the III Army Corps had retreated to the Vuoksi – Taipale line, Russian forces began to attempt crossing this waterway on 4 July 1944. The Russians managed to cross Vuosalmi and reach its south-western beaches at the narrow spot of Äyräpää and five days later they started to cross Vuosalmi with a wider front.

Ten *Mersus* of 1/HLeLv 34 led by 1Lt Väinö Pokela fought Russian planes at Nuijamaa, managing to down two Il-2s and one Yak-9 out of a mixed formation of twenty planes.

On 5 July 1944 HLeLv 34 sent Maj Luukkanen and his squadron to intercept at Uuras where they found 25 ground

Mersu MT-498 of 1/HLeLv 34 at Taipalsaari in late August 1944. 1Lt Kauko Risku stands on the wing. On 3 September 1944 WO Juutilainen claimed with MT-498 one Li-2 transport shot down as the last victory of the Continuation War, but without outside witnesses.

Mersu *MT-482 of 3/HLeLv 34 at Selänpää in mid-September 1944. It was assigned to 1Lt Lasse Lehtonen. The yellow Eastern Front markings have already been removed.*

attack planes and 20 escorting fighters. The Russians lost one *Sturmovik* and seven fighters. Flak claimed one MT of the 3rd Flight while they escorted bombers to Viipurinlahti.

On 9 July 1944 WO Urho Lehtovaara of 3/HLeLv 34 had reached the 40 aerial victory limit and was awarded the Mannerheim Cross.

Maj Eino Luukkanen of HLeLv 34 led seven MTs to Äyräpää on bomber escort, when they were attacked by ten Russian fighters. Two of these were shot down. Six planes of the 3rd Flight were led by 1Lt Myllylä flying top cover for infantry at Äyräpää, where two *Sturmoviks* assisted by ten fighters were seen to attack the troops. One Il-2 was shot down and the rest tied up in a fruitless battle.

On 11 July 1944 ten MTs of the 1st and 3rd Flight of HLeLv 34 intercepted a Russian Il-2 detachment en route to the infantry at Pölläkkälä and shot down one, turning the rest away. Russian plane formations had decreased in size and they were not as common as before.

On 15 July 1944 a swarm of HLeLv 34's 3rd Flight on reconnaissance to Tali was attacked by six La-5s, of which the Finns shot down one. The 1st Flight fought at Äyräpää, downing two Yak-9s out of a couple dozen Russian fighters. The flight was on escort for a 13-plane Junkers Ju 88 formation en route to bomb Vuosalmi. The Finns were then attacked by a 20 plane fighter detachment, but the Russians did not get at the bombers. After four Yak-9s had been shot down, the rest broke off the engagement. WO Oiva Tuominen reported the following:

"At 1650-1800 hours, altitude 3,500 m. On a bomber escort mission at least 20 fighters came on at Äyräpää. I attacked the La-5s, which in groups of six tried to get behind every Junkers and Blenheim. After wrestling with the La-5s I managed to score hits from above and behind when its tail snapped off just ahead of the stabilizers.

Then I was involved in a fight with the Yak-9s, getting to fire at one in a curve. The aileron and the wing tip broke apart and the plane went down in a spin. The I got about a dozen fighters behind my back and I broke off in a dive. My plane was MT-405."

On 18 July 1944 Maj Luukkanen of HLeLv 34 led sixteen MTs as bomber escort to Vuosalmi. In the escort area ten Russian fighters tried to block the Finnish attack, but lost one Airacobra and one Yak-9 to the Finns.

During the great offensive, in a span of forty days Messerschmitts of HLeLv 24 and HLeLv 34 were credited with 425 Russian planes shot down and 78 damaged during 355 missions and 2,168 sorties. Red fighters shot down ten *Mersus* and three went missing. Russian flak claimed three shot down and ground attack planes dispatched two more. Eight Finnish pilots were killed and three were captured.

The German II/JG 54 reported at the same time it had destroyed 126 Russian planes on 179 missions and 984 sorties. Russian losses were significant. Despite continuous replenishing, the plane complement of the once mighty 13th Air Army had been worn down to just 800 planes in the maelstrom of the Karelian Isthmus air battles.

On 25 July 1944 *Mersu* swarms of HLeLv 24 and 34 took off together to intercept a large enemy detachment attacking Hamina. The Russians lost nine planes, four thereof to HLeLv 34. Its commander Maj Eino Luukkanen described this event:

"At 1120-1235 hours. In an intercept mission south of Hamina we met about 20 Il-2s and over 10 fighters going to the south. I attacked the Il-2s and got one smoking, with the engine running idle it glided at 400 m altitude. Then the fighters attacked me and I had to leave the Il-2s.

I was able to shoot at one Yak-9 from behind, when it pulled up and left a white smoke stream. I fired another burst from obliquely behind, after which it dived in a 30 degree angle into the sea between Someri and Hailikari, leaving a long spot of oil on the water.

Being low on fuel I landed at Utti. My plane was MT-415."

As the airspace over the Karelian Isthmus was calming down, the commander of HLeLv 34 Maj Luukkanen and his pair found six Il-2 planes escorted by six Yaks while the Finns were reconnoitring Virolahti on 5 August 1944. The pair attacked and shot down one ground attack plane and two fighters near Narvi. No Russian planes were met in the air after this on patrolling missions.

On 3 September 1944 before noon a *Mersu* pair of 1/HLeLv 34 flew to the front line. WO Ilmari Juutilainen reported shooting down an Li-2 plane on the mission, but lacking corroboration it was listed as damaged. The Mustang met was more likely a Yak-9, as all earlier claims supported this. Juutilainen's combat report told this:

"At 1045-1140 hours. I was on a recce mission when I met south-east of Valkjärvi one Li-2, which was flying to east. I shot to the port engine, which caught fire. After shooting three times to the starboard engine, it too caught fire. I might have also hit

the pilot because the plane crashed from 200 m altitude on a field south of Nurmijärvi, with huge smoke and flames rising to 500-600 altitude.

A while later I met one fighter, apparently a Mustang, at which I fired twice. When I pulled up I lost it from my sight and I don't know what happened to it. 1Lt Risku met the same plane a bit earlier in the same location, south of Valkjärvi.

The flak was very fierce during the whole mission, thus 1Lt Risku had to separate from me, when doing the recce mission. My plane was MT-498."

The very last combat mission was an interception flown by a pair of the 3rd Flight *Mersus* in the afternoon.

The commander of the Air Force ordered the air regiments to tell squadrons to cease fighting on 4 September 1944 at 7 am. A ceasefire commenced and two weeks later it was confirmed by The Moscow Armistice.

Hävittäjälentolaivue 34 flew over 3,300 sorties claiming 345 enemy aircraft shot down by Messerschmitts. The squadron lost 29 fighters of which 18 were in combat. Ten pilots were killed, two captured and four bailed out.

Mersu *MT-459 of 1/HLeLv 34 running up at Utti in mid-October 1944. This fighter was assigned to 2Lt Kullervo Joutseno.*

Appendices

Appendix 1
Fighter Regiment Commanders

Air Force Commander	
Col (LtGen) Jarl Lundqvist	08.09.32–04.09.44
Chief of Staff	
LtCol Jouna Alameri	01.01.38–14.03.40
Col Yrjö Opas	14.03.40–02.04.42
LtCol Knut Ilanko	02.04.42–09.11.42
LtCol (Col) Risto Pajari	09.11.42.04.09.44
Lentorykmentti 1	
Commander	
LtCol (Col) Viljo Rekola	03.05.42–04.09.44
Chief of Staff	
LtCol Jaakko Moilanen	29.03.44–04.09.44
Fighter squadrons	
Lentolaivue 32	03.05.42–04.09.44
Lentorykmentti 2	
Commander	
LtCol (Col) Richard Lorentz	09.10.39–16.06.43
LtCol (Col) Raoul Harju-Jeanty	16.06.43–04.09.44
Chief of Staff	
Maj (LtCol) Otto Holm	07.03.44–04.09.44
Fighter squadrons	
Lentolaivue 24	30.11.39–18.07.42
Lentolaivue 26	30.11.39–16.09.41

Lentolaivue 28	08.12.39–10.06.44
Lentolaivue 22	16.01.40–01.04.40
Lentolaivue 28	24.07.44–04.09.44
Lentolaivue 26	03.08.44–04.09.44
Lentorykmentti 3	
Commander	
LtCol Einari Nuotio	27.03.40–27.05.43
LtCol (Col) Gustaf Magnusson	27.05.43–04.09.44
Chief of Staff	
Maj Eino Carlsson	02.03.44–08.08.44
Maj Lauri Bremer	08.08.44–04.09.44
Fighter squadrons	
Lentolaivue 30	25.06.41–16.11.42
Lentolaivue 32	25.05.41–03.05.42
Lentolaivue 26	16.09.41–03.08.44
Lentolaivue 24	18.07.42–04.09.44
Lentolaivue 34	23.01.43–04.09.44
Lentolaivue 28	10.06.44–24.07.44
Lentorykmentti 5	
Commander	
LtCol Knut Ilanko	03.11.42–04.09.44
Fighter squadrons	
Lentolaivue 30	16.11.42–04.09.44

Finnish Air Force top brass at a meeting on 21 May 1944. From left Col Olavi Sarko (head of war sipplies dept.) LtCol Risto Pajari (chief of staff), Col Viljo Rekola (LeR 1 CO), Col Frans Helminen (flak CO). LtGen Jarl Lundqvist (air force CO) and Col Richard Lorentz (head of operations). (SA-kuva)

From left air force HQ former chief of staff Col Yrjö Opas, present chief of staff LtCol Knut Ilanko and head of the Swdish delegation Maj Björn Bjuggren at Mikkeli villa steps on 16 May 1942. (SA-kuva)

Col Richard Lorentz, commander of Lentorykmentti 2 in 1942.

Col Viljo Rekola, commander of Lentorykmentti 1 in 1942.

Maj Raoul Harju-Jeanty in 1941, future commander of Lentorykmentti 2.

LtCol Gustaf Magnusson, commander of Lentorykmentti 3 in 1943.

Appendix 2
Lentolaivue 24 Commanders

Commanders	
Maj Richard Lorentz	10.10.34–21.10.35
Maj Niilo Jusu	21.10.35–21.11.38
Capt Gustaf Magnusson*	21.11.38–27.05.43
Capt Jorma Karhunen**	27.05.43–04.12.44
04.12.44 named HLeLv 31	

* Maj 06.12.39 and LtCol 10.11.41 ** Maj 31.08.43

Flight leaders	
1ˢᵗ Flight	
Capt Eino Carlsson	30.11.39–13.03.40
Capt Eino Luukkanen	25.06.41–10.11.42
1Lt Hans Wind	10.11.42–16.01.43
Capt Jorma Sarvanto	16.01.43–08.07.43
1Lt Lauri Nissinen †	08.07.43–17.06.44
1Lt Joel Savonen	17.06.44–03.07.44
Capt Aate Lassila	03.07.44–04.09.44
2ⁿᵈ Flight	
1Lt Jaakko Vuorela †	30.11.39–30.01.40
1Lt Jorma Karhunen	30.01.40–13.03.40
Capt Leo Ahola	25.06.41–25.06.42
Capt Pauli Ervi	25.06.42–11.02.43

1Lt Iikka Törrönen †	11.02.43–02.05.43
1Lt Aulis Lumme	02.05.43–25.08.43
Capt Jouko Myllymäki †	25.08.43–25.06.44
1Lt Aulis Lumme	25.06.44–03.07.44
1Lt Erik Teromaa	03.07.44–30.07.44
Capt Mikko Linkola	30.07.44–04.09.44
3ʳᵈ Flight	
1Lt Eino Luukkanen	30.11.30–13.03.40
1Lt Jorma Karhunen	25.06.41–27.05.43
1Lt Hans Wind	27.05.43–28.06.44
1Lt Kyösti Karhila	28.06.44–21.07.44
1Lt Väinö Suhonen	21.07.44–04.09.44
4ᵗʰ Flight	
Capt Gustaf Magnusson	30.11.39–13.03.40
1Lt Per Sovelius	25.06.41–15.02.42
1Lt Iikka Törrönen	15.02.42–11.02.43
11.02.43 attached to 2ⁿᵈ Flight	
5ᵗʰ Flight	
1Lt Leo Ahola	30.11.39–13.03.40
13.03.40 disbanded	

† killed in action

Appendix 3
Lentolaivue 26 Commanders

Commanders	
Capt Erkki Heinilä	08.10.39–06.12.39
Maj Raoul Harju-Jeanty	06.12.39–01.12.41
Capt Eino Carlsson*	01.12.41–01.03.44
Maj Lauri Larjo †	01.03.44–29.07.44
Maj Erik Metsola	30.07.44–04.12.44
04.12.44 named HLeLv 23	

*Maj 22.02.42

Flight leaders	
Osasto Heinilä	
Capt Erkki Heinilä	30.11.39–31.01.40
1ˢᵗ Flight	
1Lt Ensio Kivinen	19.01.40–13.03.40
1Lt Mikko Linkola	25.06.41–14.11.41
1Lt Aate Lassila	14.11.41–27.02.42
1Lt Mikko Linkola	27.02.42–02.07.43
1Lt Eero Enroth	02.07.43–07.10.43
Capt Juhani Ruuskanen	07.10.43–09.06.44
1Lt Erik Teromaa	09.06.44–27.06.44
1Lt Carl-Erik Bruun	27.06.44–07.07.44
Capt Aulis Bremer	07.07.44–04.09.44

2ⁿᵈ Flight	
1Lt Pekka Siiriäinen	09.02.40–13.03.40
Capt Ensio Kivinen	25.06.41–17.02.42
1Lt Lauri Hämäläinen	17.02.42–15.03.42
Capt Pentti Hyvönen	15.03.42–10.06.42
1Lt Lauri Hämäläinen	10.06.42–02.11.42
Capt Ensio Kivinen	02.11.42–11.07.43
1Lt Lauri Hämäläinen	11.07.43–25.05.44
Capt Aate Lassila	25.05.44–03.07.44
Capt Mikko Linkola	09.07.44–30.07.44
Capt Kalervo Mustonen	30.07.44–22.08.44
1Lt Erik Teromaa	22.08.44–04.09.44
3ʳᵈ Flight	
1Lt Urho Nieminen	11.02.40–05.03.40
1Lt Olli Puhakka	05.03.40–13.03.40
1Lt Urho Nieminen	25.06.41–05.12.41
1Lt Lauri Hämäläinen	05.12.41–10.06.42
1Lt Olli Puhakka	10.06.42–08.02.43
1Lt Lauri Sihvo	08.02.43–07.03.43
1Lt Aate Lassila	07.03.43–25.05.44
25.05.44 disbanded	

† killed in action

Appendix 4
Lentolaivue 28 Commanders

Commanders	
Maj Niilo Jusu	08.12.39–13.03.40
Capt Sven-Erik Siren*	25.06.41–24.08.42
Maj Auvo Maunula †	24.08.42–17.05.44
Capt Tuomo Hyrkki	17.05.44–01.06.44
Maj Per-Erik Sovelius	01.06.44–04.12.44
04.12.44 named HLeLv 21	

* Maj 16.09.41
† killed in flying accident

Flight leaders	
1st Flight	
Capt Sven-Erik Siren	08.12.39–13.03.40
Capt Timo Tanskanen	25.06.41–01.09.41
1Lt Tuomo Hyrkki	01.09.41–31.03.42
Capt Pekka Siiriäinen	31.03.42–31.10.43
Capt Reino Turkki	31.10.43–04.09.44

2nd Flight	
1Lt Reino Turkki	08.12.39–13.03.40
1Lt Reino Turkki	25.06.41–31.10.43
Capt Jaakko Puolakkainen	31.10.43–01.02.44
Capt Veikko Ala-Panula	01.02.44–10.06.44
Capt Jaakko Puolakkainen	10.06.44–04.09.44
3rd Flight	
Capt Eino Jutila	08.12.39–13.03.40
1Lt Erkki Lupari	25.06.41–26.11.42
Capt Tuomo Tyrkki	26.11.42–17.05.44
1Lt Jouko Timonen	17.05.44–17.07.44
Capt Veikko Ala-Panula	17.07.44–04.09.44
Osasto Räty	
1Lt Jussi Räty	08.03.40–13.03.40

† killed in flying accident

Appendix 5
Lentolaivue 14 Commanders

Commanders	
Capt Jaakko Moilanen*	19.02.38–18.05.40
Maj Lauri Larjo	18.05.40–09.12.41
Maj Jussi Sovio	09.12.41–18.02.42
Capt Ragnar Magnusson**	18.02.42–01.03.44
Capt Erik Kerke act.	01.03.44–03.06.44
Maj Kyösti Kurimo	03.06.44–04.09.44
04.12.44 disbanded	

*Maj 06.12.39 **Maj 26.04.42

1st Flight	
Capt Arvo Hassinen	16.06.41–09.09.41
1Lt Matti Tainio	09.09.41–01.11.41
Capt Arvo Hassinen	01.11.41.07.01.42
Capt Alpo Parviainen †	07.01.42–08.01.42

1Lt Matti Tainio	08.01.42–08.08.42
Capt Pekka Käär	08.08.42–15.11.42
1Lt Matti Tainio	15.11.42–14.06.44
Capt Esa Anttonen	14.06.44–04.09.44
2nd Flight	
1Lt Tauno Ollikainen †	02.12.39–10.03.40
1Lt Arvo Hassinen	10.03.40–13.03.40
1Lt Erik Kerke	16.06.41–23.10.42
Capt Erkki Ilveskorpi	23.10.42–17.01.43
Capt Pentti Hyvönen	17.01.43–08.09.43
Capt Arvo Hassinen	08.09.43–25.10.43
Capt Martti Kalima	25.10.43–11.06.44
Capt Toivo Vuorinen	11.06.44–04.09.44

† killed in action

Appendix 6
Lentolaivue 32 Commanders

Commanders	
Capt Erkki Heinilä	29.03.40–12.07.41
Maj Olavi Ehrnrooth	12.07.41–21.01.43
Maj Lauri Bremer	21.01.43–17.06.44
Capt Aaro Virkkunen act.	17.06.44–23.06.44
Capt Kullervo Lahtela*	23.06.44–04.12.44
04.12.44 disbanded	

*Maj 18.07.44

Flight leaders	
1st Flight	
Capt Paavo Berg	25.06.41–13.08.41
1Lt Pentti Nurminen	13.08.41–03.09.41
Capt Paavo Berg †	03.09.41–01.11.41
Capt Kullervo Lahtela	01.11.41–23.01.43
1Lt Jarl Arnkil	23.01.43–06.10.43
Capt Jorma Visapää	06.10.43–04.09.44
2nd Flight	
Capt Kullervo Lahtela	25.06.41–01.11.41

Capt Aulis Bremer	01.11.41–22.08.43
Capt Aaro Virkkunen	22.08.43–10.07.44
1Lt Ossi Marttila act.	10.07.44–26.07.44
1Lt Reino Hakulinen	26.07.44–04.09.44
3rd Flight	
1Lt Aulis Bremer	25.06.41–30.09.41
1Lt Pentti Nurminen	30.09.41–03.05.42
Capt Pauli Ervi	03.05.42–28.06.42
Capt Pentti Nurminen**	28.06.42–19.03.43
1Lt Veijo Taina	19.03.43–10.01.44
1Lt Reino Hakulinen	10.01.44–26.03.44
Capt Veikko Evinen †	26.03.44–25.06.44
1Lt Sakari Alapuro	25.06.44–04.09.44
Osasto Kalaja	
Capt Heikki Kalaja	01.07.41–20.08.41
20.08.41 attached to LLv 10	

**captured
† killed in action

Appendix 7
Lentolaivue 30 Commanders

Commanders	
Capt Lauri Bremer*	25.06.41–20.05.42
Maj Olavi Seeve	20.05.42–08.11.42
Maj Eino Luukkanen	08.11.42–27.03.43
Maj Toivo Kivilahti	27.03.43–03.11.43
Capt Arvo Hassinen**	03.11.43–22.05.44
Capt Veikko Karu***	22.05.44–29.09.44
29.09.44 attached to HLeLv 24	

*Maj 06.08.41 **Maj 14.03.44 *** Maj 18.07.44

Flight leaders	
1st Flight	
Capt Heikki Kalaja	25.06.41–01.07.41
01.07.41 attached to LLv 32	
21.10.41 ee-established	
Capt Antti Naakka †	21.10.41–17.05.42
1Lt Veikko Sauru †	17.05.42–01.10.42
Capt Per-Eric Ahonius	20.11.42–09.07.43
1Lt Tauno Kallio	09.07.43–16.02.44
Capt Kullervo Lahtela	16.02.44–22.06.44
1Lt Heikki Himmanen vs.	22.06.44–22.07.44

1Lt Tauno Kallio	22.07.44–16.08.44
Capt Toivo Puolakka	16.08.44–04.09.44
2nd Flight	
1Lt Veikko Karu	25.06.41–20.11.42
1Lt Mainio Paltila †	20.11.42–08.01.43
1Lt Tauno Kallio	08.01.43–18.05.43
Capt Lauri Karjalainen	18.05.43–06.03.44
Capt Veikko Karu	06.03.44–22.05.44
1Lt Esko Lappalainen act.	22.05.44–30.05.44
1Lt Tauno Kallio	30.05.44–22.07.44
1Lt Kyösti Karhila	22.07.44–04.09.44
3rd Flight (see Lentolaivue 10)	
1Lt Heikki Ilveskorpi	25.06.41–01.10.41
Capt Pekka Käär	01.10.41–01.08.42
1Lt Tauno Tuormaa	16.02.44–20.03.44
20.03.44 disbanded	
Lentue Käär (formally 3/LLv 30)	
Capt Pekka Käär	01.11.41–01.08.42
01.08.42 attached to LeLv 14	

† killed in action

Appendix 8
Lentolaivue 10 Commanders

Commander	
Capt Kyösti Kurimo	12.08.41–01.11.41
01.11.41 attached to LLv 14	
1st Flight	
Capt Heikki Kalaja †	12.08.41–16.09.41

Capt Erkki Ilveskorpi	18.09.41–01.10.41
Capt Pekka Käär	01.10.41–01.11.41
2nd Flight	
Capt Alpo Parviainen	12.08.41–01.11.41

† killed in action

Appendix 9
Lentolaivue 6 Commanders

Commanders	
Maj Knut Ilanko	08.04.40–12.08.41
Maj Jaakko Moilanen*	12.08.41–23.03.44

*LtCol 13.01.42

3rd Flight	
Capt Lauri Karjalainen	25.06.41–30.12.41
1Lt Per-Eric Ahonius	30.12.41–13.07.42
1Lt Mainio Paltila	13.07.42–16.11.42
16.11.42 attached to LeLv 30	

Appendix 10
Lentolaivue 34 Commanders

Commanders	
Maj Olavi Ehrnrooth †	23.01.43–27.03.43
Maj Eino Luukkanen	29.03.43–04.12.44
04.12.44 became HLeLv 33	
Flight leaders	
1st Flight	
Capt Pauli Ervi	11.02.43–07.06.43
1Lt Lauri Pekuri*	07.06.43–16.06.44
Capt Aimo Euramo	16.06.44–03.07.44
1Lt Väinö Pokela	03.07.44–04.12.44

2nd Flight	
Capt Kullervo Lahtela	23.01.43–12.02.44
1Lt Aimo Euramo	12.02.44–18.02.44
Capt Veikko Karu	18.02.44–06.03.44
06.03.44 attached to HLeLv 30	
05.09.44 re-established	
Capt Aimo Euramo	05.09.44–04.12.44
3rd Flight	
Capt Olli Puhakka	09.02.43–04.12.44

+ killed in flying accident * captured

LeLv 34 mechanics confer by a Mersu *line at Utti on 24 April 1943. The pilot at right (without hat) is MSGt Antti Tani. (SA-kuva)*

Appendix 11
Aces In Descending Order

No.	Name	LeLv	Score
1	Juutilainen, Eino Ilmari	24, 34	94
2	Wind, Hans Henrik	24	74½
3.	Luukkanen, Eino Antero	24, 34	54
4	Tuominen, Oiva Emil Kalervo	26, 30, 34	47
5	Puhakka, Risto Olli Petter	26, 34	46
6	Lehtovaara, Urho Sakari	28, 34	41½
7	Katajainen, Nils Edvard	24	34½
8	Puro Kauko, Olavi	6, 24	33
9	Karhila, Kyösti Keijo Ensio	32, 34, 30, 24	32¼
10	Karhunen, Jorma	24	31⁷/₁₂
11	Nissinen, Lauri Vilhelm	24	30⅓
12	Vesa, Emil Onerva	24	30½
13	Alakoski, Klaus Jalmari	26, 34	28
14	Järvi, Turo Tapio	24	25½
15	Saarinen, Jorma Kalevi	24	23
16	Kinnunen, Eero Aulis	24	22½
17	Myllylä, Paavo Urho Johannes	28, 34	22
18	Pyötsiä, Viktor	24	21⅓
19	Tervo, Altto Kalevi	32, 34	21¼
20	Tani, Antti Johannes	28, 34	20½
21	Suhonen, Väinö Ilmari	24	19½
22	Teromaa, Erik Uolevi	24, 26	19
23	Pekuri, Lauri Olavi	24, 34	18½
24	Turkka, Yrjö Olavi	24, 34	17¾
25	Huotari, Jouko Armas Antero	24	17½
26	Halonen, Eero Martti Olavi	24	17
27	Sarvanto, Jorma Kalevi	24	16⅚
28	Lumme, Kaarlo Aulis	24	16¾
29	Riihikallio, Eero Juhani	24	16½
30	Alho, Martti Aslak	24	15
31	Sovelius, Per Erik	24	14¾
32	Nuorala, Aaro Eerikki	30, 14, 34	14½
33	Lampi, Heimo Olavi	24	14
34	Aaltonen, Lasse Erik	26, 34	13½
35	Kokko, Pekka Johannes	24	13⅚
36	Pallasvuo, Yrjö Armas	32, 34	12¾
37	Ahokas, Leo	24	12
38	Karu, Veikko Johannes	30	11¾
39	Koskinen, Eino Eero Sakeus	32	11⅓
40	Nieminen, Urho Abraham	26	11
41	Peltola, Eino Iisakki	24, 34	11
42	Erkinheimo, Niilo Johannes	32, 34	10¾
43	Törrönen, Iikka Veikko	24	10¾
44	Berg, Paavo David	26, 32	10½
45	Kalima, Martti Tauno Johannes	30, 10, 14	10½
46	Paronen, Onni Kullervo	26, 34	10½
47	Lahtela, Kullervo	32, 34	10¼
48	Laitinen, Ahti Ilmari	24	10
49	Leino, Hemmo Kullervo	14, 34	10
50	Metsola, Johannes Kai Kalevi	24	10

No.	Name	LeLv	Score
51	Sarjamo, Urho Kaarlo	24	10
52	Kirjonen, Mauno Ilmari	32, 34	9¾
53	Pasila, Mikko	24	9
54	Savonen, Joel Adiel	24	9
55	Hillo, Jaakko Juho	32	8⅓
56	Inehmo, Martti Olavi Kalervo	28	8
57	Lyly, Erik Edvard	24, 34	8
58	Virtanen, Väinö Johannes	32	8
59	Bremer, Aulis Nathanael	32	7½
60	Jutila, Lauri Olavi	32, 34	7½
61	Kauppinen, Viljo Ilmari	24	7½
62	Huhanantti, Tatu Mauri	24	7
63	Ikonen, Heikki Sakari	24	6¾
64	Lautamäki, Lauri Johannes	26	6½
65	Porvari, Valio Valfrid	26	6½
66	Tomminen, Toivo	2	6½
67	Avikainen, Onni Ilmari	24	6
68	Durchman, Matti Ensio	34	6
69	Evinen, Veikko Arvid	32, 34	6
70	Gerdt, Aimo Emil	32, 34	6
71	Hattinen, Lars Paul Erich	28	6
72	Linnamaa, Aarre Päiviö	28	6
73	Nyman, Atte Eirik Olavi	24	6
74	Rimminen, Toivo Veikko	24	6
75	Salminen, Paul Erik	32	6
76	Trontti, Nils Rudolf	26	6
77	Virta, Toimi Kelpo Jalmari	24	6
78	Kiljunen, Aaro Jaakko	32	5⅚
79	Nurminen, Pentti Emil	32	5⅚
80	Mattila, Ture Allan Nestor	30, 34	5¾
81	Fräntilä, Mauno Mikael	24, 32, 34	5½
82	Keskinummi, Kosti Rauni	24	5½
83	Magnusson, Gustaf Erik	24	5½
84	Mellin, Paavo Kullervo	24	5½
85	Salovaara, Martti Johannes	24	5½
86	Alapuro, Veikko Sakari	32	5
87	Ehrnrooth, Erkki Olavi	KoeL, 32	5
88	Hyrkki, Tuomo Uuno Martti	28	5
89	Joensuu, Antti Ilmari	26	5
90	Kauppinen, Osmo Kalervo	24	5
91	Koskelainen, Arvo Ilmari	24	5
92	Lakio, Vilppu Mikael	24	5
93	Lindberg, Kim Konrad	24	5
94	Länsivaara, Osmo Ilmari	26, 34	5
95	Massinen, Pauli Aatos	28	5
96	Myllymäki, Jouko Jalo	28, 24	5
97	Pokela, Väinö Nikolai	24, 34	5
98	Tilli, Pentti Teodor	26	5
99	Tuomikoski, Kauko Olavi	26, 34	5
100	Vuorimaa, Toivo Olavi	24	5

Appendix 12
Almost Aces

No.	Rank	Name	Score	LeLv	Plane(s)
1	1Lt	Aalto, Mauri	4¾	32	CU-553
2	SSgt	Heinonen, Tauno	4½	24	BW-373
3.	SSgt	Hietala, Oiva	4½	12, 32, 26	FR-92, CU-505, BW-367
4	WO	Ikonen, Viljo	4½	32	CU-573, LG-1
5	SSgt	Kajanto, Jaakko	4½	32	CU-568
6	SSgt	Koskela, Paavo	4½	24	BW-379, MT-456
7	1Lt	Ruotsila, Esko	4½	32	HC452, CU-558
8	SSgt	Suikkanen, Sulo	4½	26	GL-256, FA-15
9	1Lt	Euramo, Aimo	4	32, 34	CU-571, MT-206
10	Sgt	Helava, Risto	4	24	MT-436
11	1Lt	Itävuori, Erkki	4	KoeL	FR-101
12	MSgt	Janhonen, Vesa	4	28	MS-321
13	MSgt	Jussila, Oskari	4	28	MS315
14	SSgt	Korhonen, Aarne	4	24	BW-387
15	SSgt	Lehto, Urho	4	24, 34	BW-356, MT-435
16	1Lt	Nissinen, Aarne	4	28	MS313, MS-620
17	2Lt	Nukarinen, Erkki	4	24	MT-241, MT-442
18	1Lt	Salomies, Henrik	4	34	MT-427
19	1Lt	Sartjärvi, Reino	4	26	FA-21, BW-368
20	1Lt	Sihvo, Lauri	4	26	GL-276, FA-29
21	Capt	Turkki, Reino	4	28	MS-606, MS-626
22	1Lt	Vuorela, Jaakko	4	24	FR-86

Four aces of LeLv 32 plotting the next mission at Nurmoila on 26 June 1942. From left Sgt Väinö Virtanen, Sgt Niilo Erkinheimo, WO Eino Koskinen and 2Lt Kalevi Tervo. (SA-kuva)

Appendix 13
Air Victories by Other Units

Date	Time	Location	Squadron	Rank/pilot	Plane	Victory	Unit	
20.12.39	10.15–12.05	Mantsi	46	Sgt V. Mörsky	BL-106	I-16		T
20.01.40		Pälkäne	12	SSgt O. Marttila	FR-76	SB	35 SBAP	T
01.03.40	16.00–18.50	Koivisto	46	Sgt V. Mörsky	BL-132	I-16		T
05.03.40	13.25–14.20	Äyräpää	14	SSgt M. Perälä	GL-278	I-153	68 IAP	T
07.03.40	15.15–15.25	Tervajoki	14	1Lt T. Ollikainen	GL-279	I-153	148 IAP	T
07.03.40	15.15–15.25	Tervajoki	14	2Lt R. Malinen	GL-276	I-153		T
10.03.40	11.00–12.40	Muhulahti	42	Cpl T. Hämäläinen	BL-145	I-153		T
11.03.40	07.00–08.25	Kiiskilä	42	Cpl Y. Hammaren	BL-139	I-153		T
11.03.40	07.00–08.25	Suur-Merijoki	42	Cpl Y. Hammaren	BL-139	I-153		T
26.06.41	03.00–04.05	Airisto	6	Ltm T. Hämelä	VH-14	SB		T
02.07.41	18.55–	Valkjärvi	44	Cpl M. Pohja	BL-115	I-16		T
18.07.41	17.45–17.55	Vieljärvi	42	Cpl M. Rimpivaara	BL-141	I-16	155 IAP	T
21.07.41	02.20–02.35	Petroskoi	42	Cpl R. Räty	BL-137	I-16	155 IAP	T
03.09.41	11.55–13.00	Tsalkinselkä	16	Cpl R. Pesu	LY-121	I-16	155 IAP	T
06.09.41	11.05–	Troitsankontu	12	1Lt I. Ritavuori	FR-103	I-153		T
18.09.41	11.20	Prääsä	16	2Lt T. Purhonen	GL-277	MiG-3	179 IAP	V
23.09.41	09.15–10.15	Rukajärvi	10	1Lt E. Lehtonen	FR-146	I-153		T
27.09.41	09.25–10.25	Rukajärvi	10	1Lt M. Kalima	FR-150	I-153		T
	09.25–10.25	Rukajärvi	10	2Lt L. Ukkonen	FR-153	I-153		T
	09.25–10.25	Rukajärvi	10	Sgt A. Kilpinen	FR-141	I-153		T
10.11.41	12.50	Vodlitsa	12	2Lt L. Hovilainen	FR-95	R-5		T
18.12.41	09.45	Krestnojärvi 6 km E	12	1Lt O. Tylli	FR-98	½ R-5		T
			12	Sgt S. Matilainen	FR-114	½ R-5		
08.01.42	08.45–10.15	Gora	12	2Lt O. Marttila	FR-114	MiG-3	415 IAP	T
09.01.42	12.15–12.45	Ontrosenvaara	14	Cpl H. Leino	FR-146	Hurricane		E
19.01.42	08.10–09.10	Ontajärvi	14	1Lt V. Taina	FR-156	½ R-5	669 AP	T
			14	Cpl H. Leino	FR-146	½ R-5		
19.02.42	10.30–10.40	Ojatti	12	Maj A. Maunula	FR-98	R-5		T
10.30	10.40	Ojatti	12	Sgt P. Jankko	FR-110	R-5		T
24.04.42	09.55	Alehovtshina	12	1Lt O. Tylli	FR-92	R-5	5 OAP	T
	09.55	Alehovtshina	12	Sgt V. Koskivirta	FR-110	R-5	5 OAP	T
28.04.42	16.15–17.30	Tunkua	14	Sgt A. Nuorala	FR-154	HC		T
20.06.42	09.20	Savijärvi	12	Sgt O. Hietala	FR-92	LaGG-3		E
04.07.42	22.50–23.00	Seiskari	6	Capt P-E. Ahonius	IT-16	2 x I-16		T
20.07.42	21.20–22.00	Savijärvi	12	Sgt O. Hietala	FR-92	LaGG-3		E
22.07.42	00.40–01.40	Pekonvaara	14	2Lt M. Salomaa	FR-156	R-5		T
28.07.42	00.35–02.15	Sellinjärvi	16	Capt L. Saxell	FK-78	R-5		E
04.10.42	08.10–09.25	Lavansaari	6	2Lt O. Puro	IT-18	I-153		T
05.11.42	11.55–13.40	Voijärvi	14	Sgt H. Leino	MS-313	LaGG-3		T
	11.55–13.40	Seesjärvi	14	1Lt M. Kalima	MS-326	LaGG-3		T
12.11.42	10.30.11.40	Peninsaari	6	2Lt O. Puro	IT-18	Pe-2	73 BAP	V
23.12.42	12.50–13.50	Nuokkijärvi	14	1Lt V. Hakola	FR-135	½ R-5		T
			14	2Lt R. Lilja	FR-154	½ R-5		
15.02.43	19.00–20.00	Kärkijärvi	16	1Lt H. Strömberg	GL-273	R-5		R
16.03.43	14.25–14.35	Jeljärvi 10 km NE	14	Capt M. Tainio	MS-313	I-15bis	839 IAP	T
	14.25–14.35	Jeljärvi 5 km E	14	1Lt M. Kalima	MS-326	I-15bis	839 IAP	T
	14.25–14.35	Jeljärvi 15 km N	14	2Lt K. Temmes	MS-605	I-15bis		E
	14.25–14.35	Jeljärvi	14	Sgt A. Nuorala	MS-611	I-15bis	839 IAP	E
	14.25–14.35	Jeljärvi 15 km N	14	SSgt H. Estama	MS-602	I-15bis		E

Date	Time	Location	Squadron	Rank/pilot	Plane	Victory	Unit	
16.03.43	14.25–14.35	Kotskoma 5 km SE	14	Sgt H. Leino	MS-319	2 x I-15bis		E
23.03.43	08.40–08.45	Kirasjärvi 7 km SE	14	1Lt M. Kalima	MS-326	½ I-16	197 IAP	T
			14	Sgt A. Nuorala	MS-640	½ I-16		
29.02.44	21.40–23.50	Kamenitsanjärvi	16	2Lt J. Laitinen	FK-94	U-2		T
13.04.44	13.45–14.15	Tungutjärvi	14	Capt M. Kalima	MS-622	LaGG-3		T
26.05.44	04.00–05.50	Kompakkajärvi	14	Capt M. Kalima	MS-622	LaGG-3	435 IAP	T
02.06.44	17.45–19.15	Kuutsjärvi	14	Capt E. Anttonen	MS-656	LaGG-3		T
	17.50–19.00	Kuutsjärvi 5 km S	14	Capt M. Kalima	MS-622	LaGG-3		T
	17.50–19.00	Ontajärvi	14	Capt M. Kalima	MS-622	LaGG-3		T
29.07.44	19.10–19.15	Loimola	16	Sgt V. Rinkineva	IT-31	Airacobra	773 IAP	T
30.07.44	09.20–10.45	Vegarusjärvi	14	Capt T. Vuorinen	MS-319	Airacobra	773 IAP	E
06.08.44	18.45–19.30	Tahkokoski	14	1Lt M. Niinimäki	MS-629	Airacobra		T

The list is compiled from combat records, where the pilot has claimed an aerial victory. The right hand column denotes the air force headquarters approval: T = confirmed by outside witness, E = confirmed without outside witness, R = into regiment's account without outside witness and V = damaged, but admitted loss by the Russians.

The aircraft types of a small number of claims have been amended to correspond with Russian loss records. The unit statement means a confirmed loss. The lack of it that the case is unknown, or has not been found or is missing a suitable match.

This Ilyushin Il-2 of 703 ShAP was shot down on 21 June 1944 by 1Lt Kyösti Karhila as his 19th victory. He was then a member of 3/HLeLv 34. The location is Tienhaara, north-west of Viipuri. (SA-kuva)

Appendix 14
Air Victories by Foreigners

Date	Time	Location	Squadron	Rank/pilot	Plane	Victory	Unit	
12.01.40	13.10–	Märkäjärvi	19	2Lt I. Iacobi (SE)	GL/F	I-15bis		T
17.01.40	12.00	Kuolajärvi	19	2Lt P-J. Salwén (SE)	GL/F	I-15bis		T
17.01.40	12.00	Märkäjärvi	19	2Lt R. Martin (SE)	GL/C	I-15bis		T
30.01.40	13.25	Virolahti	24	1Lt E. Frijs (DK)	FR-114	⅓ SB	50 SBAP	T
01.02.40	10.30–	Rovaniemi 70 km N	19	2Lt P-J. Salwén (SE)	GL	SB	5 OSAP	T
02.02.40	12.10	Suursaari	26	1Lt J. Ulrich (DK)	GL-259	SB	57 AP	T
12.02.40	12.15	Loimolanjärvi	26	1Lt C. Kalmberg (DK)	GL-261	SB	18 SBAP	T
13.02.40	14.15–15.05	Havuvaara	26	1Lt J. Ulrich (DK)	GL-257	2 x SB	39 SDBAP	E
17.02.40	09.20	Kämäränjärvi	24	1Lt E. Frijs (DK)	FR-100	SB	31 tai 54 SBAP	T
17.02.40	09.20	Kämäränjärvi	24	1Lt E. Frijs (DK)	FR-100	SB	31 tai 54 SBAP	V
19.02.40	15.25	Sippola	26	1Lt P. Christensen (DK)	GL-261	½ I-153	149 IAP	T
20.02.40	09.10	Vaala	19	2Lt P-J. Salwén (SE)	GL	½ SB	16 SBAP	T
			19	2Lt G. Karlsson (SE)	GL	½ SB		
	09.10	Vaala	19	2Lt P-J. Salwén (SE)	GL	SB	16 SBAP	V
21.02.40	12.00	Rovaniemi 40 km SE	19	2Lt C-O. Steninger (SE)	GL	DB-3	5 OSAP	T
	12.00	Rovaniemi	19	2Lt A. Frykholm (SE)	GL	SB	5 OSAP	T
29.02.40	12.00	Ruokolahti	26	1Lt P. Christensen (DK)	GL-261	I-16	68 IAP	V
07.03.40	14.10	Oulu 65 km SE	19	2Lt E. Tehler (SE)	GL	SB	34 DRAE	T
	14.10	Vaala	19	2Lt E. Tehler (SE)	GL	SB	34 DRAE	T
09.03.40	16.30	Viipurinlahti	28	1Lt M. Fensboe (DK)	MS-320	I-153		T
10.03.40	14.30	Kemijärvi	19	2Lt J. Karlsson (SE)	GL	TB-3	1 TAP	T

Swedish volunteer pilots of F 19 at Kauhava on 30 March 1940. From left 2Lt A. Frykholm, 2Lt H. Palme, 2Lt R. Martin, MSgt B. Bjällby, 2Lt J. Karlsson and 2Lt I. Iacobi.

Appendix 15
Destroyed Balloons

Date	Time	Location	Squadron	Rank/pilot	Plane	Victory	
20.01.40	12.05	Taipaleenjoki	24	2Lt P. Kokko	FR-78	½ Barrage balloon	T
			24	2Lt H. Ilveskorpi	FR-105	½ Barrage balloon	
09.07.41	10.15–11.25	Viipuri	24	1Lt O. Mustonen	BW-373	½ Observation balloon	T
			24	Sgt A. Vahvelainen	BW-374	½ Observation balloon	
01.08.41	13.20–15.10	Jääski	32	Capt P. Berg	CU-557	½ Observation balloon	T
			32	2Lt K. Karhila	CU-502	½ Observation balloon	
10.08.41	14.00–15.35	Mertjärvi	32	1Lt J. Arnkil	CU-566	Observation balloon	T
26.08.41	10.40–12.10	Viipuri	24	SSgt M. Alho	BW-383	Observation balloon	T
12.09.41	07.15–08.15	Näykkijärvi	14	2Lt K. Ilkama	FR-	Barrage balloon	E
	07.15–08.15	Näykkijärvi	14	WO U. Ahokas	FR-	Barrage balloon	E
30.10.41	08.55–09.15	Lempaala	26	WO O. Tuominen	FA-6	Observation balloon	T
01.11.41	14.40–15.20	Kyläjatko	26	WO L. Lautamäki	FA-6	½ Observation balloon	T
			26	Sgt P. Saarni	FA-1	½ Observation balloon	
14.05.42	15.20–16.05	Rokanjärvi	32	2Lt S. Alapuro	CU-556	½ Observation balloon	T
			32	Kers E. Emaus	CU-	½ Observation balloon	
12.10.42	12.50–13.55	Ojatin suu	32	WO V. Ikonen	CU-559	½ Observation balloon	T
			32	Sgt L. Mäittälä	CU-560	½ Observation balloon	
15.02.43	15.50	Taipale	26	Sgt K. Alakoski	FA-15	Observation balloon	T
14.06.44	10.50–11.50	Tyrisevä	34	Maj E. Luukkanen	MT-422	Observation balloon	T
	10.55–12.15	Termola	34	Capt O. Puhakka	MT-419	Observation balloon	T
	10.55–12.15	Terijoki	34	Capt O. Puhakka	MT-419	Observation balloon	T
	13.35–14.40	Kivennapa	34	SSgt K. Alakoski	MT-415	Observation balloon	T
19.06.44	10.25–10.55	Perkjärvi	34	Maj E. Luukkanen	MT-417	Observation balloon	T
21.06.44	13.10–13.40	Lotinanpelto	32	Sgt O. Ojapalo	CU-553	Observation balloon	T
22.06.44	17.10–18.10	Heinjoki	24	1Lt A. Laitinen	MT-441	Observation balloon	T
23.06.44	08.30–09.40	Tali	24	Sgt R. Helava	MT-436	Observation balloon	T
26.06.44	07.20–08.40	Lahdenperä	24	Capt H. Wind	MT-439	Observation balloon	T
	07.20–08.40	Heinjoki	24	Sgt R. Helava	MT-437	Observation balloon	T
28.06.44	09.00–10.00	Viipuri	24	Capt H. Wind	MT-439	Observation balloon	T
	09.00–10.00	Peippola	24	1Lt A. Laitinen	MT-441	Observation balloon	E
	10.25–11.05	Tali	24	MSgt N. Katajainen	MT-436	Observation balloon	T
01.07.44	10.55–12.35	Vitele	32	1Lt S. Alapuro	CU-574	Observation balloon	T
03.07.44	12.00	Rajakontu	32	1Lt E. Eerola	CU-	Observation balloon	T
06.07.44	16.40–18.20	Salmi	32	1Lt E. Eerola	CU-578	Observation balloon	T
09.07.44	03.00–04.30	Käsnäselkä	32	1Lt S. Alapuro	CU-587	Observation balloon	T
16.07.44	09.50–10.40	Pitkäranta	32	1Lt S. Alapuro	CU-587	Observation balloon	T

Appendix 16
Destroyed on the Ground/Water

Date	Time	Location	Squadron	Rank/pilot	Plane	Victory	
12.09.41	16.05–16.25	Pyhäjärvi	28	1Lt A. Nissinen	MS-313	½ MBR	T
			28	1Lt J. Ruuskanen	MS-329	½ MBR	
	16.05–16.25	Pyhäjärvi	28	MSgt U. Karhumäki	MS-324	½ MBR	T
			28	SSgt M. Laitinen	MS-603	½ MBR	
19.01.42	08.10–09.10	Ontajärvi	30	5 pilots	FR-	R-5	T
01.02.42		Kolmajärvi	14	1Lt V. Hakola	FR-105	½ R-5	T
			14	2Lt A. Parviainen	FR-106	½ R-5	
		Kolmajärvi	14	MSgt M. Perälä	FR-109	½ R-5	T
			14	Sgt H. Estama	FR-90	½ R-5	
04.03.42	07.45–08.00	Suikujärvi	24	MSgt L. Nissinen	BW-384	I-16	T
	07.45–08.00	Suikujärvi	24	2Lt V. Pokela	BW-381	DB-3	T
	07.45–08.00	Suikujärvi	24	Sgt E. Peltola	BW-356	½ DB-3	T
			24	Sgt U. Lehto	BW-372	½ DB-3	
14.04.42		Tiiksjärvi	30	2 pilots	FR-	U-2	T
08.06.42	01.00–02.50	Suunujärvi	24	1Lt I. Törrönen	BW-380	½ LaGG-3	T
			24	1Lt H. Wind	BW-378	½ LaGG-3	
14.11.42	10.15–12.15	Monastirskaya	32	2 pilots	CU-	MBR	T
30.12.42	11.15–12.15	Shotkusha	32	2Lt Y. Pallasvuo	CU-552	U-2	T
07.01.43		Petrovskiy Jam	16	1Lt M. Somppi	GL-270	½ R-5	T
			16	2Lt V. Koski	GL-272	½ R-5	
12.01.43	10.00–11.20	Mikkola	14	1Lt M. Nysten	FR-	R-5	T
	10.00–11.20	Mikkola	14	2Lt E. Hyvärinen	FR-	R-5	T
	10.00–11.20	Mikkola	14	Sgt S. Valkeiskangas	FR-	R-5	T
	12.00–13.50	Mikkola	14	5 pilots	MS-	R-5	T
	13.00	Mikkola	14	1Lt M. Nysten	FR-	R-5	T
	13.00	Mikkola	14	2Lt E. Hyvärinen	FR-	R-5	T
	13.00	Mikkola	14	Sgt S. Valkeiskangas	FR-	R-5	T
09.03.44	13.20–14.20	Kordoijärvi	14	4 pilots	MS-	DB-3	T
26.03.44	14.40	Tsolmutsanlahti	16	Sgt S. Peijari	FK-79	DB-3	T

This Yakovlev Yak-9 (red 38) was damaged in a combat, making a forced landing to a rye field on the Karelian Isthmus on 5 July 1944. (SA-kuva)

Appendix 17
Bases

Lentolaivue 6 Bases

Headquarters	
Turku	18.06.41–26.06.41
Nummela	26.06.41–06.12.41
Malmi	06.12.41–17.12.41
Santahamina	17.12.41–23.02.44

3rd Flight	
Turku	18.06.41–26.06.41
Nummela	26.06.41–12.12.41
Malmi	12.12.41–27.05.42
Römpötti	27.05.42–16.11.42
16.11.42 attached to LeLv 30	

Lentolaivue 14 Bases (landing grounds)

Headquarters	
Laikko	30.11.39–17.03.4
Utti	17.03.40–19.06.41
Padasjoki	19.06.41–07.07.41
Utti	07.07.41–10.08.41
Lappeenranta	10.08.41–25.09.41
Suulajärvi	25.09.41–23.10.41
Tiiksjärvi	23.10.41–31.08.44
Kolvasjärvi (Tiiksjärvi)	31.08.44–21.09.44
Paltamo (Vaala)	21.09.44–15.10.44
Pudasjärvi	15.10.44–07.11.44
Kemi	07.11.44–23.11.44
Rissala	23.11.44–04.12.44
04.12.44 disbanded	

1st Flight	
Utti	25.06.41–10.08.41
Lappeenranta	10.08.41–02.09.41
Suur-Merijoki	02.09.41–14.09.41
Suulajärvi	14.09.41–27.10.41
Tiiksjärvi	27.10.41–05.09.44
Paltamo (Vaala)	05.09.41–15.10.44
Pudasjärvi	15.10.44–07.11.44
Kemi	07.11.44–23.11.44
Rissala	23.11.44–04.12.44

2nd Flight	
Laikko	30.11.39–13.03.40
Selänpää	25.06.41–07.07.41

Utti	07.07.41–31.07.41
Lappeenranta	31.07.41–20.09.41
Suulajärvi	20.09.41–27.10.41
Tiiksjärvi	27.10.41–15.09.44
Onttola	15.09.44–21.09.44
Paltamo (Vaala)	21.09.44–15.10.44
Pudasjärvi	15.10.44–07.11.44
Kemi	07.11.44–23.11.44
Rissala	23.11.44–04.12.44

Utti air base, east of Kouvola, seen from the east in June 1937. (Finnish Air Force)

Lentolaivue 24 Bases (landing grounds)

Headquarters	
Immola	12.10.39–28.12.39
Joutseno	28.12.39–01.01.40
Utti	01.01.40–08.01.40
Joutseno	08.01.40–25.01.40
Immola	25.01.40–01.03.40
Lemi	01.03.40–10.03.40
Ristiina	10.03.40–18.03.40
Joroinen	18.03.40–17.04.40
Malmi	17.04.40–13.08.40
Vesivehmaa	13.08.40–02.07.41
Rantasalmi	02.07.41–19.08.41
Immola	19.08.41–16.09.41
Lunkula	16.09.41–23.12.41
Kontupohja	23.12.41–14.04.42
Hirvas	14.04.42–01.08.42
Römpötti	01.08.42–11.11.42
Suulajärvi	11.11.42–11.06.44
Immola	11.06.44–15.06.44
Lappeenranta	15.06.44–12.09.44
Utti	12.09.44–23.11.44
Vesivehmaa	23.11.44–04.12.44

4.12.44 named HLeLv 31

1st Flight	
Immola	12.10.39–28.12.39
Joutseno	28.12.39–01.01.40
Utti	01.01.40–08.01.40
Joutseno	08.01.40–25.01.40
Immola	25.01.40–01.03.40
Lemi	01.03.40–10.03.40
Ristiina	10.03.40–18.03.40
Joroinen	18.03.40–17.04.40
Malmi	17.04.40–13.08.40
Vesivehmaa	13.08.40–02.07.41
Rantasalmi (Mikkeli)	02.07.41–19.08.41
Immola	19.08.41–13.09.41
Nurmoila	13.09.41–30.05.42
Hirvas	30.05.42–10.06.4
Suulajärvi (Mensuvaara)	10.06.42–07.08.42
Römpötti	07.08.42–16.09.42
Hirvas	16.09.42–01.10.42
Römpötti	01.10.42–15.11.42
Suulajärvi	15.11.42–11.05.44
Nurmoila	11.05.44–03.06.44
Suulajärvi	03.06.44–11.06.44
Immola	11.06.44–15.06.44
Lappeenranta	15.06.44–12.09.44

2nd Flight	
Suur-Merijoki	12.10.39–09.12.39
Lappeenranta	09.12.39–20.12.39
Ruokolahti	20.12.39–02.02.40
Turku	02.02.40–25.02.40
Naantali	25.02.40–03.03.40
Lemi	03.03.40–10.03.40
Ristiina	10.03.40–18.03.40
Joroinen	18.03.40–17.04.40
Lemi	01.03.40–10.03.40
Ristiina	10.03.40–18.03.40
Joroinen	18.03.40–17.04.40
Malmi	17.04.40–13.08.40
Vesivehmaa	13.08.40–02.07.41
Rantasalmi	03.07.41–16.08.41
Lappeenranta	16.08.41–02.09.41
Immola	02.09.41–17.09.41
Mantsinsaari	17.09.41–11.11.41
Lunkula	11.11.41–06.12.41
Äänislinna	06.12.41–11.12.41
Kontupohja	11.12.41–23.03.42
Immola	23.03.42–31.03.42
Hirvas	14.04.42–01.08.42
Römpötti	01.08.42–14.11.42
Suulajärvi	14.11.42–11.06.44
Immola	11.06.44–15.06.4
Lappeenranta	15.06.44–12.09.44

3rd Flight	
Immola	12.10.39–25.12.39
Värtsilä (Suistamo)	25.12.39–04.02.40
Ruokolahti	04.02.40–01.03.40
Lemi	01.03.40–10.03.40
Ristiina	10.03.40–18.03.40
Joroinen	18.03.40–17.04.40
Malmi	17.04.40–13.08.40
Vesivehmaa	13.08.40–02.07.41
Rantasalmi	03.07.41–16.08.41
Lappeenranta	16.08.41–02.09.41
Immola	02.09.41–17.09.41
Mantsinsaari	17.09.41–11.11.41
Lunkula	11.11.41–06.12.41
Äänislinna	06.12.41–11.12.41
Kontupohja	11.12.41–23.03.42
Immola	23.03.42–31.03.42
Hirvas	14.04.42–01.08.42
Römpötti	01.08.42–14.11.42
Suulajärvi	14.11.42–11.06.44
Immola	11.06.44–15.06.44
Lappeenranta	15.06.44–12.09.44

4th Flight	
Immola	12.10.39–28.12.39
Joutseno	28.12.39–01.01.40
Utti	01.01.40–08.01.40
Joutseno	08.01.40–25.01.40
Immola	25.01.40–01.03.40
Lemi	01.03.40–10.03.40
Ristiina	10.03.40–18.03.40
Joroinen	18.03.40–17.04.40
Malmi	17.04.40–13.08.40
Vesivehmaa	13.08.40–02.07.41
Rantasalmi	02.07.41–19.08.41
Immola	19.08.41–16.09.41
Lunkula	16.09.41–23.12.41
Kontupohja	23.12.41–14.04.42
Hirvas	14.04.42–01.08.42
Römpötti	01.08.42–11.11.42
Suulajärvi	11.11.42–11.02.43

11.2.43 attached to 2nd Flight

5th Flight	
Immola	23.10.39–09.12.39
Lappeenranta	09.12.39–20.12.39
Ruokolahti	20.12.39–02.02.40
Turku	02.02.40–25.02.40
Naantali	25.02.40–03.03.40
Lemi	03.03.40–10.03.40
Ristiina	10.03.40–18.03.40

18.3.40 disbanded

Lentolaivue 26 Bases (landing grounds)

Lentolaivue 26 Bases (landing grounds)	
Headquarters	
Utti	*30.11.39–30.05.40*
Joroinen	*30.05.40–06.07.41*
Joensuu	*06.07.41–30.07.41*
Lunkula	*30.07.41–16.09.41*
Immola	*16.09.41–20.06.42*
Kilpasilta	*20.06.42–08.05.44*
Heinjoki	*08.05.44–14.06.44*
Immola	*14.06.44–16.06.44*
Käkisalmi	*16.06.44–06.07.44*
Mensuvaara	*06.07.44–27.07.44*
Värtsilä	*27.07.44–05.08.44*
Sortavala	*05.08.44–08.09.44*
Onttola	*08.09.44–02.10.44*
Vaala	*02.10.44–17.10.44*
Kemi	*17.10.44–04.12.44*

Osasto Heinilä	
Heinjoki	*30.11.39–19.12.39*
Käkisalmi	*19.12.39–31.12.39*
Littoinen	*31.12.39–31.01.40*

31.01.40 disbanded

1ˢᵗ Flight	
Utti	*21.01.40–09.02.40*
Kuluntalahti	*09.02.40–15.02.40*
Joutseno	*15.02.40–25.02.40*
Ruokolahti	*25.02.40–01.03.40*
Pyhäniemi	*01.03.40–10.03.40*
Haukkajärvi	*10.03.40–15.03.40*
Vuohijärvi	*15.03.40–05.04.40*
Utti	*05.04.40–30.05.40*
Joroinen	*30.05.40–01.07.41*

Joensuu	*01.07.41–29.07.41*
Lunkula	*29.07.41–16.09.41*
Immola	*16.09.41–17.12.41*
Malmi	*17.12.41–15.01.42*
Immola	*15.01.42–15.02.42*
Malmi	*15.02.42–15.03.42*
Immola	*15.03.42–15.04.42*
Malmi	*15.04.42–15.07.42*
Kilpasilta	*15.07.42–08.05.44*
Heinjoki	*08.05.44–14.06.44*
Immola	*14.06.44–16.06.44*
Käkisalmi	*14.06.44–06.07.44*
Mensuvaara	*06.07.44–05.08.44*
Värtsilä	*05.08.44–08.09.44*

2ⁿᵈ Flight	
Utti	*21.01.40–05.02.40*
Mensuvaara	*05.02.40–07.02.40*
Värtsilä	*07.02.40–14.02.40*
Mensuvaara	*14.02.40–21.02.40*
Ruokolahti	*21.02.40–29.02.40*
Haukkajärvi	*29.02.40–05.04.40*
Utti	*05.04.40–29.05.40*
Joroinen	*30.05.40–28.06.41*
Rantasalmi	*28.06.41–06.07.41*
Joensuu (Värtsilä)	*06.07.41–02.08.41*
Lunkula	*02.08.41–15.08.41*
Vitele	*15.08.41–16.09.41*
Immola	*16.09.41–21.09.41*
Sakkola	*21.09.41–23.10.41*
Immola	*23.10.41–22.11.41*
Sakkola	*22.11.41–14.12.41*
Immola	*14.12.41–15.01.42*

Malmi	*15.01.42–15.02.42*
Immola	*15.02.42–15.03.42*
Malmi	*15.03.42–15.04.42*
Immola	*15.04.42–29.06.42*
Kilpasilta	*29.06.42–15.07.42*
Malmi	*15.07.42–18.03.43*
Kilpasilta	*18.03.43–14.07.43*

14.7.43 without planes
25.5.44 re-established

Heinjoki	*25.05.44–14.06.44*
Immola	*14.06.44–16.06.44*
Käkisalmi	*14.06.44–25.07.44*
Mensuvaara	*25.07.44–27.07.44*
Värtsilä	*27.07.44–08.09.44*

3ʳᵈ Flight	
Haukkajärvi	*14.02.40–01.03.40*
Pyhäniemi	*01.03.40–09.03.40*
Haukkajärvi	*09.03.40–15.03.40*
Vuohijärvi	*15.03.40–05.04.40*
Utti	*05.04.40–29.05.40*
Joroinen	*29.05.40–03.07.41*
Joensuu (Värtsilä)	*03.07.41–30.07.41*
Lunkula	*30.07.41–16.09.41*
Immola	*16.09.41–23.10.41*
Sakkola	*23.10.41–22.11.41*
Immola	*22.11.41–07.12.41*

07.12.41 disbanded
23.04.42 re-established

Sakkola	*28.04.42–07.06.42*
Petäjärvi	*07.06.42–29.06.42*
Kilpasilta	*29.06.42–08.05.44*
Heinjoki	*08.05.44–25.05.44*

25.5.44 disbanded

Lappeenranta airfield in June 1940. North is at 12 o'clock. (Finnish AIr Force)

Lentolaivue 28 Bases (landing grounds)

Lentolaivue 28 Bases (landing grounds) Headquarters	
Säkylä	06.02.40–04.04.40
Turku	04.04.40–06.08.40
Naarajärvi	06.08.40–04.07.41
Joroinen	04.07.41–17.07.41
Joensuu	17.07.41–31.07.41
Joroinen	31.07.41–19.08.41
Karkunranta	19.08.41–24.10.41
Äänislinna	24.10.41–25.06.42
Viitana	25.06.42–03.08.42
Hirvas	03.08.42–10.06.44
Lappeenranta	10.06.44–23.07.44
Värtsilä	23.07.44–09.09.44
Onttola	09.09.44–02.10.44
Paltamo	02.10.44–15.10.44
Pudasjärvi	15.10.44–08.11.44
Kemi	08.11.44–27.11.44
Rissala	27.11.44–04.12.44

1st Flight	
Säkylä	06.02.40–07.03.40
Pyhäniemi	07.03.40–15.03.40
Säkylä	15.03.40–04.04.40
Turku	04.04.40–06.08.40
Naarajärvi	25.06.41–25.06.41
Joroinen	25.06.41–11.07.41
Joensuu	11.07.41–30.07.41
Läskelä	30.07.41–06.08.41
Lunkula	06.08.41–19.10.41

Viitana	19.10.41–21.04.42
Äänislinna	21.04.42–25.06.42
Viitana	25.06.42–10.07.42
Latva	10.07.42–07.08.42
Äänislinna	07.08.42–21.06.43
Latva	21.06.43–20.09.43
Äänislinna (Kuutamolahti)	20.09.43–03.07.44
Värtsilä (Tiiksjärvi)	03.07.44–09.09.44
Onttola	09.09.44–02.10.44
Paltamo	02.10.44–15.10.44
Pudasjärvi	15.10.44–08.11.44
Kemi	08.11.44–27.11.44
Rissala	27.11.44–04.12.44

2nd Flight	
Säkylä	06.02.40–25.02.40
Turku	25.02.40–16.03.40
Säkylä	16.03.40–04.04.40
Turku	04.04.40–06.08.40
Naarajärvi	06.08.40–04.07.41
Joroinen	04.07.41–17.07.41
Joensuu	17.07.41–31.07.41
Joroinen	31.07.41–20.08.41
Karkunranta	20.08.41–19.10.41
Viitana	19.10.41–21.04.42
Äänislinna	21.04.42–25.06.42
Viitana	25.06.42–23.07.42

23.07.42 disbanded
10.10.42 re-estrablished

Hirvas (Pintuinen)	10.10.42–12.06.44
Lappeenranta	12.06.44–16.06.44
Utti	16.06.44–23.07.44
Värtsilä	23.07.44–09.09.44
Onttola	09.09.44–27.11.44
Rissala	27.11.44–04.12.44

3rd Flight	
Säkylä	06.02.40–07.03.40
Haukkajärvi	07.03.40–11.03.40
Pyhäniemi	11.03.40–29.03.40
Säkylä	20.03.40–04.04.40
Turku	04.04.40–06.08.40
Naarajärvi	06.08.40–04.07.41
Joroinen	04.07.41–17.07.41
Joensuu	17.07.41–31.07.41
Joroinen	31.07.41–19.08.41
Karkunranta	19.08.41–12.10.41
Äänislinna	12.10.41–25.06.42
Viitana	25.06.42–03.08.42
Hirvas (Torasjärvi)	03.08.42–12.06.44
Lappeenranta	12.06.44–16.06.44
Utti	16.06.44–23.07.44
Värtsilä	23.07.44–09.09.44
Onttola	09.09.44–27.11.44
Rissala	27.11.44–04.12.44

Osasto Räty	
Säkylä	01.03.40–04.04.40
Turku	27.05.40–29.08.40

Quickly built servicing hangar at Lunkula, south-west coast of Lake Ladoga in October 1941. Inside is a Morane of LLv 28. (SA-kuva)

Lentolaivue 30 Bases (landing grounds)

Lentolaivue 30 Bases (landing grounds) Headquarters	
Pori	29.03.40–03.07.41
Hyvinkää	03.07.41–02.09.41
Utti	02.09.41–15.01.42
Suur-Merijoki	15.01.42–20.05.42
Suulajärvi	20.05.42–11.11.42
Römpötti	11.11.42–22.02.44
Utti	22.02.44–22.05.44
Malmi	22.05.44–21.09.44
Hyvinkää	21.09.44–29.09.44

29.09.44 attached to HLeLv 24

1st Flight	
Hollola	25.06.41–01.07.41

01.07.41 attached to LLv 32
20.10.41 re-established

Utti	20.10.41–15.01.42
Suur-Merijoki (Hamina)	15.01.42–20.05.42
Suulajärvi	20.05.42–11.11.42
Römpötti	11.11.42–22.02.44
Utti	22.02.44–11.06.44
Kymi	11.06.44–03.08.44
Malmi	03.08.44–21.09.44

2nd Flight	
Pori	25.06.41–03.07.41
Hyvinkää (Malmi)	03.07.41–02.09.41
Utti	02.09.41–25.10.41
Suulajärvi (Hamina)	25.10.41–11.11.42
Römpötti	11.11.42–05.03.44
Malmi	05.03.44–05.08.44
Kymi	05.08.44–13.09.44

3rd Flight	
Turku	25.06.41–15.09.41
Utti	15.09.41–21.09.41

21.09.41 attached to LLv 10
16.02.44 re-established for three weeks

Römpötti	16.02.44–06.03.44

Lentolaivue 10 Bases

Headquarters	
Tiiksjärvi	20.08.41–01.11.41

01.11.41 attached to LLv 14

1st Flight	
Tiiksjärvi	20.08.41–01.11.41

01.11.41 re-named Osasto Käär (formally 3/LLv 30)

Osasto Käär	01.11.41–01.08.42

01.08.42 attached to LeLv 14

2nd Flight	
Tiiksjärvi	20.08.41–01.11.41

01.11.41 attached to LLv 14

Helsinki Malmi airport in September 1943. North is at 10 o'clock. (Finnish AIr Force)

Lentolaivue 32 Bases (landing grounds)

Headquarters	
Siikakangas	08.05.40–21.06.41
Hyvinkää	21.06.41–04.07.41
Utti	04.07.41–30.07.41
Lappeenranta	30.07.41–23.09.41
Suulajärvi	23.09.41–19.05.42
Immola	19.05.42–30.05.42
Nurmoila	30.05.42–23.06.44
Uomaa	23.06.44–04.07.44
Mensuvaara	04.07.44–20.09.44
Mikkeli	20.09.44–24.10.44
Vesivehmaa	24.10.44–23.11.44
Pori	23.11.44–04.12.44

4.12.44 named HLeLv 13

1st Flight	
Hyvinkää	21.06.41–02.07.41
Utti	02.07.41–30.07.41
Lappeenranta	30.07.41–23.09.41
Suulajärvi	23.09.41–30.10.41

Nummela	30.10.41–07.12.41
Suulajävi	07.12.41–19.05.42
Immola	19.05.42–30.05.42
Nurmoila (Immola)	30.05.42–14.04.44
Onttola	14.04.44–07.05.44
Nurmoila	07.05.44–23.06.44
Uomaa	23.06.44–04.07.44
Mensuvaara	04.07.44–20.09.44

2nd Flight	
Hyvinkää	21.06.41–04.07.41
Utti	04.07.41–30.07.41
Lappeenranta	30.07.41–23.09.41
Suulajärvi	07.12.41–19.05.42
Immola	19.05.42–30.05.42
Nurmoila (Immola)	30.05.42–01.04.44
Koveri	01.04.44–23.06.44
Uomaa	23.06.44–04.07.44
Mensuvaara	04.07.44–20.09.44

3rd Flight	
Siikakangas	24.06.41–29.06.41
Hyvinkää	29.06.41–03.07.41
Vesivehmaa	03.07.41–10.07.41
Utti	10.07.41–30.07.41
Lappeenranta	30.07.41–23.09.41
Suulajärvi	23.09.41–19.05.42
Immola	19.05.42–30.05.42
Nurmoila	30.05.42–06.06.42
Lunkula	06.06.42–19.06.42
Nurmoila (Immola)	19.06.42–23.06.44
Uomaa	23.06.44–04.07.44
Mensuvaara	04.07.44–20.09.44

Osasto Kalaja	
Utti	01.07.41–20.08.41

20.08.41 attached to LLv 10

Suulajärvi airfield in June 1942. North is at 2 o'clock. (Finnish AIr Force)

Lentolaivue 34 Bases (landing grounds)

Lentolaivue 34 Bases (landing grounds)	
Headquarters	
Utti	17.03.43–02.08.43
Kymi	02.08.43–12.06.44
Immola	12.06.44–16.06.44
Lappeenranta	16.06.44–23.06.44
Taipalsaari	23.06.44–07.09.44
Selänpää	07.09.44–14.10.44
Utti	14.10.44–23.11.44
Vesivehmaa	23.11.44–04.12.44

04.12.44 became HLeLv 33

1ˢᵗ Flight	
Malmi	17.03.43–01.06.43
Utti	01.06.43.16.07.43
Suulajärvi	16.07.43–01.05.44

Kymi	01.05.44–31.05.44
Suulajärvi	31.05.44–12.06.44
Immola	12.06.44–16.06.44
Lappeenranta	16.06.44–23.06.44
Taipalsaari (Utti)	23.06.44–05.09.44
Selänpää	05.09.44–22.10.44
Utti	22.10.44–23.11.44
Vesivehmaa	23.11.44–04.12.44

2ⁿᵈ Flight	
Utti	20.03.43–02.08.43
Kymi	02.08.43–19.08.43
Malmi	19.08.43–06.03.44

06.03.44 attached to HLeLv 30
05.09.44 re-established

Taipalsaari	05.09.44–23.09.44

Selänpää	23.09.44–14.10.44
Utti	14.10.44–23.11.44
Vesivehmaa	23.11.44–04.12.44

3ʳᵈ Flight	
Utti	20.03.43–01.06.43
Malmi	01.06.43–19.08.43
Kymi	19.08.43–12.06.44
Immola	12.06.44–16.06.44
Lappeenranta	16.06.44–23.06.44
Taipalsaari (Utti)	23.06.44–21.08.44
Immola	21.08.44–05.09.44
Selänpää	05.09.44–14.10.44
Utti	14.10.44–23.11.44
Vesivehmaa	23.11.44–04.12.44

Solomanni airfield hangar seen on 24 October 1941, with Moranes of 3/LLv 28 inside, closest MS-603. When the Finnish troops occupied Petrozavodsk three weeks earlier, they found the town airfield hangar intact and still under construction. (SA-kuva)

Servicing hangar at Kymi airfield in May 1944. Inside is a Bf 109 G-6 of HLeLv 34.

Mersu MT-207 of 2/LeLv 34 parked in front of the hangar at Helsinki Malmi in early May 1943. At right is the passenger terminal.

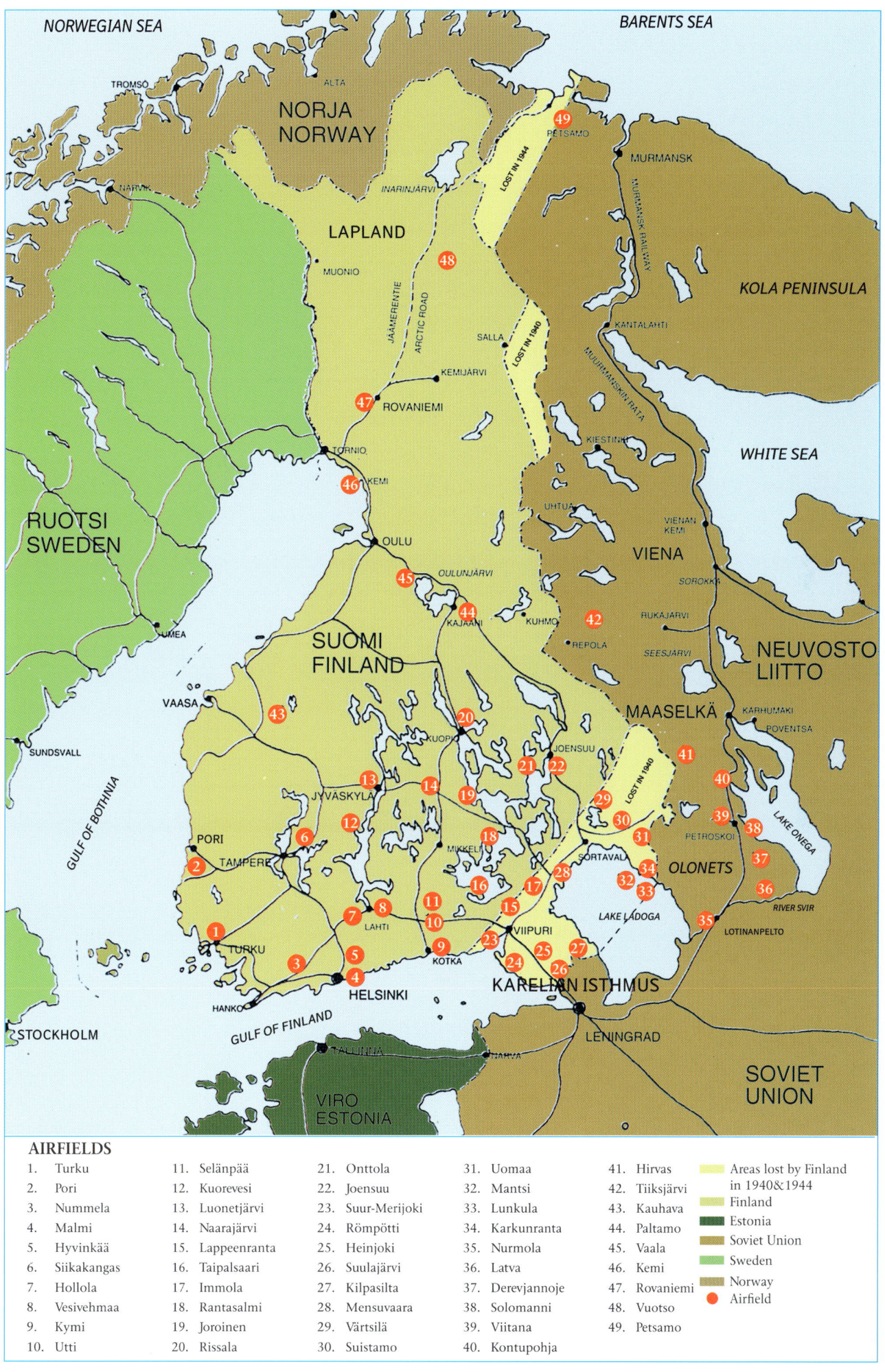

AIRFIELDS

1. Turku	11. Selänpää	21. Onttola	31. Uomaa	41. Hirvas
2. Pori	12. Kuorevesi	22. Joensuu	32. Mantsi	42. Tiiksjärvi
3. Nummela	13. Luonetjärvi	23. Suur-Merijoki	33. Lunkula	43. Kauhava
4. Malmi	14. Naarajärvi	24. Römpötti	34. Karkunranta	44. Paltamo
5. Hyvinkää	15. Lappeenranta	25. Heinjoki	35. Nurmola	45. Vaala
6. Siikakangas	16. Taipalsaari	26. Suulajärvi	36. Latva	46. Kemi
7. Hollola	17. Immola	27. Kilpasilta	37. Derevjannoje	47. Rovaniemi
8. Vesivehmaa	18. Rantasalmi	28. Mensuvaara	38. Solomanni	48. Vuotso
9. Kymi	19. Joroinen	29. Värtsilä	39. Viitana	49. Petsamo
10. Utti	20. Rissala	30. Suistamo	40. Kontupohja	

Areas lost by Finland in 1940&1944
Finland
Estonia
Soviet Union
Sweden
Norway
● Airfield

Month		1.01.	1.02.	1.03.	1.04.	1.05.	1.06.	1.07.	1.08.	1.09.	1.10.	1.11.	1.12.
Squadron	*Type*												
LeLv 24	FRm												35+0
1940													
LeLv 22	BW			4+0									
LeLv 24	FRm	27+3	27+1	25+3	24+0								
	BW					32+5	28+9	32+0	31+0	31+0	30+0	29+1	30+1
LeLv 26	BU	8+0	8+2										
	GL		11+1	5+0									
	FA			13+0	20+3	20+4	19+6	15+10	18+7	19+7	17+10	18+9	15+8
	BW										1+0	1+0	1+0
LeLv 28	MS			21+8	19+9	19+2	16+5	14+2	14+2	20+3	20+4	21+3	20+3
	HC				8+2	8+2	7+2	6+2	4+3				
LeLv 30	HC									5+3	5+3	5+3	5+3
LeLv 32	BW				14+3								
	FRm					18+5	10+1	18+1	16+2	12+3	14+1	11+4	12+2
1941													
LeLv 6	IT					4+1	3+0	3+3	3+1	4+0	3+1	1+2	2+1
LeLv 10	HC									3+0			
	FRw										6+0	7+0	
LeLv 24	BW	30+10	27+13	33+7	26+13	38+3	39+3	39+2	28+7	34+1	33+2	35+2	35+2
LeLv 26	FA	13+13	14+11	16+10	17+9	20+6	26+3	20+9	11+9	5+10	6+10	5+11	6+12
	BW	1+0	1+0	1+0	1+0					1+0			
LeLv 28	MS	20+2	17+5	17+4	23+1	22+9	24+7	26+5	23+7	16+11	20+6	21+3	20+5
LeLv 30	HC	5+3	6+3	6+3	6+3	6+3	7+1						
	FRw					12+0	24+0	19+2	13+8	25+11	15+10	18+7	20+5
LeLv 32	FRm	12+1	13+1	13+1	14+2	18+1	17+1	15+0					
	FRw					12+0	10+1	15+5	18+1	4+0	4+0		
	HC							4+0	4+1		2+1	1+2	0+2
	CU								8+1	12+3	15+2	11+3	11+3
1942													
LeLv 6	IT	3+2	2+2	3+2	4+2	2+4	3+1	2+3	2+2	3+1	4+1	4+1	
LeLv 14	MS									4+1	7+0	9+1	9+0
LeLv 24	BW	33+3	32+1	28+3	29+2	30+1	30+1	28+1	27+2	26+4	24+6	21+6	23+7
LeLv 26	FA	7+11	7+10	6+10	9+7	13+3	15+4	14+5	16+5	16+5	18+3	18+3	17+5
	HC							2+0	1+1	0+2	0+2	1+1	0+2
LeLv 28	MS	12+9	12+10	9+9	11+6	10+3	11+5	15+4	18+3	8+11	6+13	11+11 / 12+11	
LeLv 30	FRw	17+5	19+1	18+1	18+1	20+0	15+2	16+0	16+1	13+0	13+0	6+2	4+0
	IT												4+0
LeLv 32	CU	14+1	15+1	16+0	15+1	14+1	13+1	10+4	11+1	9+3	9+2	9+3	9+1
	HC	2+1	2+1	0+3	0+3	0+3	0+3						
1943													
LeLv 14	MS	7+2	6+4	11+1	7+4	8+3	8+5	7+6	8+9	9+9	12+6	15+3	15+3
LeLv 24	BW	23+7	24+6	25+5	27+3	25+3	23+3	23+3	18+5	18+3	20+0	17+5	15+0
LeLv 26	FA	20+2	19+2	17+3	19+1	15+6	5+15	5+15	8+12	9+10	13+7	12+7	16+3
	HC	2+0	1+1	1+1	1+2	1+2	1+1	1+0					
LeLv 28	MS	16+6	17+9	21+5	16+8	21+5	20+10	21+8	23+7	27+5	26+6	24+9	25+8
	GL			1+0									
LeLv 30	FRw	4+0	4+0	4+0									

Month Squadron	Type	1.01.	1.02.	1.03.	1.04.	1.05.	1.06.	1.07.	1.08.	1.09.	1.10.	1.11.	1.12.
LeLv 32	IT	3+1	3+0	3+0	3+0	5+2	6+0	6+1	8+0	7+0	5+2	5+0	5+0
	CU	9+2	5+5	8+4	9+1	9+0	10+1	6+2	7+2	11+2	15+2	16+3	20+1
	LG				1+0	1+0	1+0	1+0	0+1	0+1	2+0	2+0	1+1
LeLv 34	MT				15+1	13+3	19+6	18+4	15+3	17+1	16+1	15+2	16+2
1944													
LeLv 14	MS	15+4	12+5	10+7	9+7	8+8	7+6	7+4	4+3	7+4	10+0	6+0	6+0
LeLv 24	BW	16+3	14+0	15+1	13+3	12+4							
	MT					13+2	13+1	29+11	25+4	23+3	28+0	28+0	28+0
LeLv 26	FA	6+13	6+13	4+3	3+3	4+1							
	BW						16+0	13+3	2+7	4+5	10+0	4+1	4+1
LeLv 28	MS	24+9	22+11	18+15	22+9	24+7	18+3	14+4	6+4	4+2			
	MSv								0+1	2+0	5+0		
	MT								5+2	9+1	9+3	10+0	10+0
	FRw											5+0	5+0
	IT											8+0	8+0
LeLv 30	IT	5+1	6+0	6+1									
	FA			3+4	6+5	6+5	11+1						
	MT				7+4	9+1	6+4	3+2	8+1	11+1	13+0	13+0	13+0
LeLv 32	CU	18+3	18+1	15+4	16+2	18+1	14+4	13+1	11+4	13+1	12+2	13+1	13+0
	LG	0+1	1+0	1+0	2+0	2+0	2+0	1+1	2+0	0+2	0+2	0+2	0+2
	KH								1+0	0+1	0+1	1+0	1+0
LeLv 34	MT	16+1	20+0	19+3	17+5	22+0	15+1	28+4	22+3	27+1	30+0	30+0	30+0

The numbers inform serviceable + overhaul.

Morane MS-613 after a thorough overhaul at the State Aircraft Factory at Tampere. On 26 Juner 1942 the plane was flown to Hirvas for 2/LeLv 28.
(Finnish Air Force)

Appendix 19
Monthly Missions of the Squadrons

	January	February	March	April	May	June	July	August	September	October	November	December	Total
1939													
LeLv 24												506	506
LeLv 26												172	172
1940													
LeLv 24	564	1090	228										1882
LeLv 26	222	707	69										998
LeLv 28		150	166										316
1941													
LeLv 24						298	881	840	584	350	326	159	3438
LeLv 26						237	490	430	103	129	45	36	1470
LeLv 28						250*	500*	350*	342	168	190	133	1950*
LeLv 30						158	553	542	393	230	326	123	2405
LeLv 32						102	587	391	537	209	161	123	2310
1942													
LeLv 24	238	359	451	134	343	405	545	523	325	244	160	23	3750
LeLv 26	12	28	40	8	43	140	220	365	217	224	150	39	1486
LeLv 28	102	93	125	43	119	221	169	209	148	34	92	19	1380
LeLv 30	254	247	477	325	208	167	202	222	87	92	58	39	2378
LeLv 32	84	155	358	82	170	515	246	230	101	57	76	21	2095
1943													
LeLv 24	136	163	238	236	434	194	218	243	191	78	102	84	2317
LeLv 26	93	85	184	77	262	51	92	59	46	58	18	6	1031
LeLv 28	76	106	195	36	372	234	103	132	44	53	13	16	1380
LeLv 30	65	53	77	22	109	89	57	44	32	42	18	10	618
LeLv 32	65	49	150	4	215	242	184	150	143	129	199	16	1447
LeLv 34			24	161	203	166	253	272	143	197	126	103	1658
1944													
LeLv 24	34	104	184	194	431	648	711	102	2				2410
LeLv 26	2	12	37	11	88	292	421	84					947
LeLv 28	14	144	136	41	103	303	485	129	4				1359
LeLv 30	10	6	30	38	104	118	87	66	4				457
LeLv 32	23	144	190	54	286	659	581	199	-				2136
LeLv 34	33	141	170	185	155	538	582	82	9				1895

* estimated

Appendix 20
Victory Markings

As with other individual marking the victory bars or similar are very rare on the Finnish fighters during the Winter War. There are actually only two known cases, a silver star on the rudder of Morane MS318 and a row of seven white kill bars on the fin of Fokker FR-110.

In the Continuation War the situation changed right from the beginning, as the first applications of victory markings were made already on the first day of the conflict, 25 June 1941. At first they were simple bars, white or yellow, vertical or horizontal, some with dates. The frontal silhouette was more sophisticated as it could depict the type shot down. The earliest know silhouettes appeared in September 1941. There were two types of silhouettes, one distinguished only single- and twin-engined aircraft while the other all types.

Main users were of course the fighters of *Lentorykmentti* 2 (LLv 24 Brewsters, LLv 26 FIATs and LLv 28 Moranes) and *Lentorykmentti* 3 (LLv 30 Fokkers). *Lentolaivue* 32 and *Lentolaivue* 34 later were exceptions to the rule. Until now no kill markings have been observed on their Curtiss Hawks or Messerschmitt 109s.

Out of the reconnaissance or other squadrons flying fighters (Polikarpov I-153, Fokker D.XXI and Morane 406), only *Lentolaivue* 6, 10 and 14 applied victory markings, the latter both air and ground.

By the Soviet major offensive in June 1944 all victory markings had disappeared from the fighters and the situation remained so until the end of the hostilities in September 1944. But in autumn 1945, in the peace time, the victory markings appeared to a couple of FIAT's.

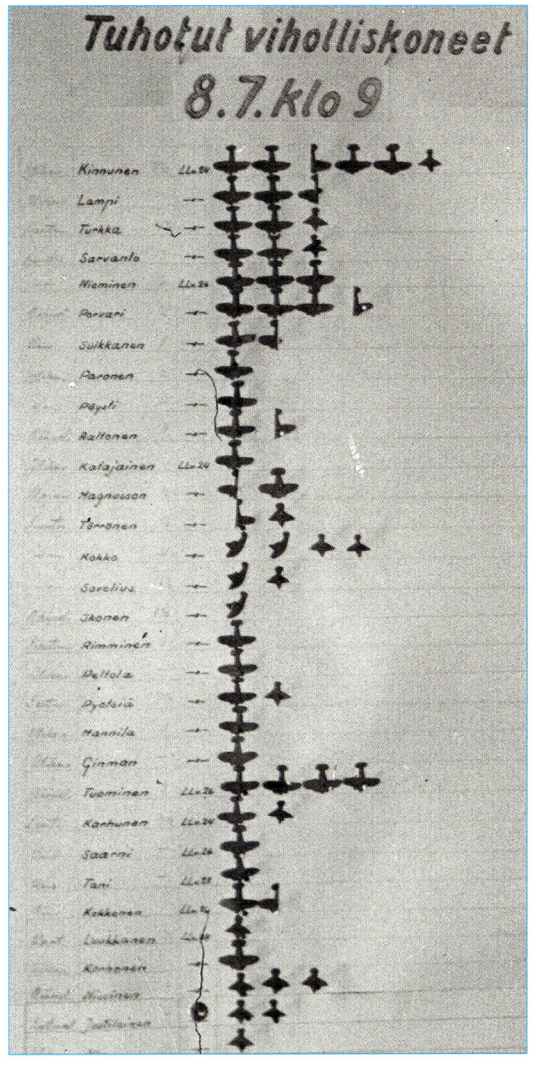

LeR 2 (LLv 24, 26 and 28) scoreboard after two weeks of war, on 8 July 1941 at 9 am. SSgt Eero Kinnunen is leading with 5½ confirmed victories followed by 1Lt Pekka Kokko and MSgt Oiva Tuominen both with 4. (SA-kuva)

Chaika coded IT-16 of 3/LeLv 6 leader Capt Pehr-Eric Ahonius (right) at Römpötti on 9 July 1941 showing both air and sea victories. (SA-kuva)

2/TLeLv 14 leader Capt Martti Kalima (left) at his Morane MS-622 at Tiiksjärvi ion early June 1944, showing 11 kill bars.

Fokker FR-110 of 3/LLv 24 leaning on its wing at Joroinen on 8 April 1940, showing the kill bars of WO Viktor Pyötsiä.

Also the shot down observation or barrage balloons were sometimes marked on the planes. Here on Brewster BW-373 of 1/LLv 24 on 24 October 1941. (SA-kuva)

SSgt Eero Kinnunen (middle) of 2/LLv 24 had at first in June 1941 small victory bars on the fin of his Brewster BW-352. Kinnunen (middle)

1Lt Jorma Sarvanto of 2/LLv 24 had the kill bars on the rudder top in July 1941, white for Winter War and yellow for Continuation War.

4/LLv 24 leader Capt Per Sovelius in front of his Brewster BW-378 at Lunkula in October 1941.

A mechanic sitting on the tail of Brewster BW-354 of 2/LLv 24 at Tiiksjärvi. The victories belong to SSgt Heimo Lampi, last two gained on 30 March 1942.

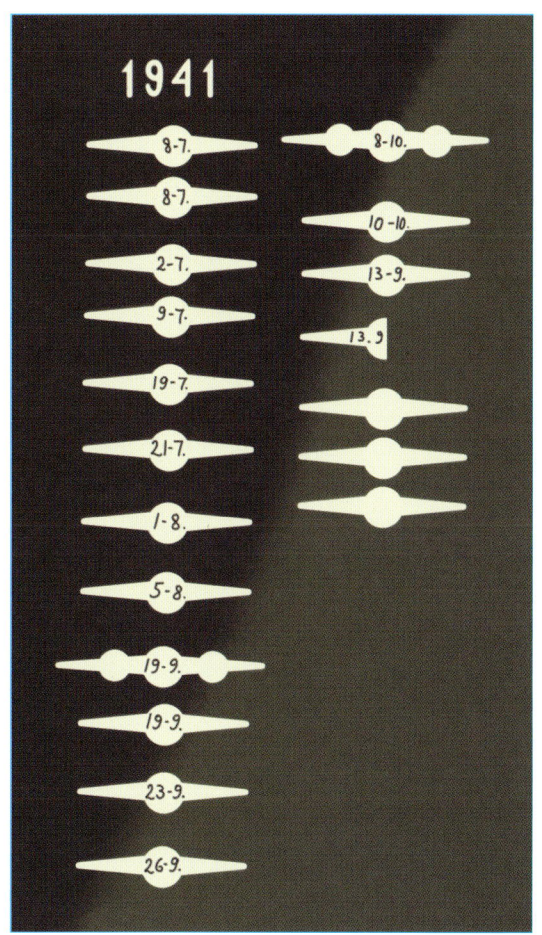

The victory silhouettes of 2Lt Lauri Nissinen. He is at his assigned plane BW-384 of 2/LeLv 24 at Tiiksjärvi on 10 April 1942. (SA-kuva)

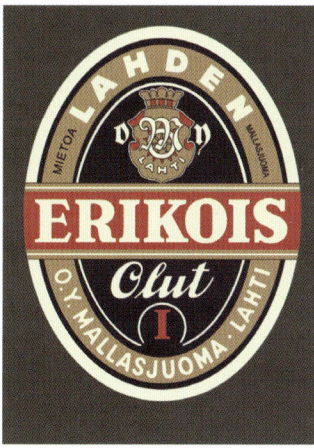

Another kind of demonstrating the air victories. 1/LeLv 24 leader Capt Eino Luukkanen glued beer bottle labels on both sides of the fin of his Brewster BW-393, in Römpötti on 1 November 1942.

WO Ilmari Juutilainen of 3/LLv 24 showing his 20 kill bars on BW-364 at Kontupohja in early April 1942.

WO Oiva Tuominen of 1/LLv 26 at the tail of his FIAT FA-26 at Vitele on 1 September 1941. (SA-kuva)

A mechanic at the tail of FIAT FA-11 of 3/LLv 26 leader 1Lt Urho Nieminen. The last kill bar came on 5 August 1941.

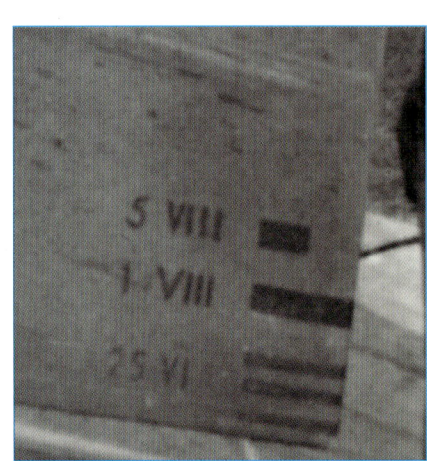

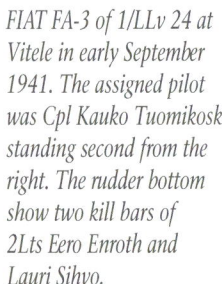

3/LeLv 24 leader 1Lt Hans Wind at the tail of his BW-393 at Suulajärvi on 12 September 1943. (SA-kuva)

FIAT FA-3 of 1/LLv 24 at Vitele in early September 1941. The assigned pilot was Cpl Kauko Tuomikoski standing second from the right. The rudder bottom show two kill bars of 2Lts Eero Enroth and Lauri Sihvo.

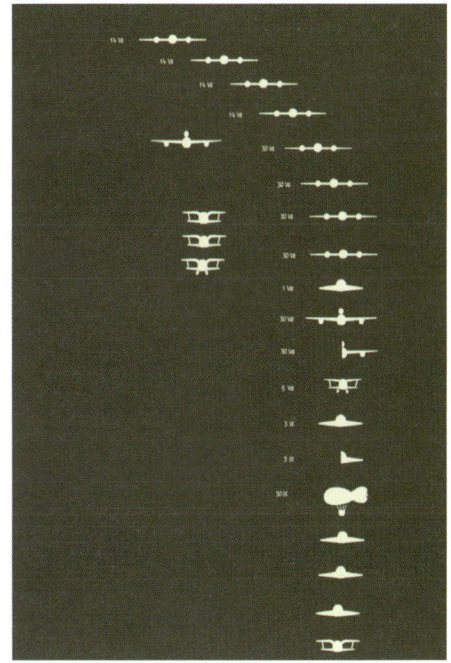

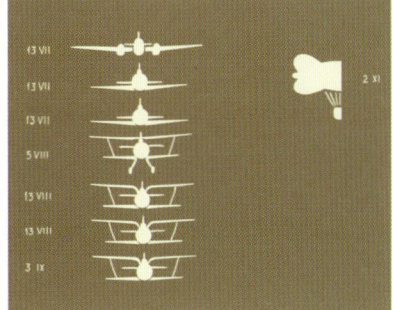

3/LLv 26 pilot 1Lt Olli Puhakka's kill markings on the rudder of his FIAT FA-1 in March 1942. Sgt Paavo Saarni shared the balloon on 1 November 1941.

1/LLv 26 pilot WO Oiva Tuominen's impressive row of kill markings on the rudder of his FIAT FA-26 in March 1942 (photo) and October 1942.

Hurricane HC-452 of 2/LeLv 26 at Kilpasilta in June 1943. It kept the victory bars gained two years earlier by 2Lt Esko Ruotsila with LLv 32.

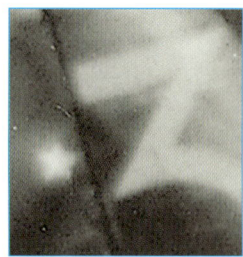

Morane MS318 of 2/LLv 28 at Säkylä. On 2 March 1940 2Lt Pauli Massinen shot down one DB-3 bomber and marked it on the rudder with a silver star.

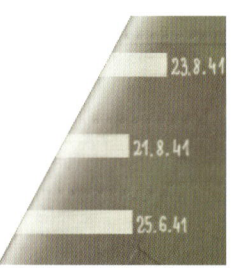

Two mechanics at the tail of 1/LLv 28 pilot Antti Tani's Morane MS-311 at Lunkula in September 1941. Normally LLv 28 did not mark the victory dates, but Tani did.

Morane MS-317 of 1/LLv 28 at Solomanni in summer 1942. The 4½ victory bars denote scored by three different pilots, but not these, from left Sgt Urho Jääskeläinen, 2Lt Yrjö Pulliainen and SSgt Vesa Janhonen.

2/LLv 30 pilot 1Lt Teuvo Ruohola at the tail of his Fokker FR-119 at Suulajärvi on 20 November 1941. The unit practice was applying both air and sea victories to the planes. (SA-kuva)

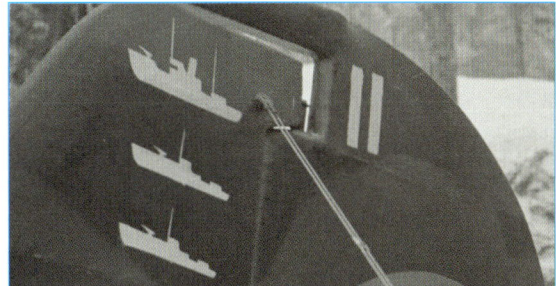

Lentue Käär/LLv 14 pilot 1Lt Martti Kalima at the tail of his Fokker FR-148 at Tiiksjärvi in March 1942. The markings contain both air and ground victories. Plan view silhouette practice was initiated by Lentue Käär predecessor LLv 10.

Not a victory marking, but showing 1500 reconnaissance sorties flown to the east from the front, which is shown by a red wavy line. LeLv 12 flew this sortie on 6 May 1942 and Fokker FR-103 got the honour to carry the marking.

Appendix 21
Tactical Markings

The first tactical markings were introduced in October 1939 during the Winter War mobilization. All modern fighters (Fokker D.XXI) were concentrated in *Lentolaivue* 24, which had five flights. All planes received a tactical number on its rudder, bu using existing standard colours: White, Black, Blue and Green. The original application was:

- 1st Flight - Black number on Green rudder
- 2nd Flight - White number on a Black rudder
- 3rd Flight - Blue number with white outline on Green rudder
- 4th Flight - Black number on White rudder
- 5th Flight - White number on Green rudder

Lentolaivue 28 was the only other squadron during the Winter War applying tactical numbers on the fin of their Moranes. 2/LLv 28 had Yellow numbers and 3/LLv 28 White.

Within *Lentorykmentti* 2 squadrons (LLv 24, 26 and 28) the tactical markings were carried on the planes and applied to the new planes through the intermediary peace.

Within *Lentorykmentti* 3 LLv 32 began applying gradually their own numbers in June 1940 and LLv 30 followed in May 1941.

When the Continuation began all Lentorykmentti 2 and 3 squadrons had tactical numbers, typically a different colour for every flight. Basically this type of tactical numbering remained in use throughout the Continuation war. This system is lavishly illustrated in the colour section below.

Lentolaivue 32 Curtiss Hawks had a different system through the war. White tail number was issued to aircraft in serial block CU-551–559, Yellow number to CU-560–569 and a Blue with White Border to CU-571–, being the same as the last digit of the serial. This practice continued with aircraft arriving later, but no tie to the serial blocks.

The same system was used on the Messerschmitts of *Lentolaivue* 34. White number for block MT-201–210, Red with white border for MT-211–220 and Yellow for MT-221–230. But now the numbers were on the nose.

Lentorykmentti 3 (HLeLv 24, 26 and 34) introduced a new style of tactical numbers on 22 May 1944. The first flight in every squadron had a Yellow number in the forward fuselage, the second flight had the number in the centre fuselage and the third flight on the fin. 1/HLeLv 34 was an exception in using White numbers.

Though this type of tactical numbering concerned only *Lentorykmentti* 3 squadrons, photographic evidence suggests that it was used to some extent also by *Lentorykmentti* 2 and 5 squadrons.

3/LLv 6
June 1941 – November 1942
Here on IT-16 in July 1942

1/LLv 10
September 1941 – October 1941
Here on FR-154 in September 1941

1/LeLv 12
April 1943 – May 1944
Here on FR-100 in September 1943

2/LeLv 12
April 1943 – May 1944
Here on FR-142 in August 1943

2/TLeLv 12
August 1944 – November 1944
Here on MY-16 in September 1944

Lentue Kään/LLv 14
November 1941 – August 1942
Here on FR-148 in June 1942

1/LeLv 14
September 1942 – September 1944
Here on MS-311 in July 1943

2/LeLv 14
September 1942 – December 1943
Here on FR-140 in October 1942

2/TLeLv 14
January 1944 – September 1944
Here on MS-622 in June 1944

2/LeLv 16
January 1943 – July 1944
Here on GL-276 in September 1943

F 19 Fighter squadron
January 1940 – March 1940
Here on Gladiator H in January 1940

1/LLv 24
October 1939 – April 1940
Here on FR-81 in January 1940

2/LLv 24
October 1939 – April 1940
Here on FR-97 in January 1940

3/LLv 24
October 1939 – April 1940
Here on FR-110 in April 1940

4/LLv 24
October 1939 – April 1940
Here on FR-99 in January 1940

5/LLv 24
October 1939 – April 1940
Here on FR-105 in April 1940

1/LLv 24
September 1940 – April 1944
Here on BW-373 in July 1941

2/LLv 24
September 1940 – February 1943
Here on BW-357 in July 1941

3/LLv 24
September 1940 – April 1944
Here on BW-393 in September 1943

4/LLv 24
September 1940 – February 1943
Here on BW-370 in August 1942

1/HLeLv 24
May 1944 – September 1944
Here on MT-456 in July 1944

2/HLeLv 24
May 1944 – September 1944
Here on MT-449 in June 1944

3/HLeLv 24
May 1944 – September 1944
Here on MT-476 in July 1944

1/LLv 26
June 1941 – December 1941
Here on FA-26 in September 1941

2/LLv 26
June 1941 – December 1941
Here on FA-22 in June 1941

3/LLv 26
June 1941 – May 1944
Here on FA-6 in September 1941

1/LeLv 26
July 1942 – May 1944
Here on FA-26 in April 1943

2/LeLv 26
March 1943 – May 1944
Here on FA-17 in March 1943

1/HLeLv 26
May 1944 – December 1944
Here on BW-375 in June 1944

2/HLeLv 26
May 1944 – December 1944
Here on BW-368 in June 1944

2/LLv 28
February 1940 – May 1940
Here on MS-319 in March 1940

3/LLv 28
February 1940 – May 1940
Here on MS-305 in March 1940

1/LLv 28
June 1941 – July 1944
Here MS-328 in September 1942

2/LLv 28
June 1941 – July 1944
Here on MS-327 in September 1941

3/LLv 28
June 1941 – July 1944
Here on MS-623 in September 1942

LLv 30 Commander
June 1941 – March 1942
Here on FR-157 in September 1941

1/LLv 30
May 1941 – June 1941
Here on HC-456 in June 1941

2/LLv 30
May 1941 – November 1942
Here on FR-129 in October 1941

3/LLv 30
May 1941 – September 1941
Here on FR-154 in September 1941

1/LLv 30
October 1941 – November 1942
Here on FR-125 in November 1941

1/LeLv 30
November 1942 – March 1944
Here on IT-11 in April 1943

2/LeLv 30
November 1942 – March 1944
Here on IT-15 in May 1943

LLv 32 Commander
June 1941 – July 1941
Here on FR-109 in July 1941

1/LLv 32
May 1941 – July 1941
Here on FR-112 in June 1941

2/LLv 32
May 1941 – July 1941
Here on FR-113 in June 1941

3/LLv 32
June 1941 – August 1941
Here on FR-160 in June 1941

LLv 32
July 1941 – September 1944
Here on CU-552 in June 1942

LLv 32
July 1941 – September 1944
Here on CU-560 in May 1942

LLv 32
July 1941 – September 1944
Here on CU-581 in October 1943

LeLv 34
May 1943 – April 1944
Here on MT-201 in June 1943

LeLv 34
May 1943 – April 1944
Here on MT-215 in June 1943

LeLv 34
May 1943 – April 1944
Here on MT-227 in May 1943

1/HLeLv 34
May 1944 – September 1944
Here on MT-426 in July 1944

3/HLeLv 34
May 1944 – September 1944
Here on MT-415 in June 1944

Appendix 22
Unit Emblems

This chapter deals with the unit emblems on aircraft classified as fighters in at least one point of their career. Some of the emblems are those belonging to a reconnaissance squadron.

Only one fighter outfit, *Lentolaivue* 24 had an early squadron emblem, the striking lynx, which was first applied in early July 1941 to the unit's Brewsters. By 1944 the lynxes were all painted out.

On 7 June 1944 *Lentorykmentti* 3 introduced new squadron emblems to its three squadrons:
- HLeLv 24 - Lynx's head
- HLeLv 26 - Snake in upward circle
- HLeLv 34 - Eagle Fledgling

All these wereto be painted on the tails of the unit's planes, but photographic evidence confirms that only four planes ever got it. HLeLv 24 on MT-437 only, HLeLv 26 on BW-375 only and HLeLv 34 on MT-423 and MT-451 only. These were carried as long as the respective plane existed or was re-painted.

Within *Lentolaivue* 24 are the two known cases when a fighter flight introduced its own emblem. 2/LeLv 24 had the farting elk and 4/LeLv 24 the osprey, both appearing in summer 1942 and the former remaining on the planes to February 1943 and the latter until summer 1943.

Out of the reconnaissance squadrons flying fighters, only Lentolaivue 12 applied flight emblems. Stalin the main devil decorated 1/LLv 12 Curtiss Hawks in June and July 1941.

When LLv 12 received Fokker D.XXIs in July 1941, the flight emblems appeared after five months: 1/LLv 12 got the butting ram and 2/LLv 12 the cheerful donkey. They lasted on the planes until the end of 1942.

Finally, when 2/TLeLv 12 got the domestic Myrsky fighters, they introduced the lightning emblem in either August or September 1944, lasting only for a couple of months.

1/LLv 12
July 1941
Here on CU-502 in July 1941

1/LLv 12
November 1941 – November 1942
Here on FR-98 in May 1942

2/LLv 12
November 1941 – November 1942
Here on FR-83 in February 1942

2/TLeLv 12
September 1944 – December 1944
Here on MY-23 in September 1944

LeLv 24
July 1941 – December 1943
Here on BW-366 in August 1941

2/LeLv 24
June 1942 – February 1943
Here on BW-357 in July 1942

4/LeLv 24
August 1942 – February 1943
Here on BW-370 in October 1942

HLeLv 24
June 1944 – September 1944
Here on MT-437 in June 1944

HLeLv 26
April 1943 – November 1943
Here on FA-19 in May 1943

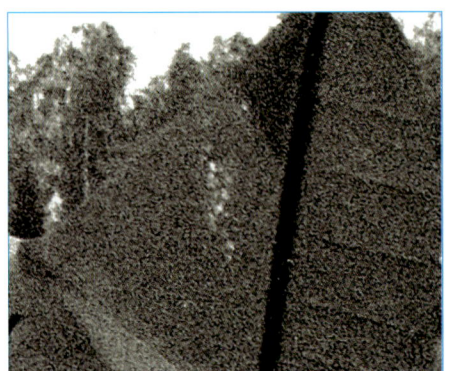

HLeLv 26
June 1944 – September 1944
Here on BW-375 in June 1944

HLeLv 34
June 1944 – September 1944
Here on MT-423 in June 1944

Appendix 23
Individual Markings

In the Winter War this kind of markings were basically prohibited on the Finnish Air Force aircraft. Actually there is only one known example from this conflict. A large skull and bomb on the rudder on Blenheim BL-117.

Though the practice in the Continuation War became more liberal, individual or personal emblems or inscriptions were not overly popular, many being also short-lived. The selection below shows most but not all markings, some photos being too poor to reproduce or just showing part of it.

VH-17 Olive
3/LLv 6 leader Capt Lauri Karjalainen
March 1942

FR-100 X
1/LeLv 14 leader 1Lt Matti Tainio
July 1942

FR-148 Pinocchio
Lentue Kään *1Lt Martti Kalima*
July 1942

FR-144 A
Lentue Kään *leader Capt Pekka Kään*
July 1942

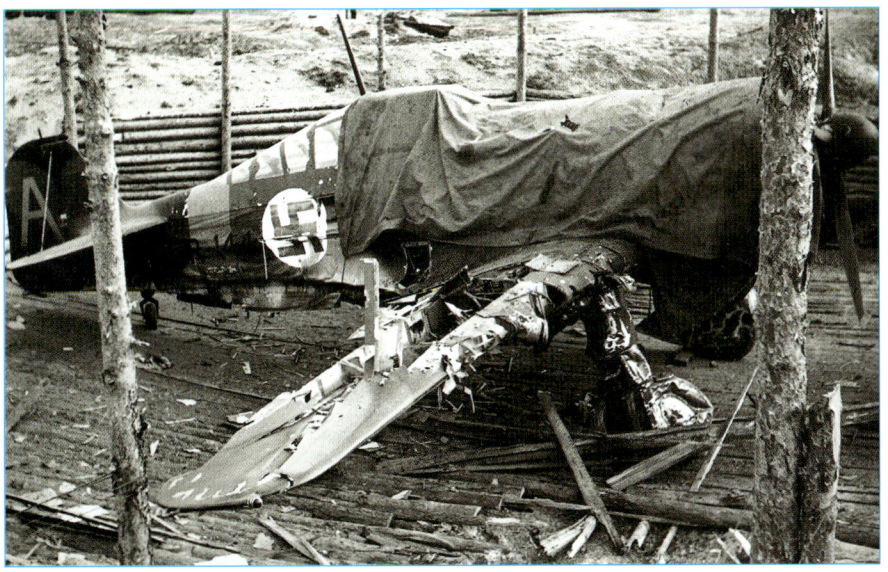

MS-326 Sea horse
1/LeLv 14 1Lt Martti Kalima
July 1943

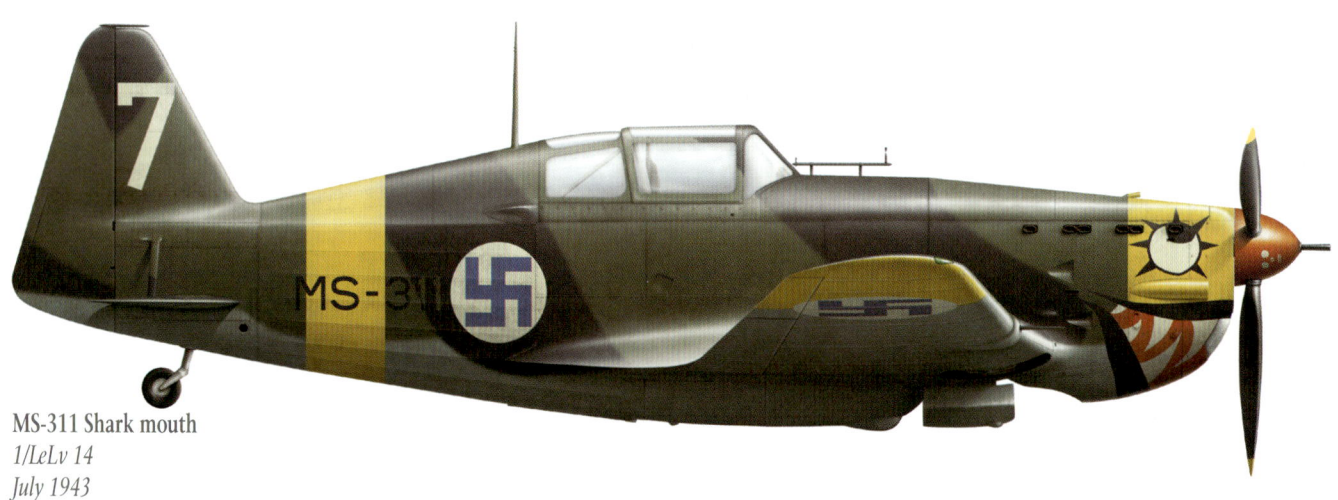

MS-311 Shark mouth
1/LeLv 14
July 1943

GL-264 Alligator
2/LeLv 16
June 1942

GL-265 Skull
2/LeLv 16
June 1943

„TRE BRÖDER"

BW-367 "TRE BRÖDER"
2/LLv 24
September 1940

BW-379 Pekka
3/LLv 24 1Lt Pekka Kokko
September 1941

BW-355 NOKA
3/LLv 24
July 1941

OTTO WREDE

BW-378 OTTO WREDE
4/LLv 24 leader Capt Per Sovelius
October 1941

BW-364 Skull
3/LeLv 24
April 1943

FA-1 Skull
3/LeLv 26
March 1943

MS-615 X
LeLv 28 commander Maj Auvo Maunula
March 1943

FR-154 Cartoons
3/LLv 30
July 1941

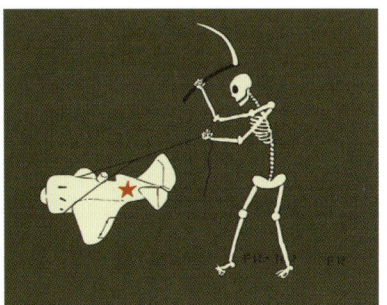

FR-129 Devil chasing I-16
2/LLv 30 leader Capt Veikko Karu
September 1941

FR-142 Death spearing I-16
2/LLv 30
September 1941

FR-157 Russkie
1/LeLv 30
February 1943

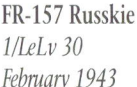

CU-557 Teeth
LLv 32 commander Maj Olavi Ehrnrooth
August 1941

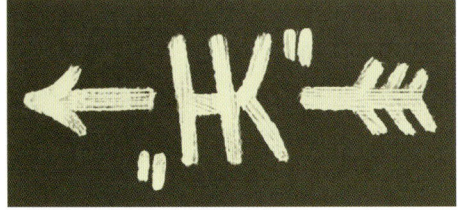

HC-456 <HK<
Os Kalaja/LLv 32
leader Capt Heikki Kalaja
August 1941

Index
of Personal Names

Bibliography

Juutilainen, Ilmari, *Double Fighter Knight*, Apali, Finland 1996

Keskinen, Kalevi, Stenman, Kari, *Finnish Air Force 1939-1945*, Squadron/Signal, USA 1998

Keskinen, Kalevi, Stenman, Kari, *Finnish Air Force History 1, 4, 5, 7–10, 17–28*, Kari Stenman Publishing, Finland 2001–2008

Keskinen, Kalevi, Stenman, Kari, *Finnish Air Force History 2, 3, 6*, Hobby-Kustannus, Finland 1999–2004

Keskinen, Kalevi, Stenman, Kari, *Finnish Air Force I–VI*, Kari Stenman Publishing, Finland 2001-2008

Luukkanen, Eino, *Fighter over Finland*, Macdonald, England 1963

Persyn, Lionel, Stenman, Kari, Thomas, Andrew, *Aircraft of the Aces 86 – P-36 Hawk Aces of World War 2*, Osprey Publishing, England 2009

Stenman, Kari, *Air Enthusiast 23, 46, 50, 66, 88, 120*, Key Publishing, England 1984–2005

Stenman, Kari, de Jong, Peter, *Aircraft of the Aces 112 – Fokker D.XXI Aces of World War 2*, Osprey Publishing, England 2013

Stenman, Kari, Ehrengardt, Christian-Jacques, *Aircraft of the Aces 121 – Morane-Saulnier MS. 406 Aces*, Osprey Publishing, England 2014

Stenman, Kari, Keskinen, Kalevi, *Aircraft of the Aces 23 – Finnish Fighter Aces of World War 2*, Osprey Publishing, England 1998

Stenman, Kari, Thomas, Andrew, *Aircraft of the Aces 91 – Brewster F2A Buffalo Aces of World War 2*, Osprey Publishing, England 2010

Stenman, Kari, Keskinen, Kalevi, *Aviation Elite 4 – Lentolaivue 24*, Osprey Publishing, England 2001

Stenman, Kari, Hołda, Karolina, *Finnish Fighter Colours*, vol 1, Stratus, Poland 2014

Stenman, Kari, Hołda, Karolina, *Finnish Fighter Colours*, vol 2, Stratus, Poland 2015

Official sources

Finnish National Archives, War Archives Branch, Helsinki
Air Force HQ permanent orders
Individual aircraft files
Squadron operational records and logbooks
Combat reports
Personnel files

Central Archive of Ministry of Defence (TsAMO), Podolsk, Russia

Russian State Military Archive (RGVA), Moscow, Russia

Central Naval Archive (TsVMA), Gatchina, Russia

Air war cadets at Kauhava in April 1930. Rear row from left: Poimio, Hytönen, Savolainen, Kalervo, Heinilä (future LLv 26 CO) and Säilä. Front row from left: Ehrnrooth (future LLv 32 CO), Sirén (future LLv 28 CO), Appelelroth, Bremer (future LLv 30 CO) and Pajari (future LLv 16 CO). (Finnish Air Force)